Introduction to Time Series Using Stata

Stata

Revised Edition

SEAN BECKETTI

A Stata Press Publication
StataCorp LLC
College Station, Texas

Published by Stata Press, 4905 Lakeway Drive, College Station, Texas 77845
Typeset in LaTeX 2_ε
Printed in the United States of America

10 9 8 7 6 5 4 3 2 1

Print ISBN-10: 1-59718-306-7
Print ISBN-13: 978-1-59718-306-2
ePub ISBN-10: 1-59718-307-5
ePub ISBN-13: 978-1-59718-307-9
Mobi ISBN-10: 1-59718-308-3
Mobi ISBN-13: 978-1-59718-308-6

Library of Congress Control Number: 2020932011

Introduction to Time Series Using Stata

Revised Edition

Contents

List of tables   xiii

List of figures   xv

Preface   xxi

Acknowledgments   xxvii

1   Just enough Stata   1

   1.1   Getting started   2

       1.1.1   Action first, explanation later   2

       1.1.2   Now some explanation   6

       1.1.3   Navigating the interface   7

       1.1.4   The gestalt of Stata   13

       1.1.5   The parts of Stata speech   15

   1.2   All about data   20

   1.3   Looking at data   29

   1.4   Statistics   49

       1.4.1   Basics   49

       1.4.2   Estimation   53

   1.5   Odds and ends   60

   1.6   Making a date   62

       1.6.1   How to look good   63

       1.6.2   Transformers   65

   1.7   Typing dates and date variables   68

   1.8   Looking ahead   69

2   Just enough statistics   71

   2.1   Random variables and their moments   72

2.2   Hypothesis tests   73

2.3   Linear regression   74

    2.3.1   Ordinary least squares   74

    2.3.2   Instrumental variables   77

    2.3.3   FGLS   77

2.4   Multiple-equation models   78

2.5   Time series   79

    2.5.1   White noise, autocorrelation, and stationarity   80

    2.5.2   ARMA models   82

3   Filtering time-series data   **85**

3.1   Preparing to analyze a time series   87

    3.1.1   Questions for all types of data   87

        How are the variables defined?   87

        What is the relationship between the data and the phe-
nomenon of interest?   88

        Who compiled the data?   90

        What processes generated the data?   90

    3.1.2   Questions specifically for time-series data   91

        What is the frequency of measurement?   91

        Are the data seasonally adjusted?   91

        Are the data revised?   92

3.2   The four components of a time series   92

        Trend   93

        Cycle   95

        Seasonal   98

3.3   Some simple filters   100

    3.3.1   Smoothing a trend   103

    3.3.2   Smoothing a cycle   109

    3.3.3   Smoothing a seasonal pattern   114

    3.3.4   Smoothing real data   115

3.4   Additional filters   121

    3.4.1   ma: Weighted moving averages   123

    3.4.2   EWMAs   125

        exponential: EWMAs   126

        dexponential: Double-exponential moving averages   130

    3.4.3   Holt–Winters smoothers   131

        hwinters: Holt–Winters smoothers without a seasonal
component   131

        shwinters: Holt–Winters smoothers including a seasonal
component   137

3.5   Points to remember   138

4   A first pass at forecasting   **141**

4.1   Forecast fundamentals   141

    4.1.1   Types of forecasts   142

    4.1.2   Measuring the quality of a forecast   144

    4.1.3   Elements of a forecast   144

4.2   Filters that forecast   146

    4.2.1   Forecasts based on EWMAs   148

    4.2.2   Forecasting a trending series with a seasonal component . .   159

4.3   Points to remember   165

4.4   Looking ahead   166

5   Autocorrelated disturbances   **167**

5.1   Autocorrelation   168

    5.1.1   Example: Mortgage rates   169

5.2   Regression models with autocorrelated disturbances   172

    5.2.1   First-order autocorrelation   173

    5.2.2   Example: Mortgage rates (cont.)   175

5.3   Testing for autocorrelation   176

    5.3.1   Other tests   177

5.4 Estimation with first-order autocorrelated data 178

   5.4.1 Model 1: Strictly exogenous regressors and autocorre-
           lated disturbances 179

          The OLS strategy 182

          The transformation strategy 183

          The FGLS strategy 186

          Comparison of estimates of model 1 188

   5.4.2 Model 2: A lagged dependent variable and i.i.d. errors . . . 189

   5.4.3 Model 3: A lagged dependent variable with AR(1) errors . . 193

          The transformation strategy 194

          The IV strategy 196

5.5 Estimating the mortgage rate equation 197

5.6 Points to remember 199

6   Univariate time-series models   **201**

6.1 The general linear process 202

6.2 Lag polynomials: Notation or prestidigitation? 203

6.3 The ARMA model 205

6.4 Stationarity and invertibility 208

6.5 What can ARMA models do? 210

6.6 Points to remember 214

6.7 Looking ahead 215

7   Modeling a real-world time series   **217**

7.1 Getting ready to model a time series 218

7.2 The Box–Jenkins approach 226

7.3 Specifying an ARMA model 228

   7.3.1 Step 1: Induce stationarity (ARMA becomes ARIMA) . . . 228

   7.3.2 Step 2: Mind your p's and q's 233

7.4 Estimation 243

7.5   Looking for trouble: Model diagnostic checking 253

     7.5.1   Overfitting 253

     7.5.2   Tests of the residuals 254

7.6   Forecasting with ARIMA models 257

7.7   Comparing forecasts 262

7.8   Points to remember 266

7.9   What have we learned so far? 267

7.10   Looking ahead 269

8   Time-varying volatility   **271**

8.1   Examples of time-varying volatility 272

8.2   ARCH: A model of time-varying volatility 277

8.3   Extensions to the ARCH model 285

     8.3.1   GARCH: Limiting the order of the model 286

     8.3.2   Other extensions 292

          Asymmetric responses to "news" 293

          Variations in volatility affect the mean of the observable series 295

          Nonnormal errors 296

          Odds and ends 296

8.4   Points to remember 298

9   Models of multiple time series   **299**

9.1   Vector autoregressions 300

     9.1.1   Three types of VARs 302

9.2   A VAR of the U.S. macroeconomy 303

     9.2.1   Using Stata to estimate a reduced-form VAR 305

     9.2.2   Testing a VAR for stationarity 309

          Other tests 312

     9.2.3   Forecasting 316

          Evaluating a VAR forecast 325

9.3   Who's on first?   329

        9.3.1   Cross correlations   330

        9.3.2   Summarizing temporal relationships in a VAR   335

                 Granger causality   336

                 How to impose order   339

                 FEVDs   343

                 Using Stata to calculate IRFs and FEVDs   344

9.4   SVARs   358

        9.4.1   Examples of a short-run SVAR   361

        9.4.2   Examples of a long-run SVAR   370

9.5   Points to remember   373

9.6   Looking ahead   374

10   Models of nonstationary time series   377

10.1   Trends and unit roots   378

10.2   Testing for unit roots   382

10.3   Cointegration: Looking for a long-term relationship   387

10.4   Cointegrating relationships and VECMs   389

        10.4.1   Deterministic components in the VECM   393

10.5   From intuition to VECM: An example   394

                 Step 1: Confirm the unit root   399

                 Step 2: Identify the number of lags   401

                 Step 3: Identify the number of cointegrating relationships .   402

                 Step 4: Fit a VECM   406

                 Step 5: Test for stability and white-noise residuals   416

                 Step 6: Review the model implications for reasonableness .   417

10.6   Points to remember   424

10.7   Looking ahead   424

11   Closing observations   427

11.1   Making sense of it all   427

11.2   What did we miss?   428

    11.2.1   Advanced time-series topics   429

    11.2.2   Additional Stata time-series features   431

        Data management tools and utilities   431

        Univariate models   432

        Multivariate models   433

11.3   Farewell   433

References   **435**

Author index   **439**

Subject index   **441**

Tables

1.1   Stata operators 18

1.2   Importing and exporting data 24

1.3   Date and time formats 64

3.1   Questions to answer prior to data analysis 88

4.1   Forecasting with `tssmooth` 147

4.2   Comparison of forecasts 155

4.3   Share of "respectable" forecasts 157

6.1   Dynamic behavior of an AR(2) model 213

7.1   Indicators of p, d, and q 242

8.1   Extensions to the ARCH model and `arch` command 297

9.1   FEVD 351

Figures

1.1 The relationship between automobile mileage and automobile weight   6

1.2 Stata for Windows opening screen   7

1.3 Viewer window displaying `search` results   9

1.4 **Data** menu 11

1.5 The `describe` dialog box 12

1.6 A histogram of GNI per capita, Atlas $ 36

1.7 A histogram of GNI per capita, weighted by population 37

1.8 The ratio of girls to boys in primary and secondary school 39

1.9 Income per capita and the ratio of girls to boys in school 40

1.10 Log income per capita and the ratio of girls to boys in school 41

1.11 Log income per capita and the ratio of girls to boys in school 42

1.12 Population-weighted relationship between income and female
education 43

1.13 U.S. GDP, 1990–2010 44

1.14 CO$_2$ emissions in the United States, 1990–2008 45

1.15 CO$_2$ emissions and GDP in the United States 46

1.16 CO$_2$ emissions and GDP in the United States 47

1.17 Energy use per capita in the United States (kg of oil equivalent) . . 48

1.18 U.S. GDP, 1990–2010 49

1.19 Fitted and actual values of CO$_2$ emissions 56

1.20 Residuals from the CO$_2$ regression 57

3.1 The relationship between the unemployment rate and nonfarm
payroll employment, January 1950 to January 2012 89

3.2 U.S. GDP, billions of dollars, 1947:1 through 2012:1 94

3.3 Log of U.S. GDP, trillions of dollars, 1947:1 through 2012:1 95

3.4 U.S. civilian unemployment rate, January 1948 through March 2012 97

3.5 Average monthly prepayment rates (annualized) on seasoned Federal National Mortgage Association 30-year discount mortgages, 2000–2007 99

3.6 Trend (unobserved) and Trend+Residual (observed) 104

3.7 A span-3 median smoother and a Hanning smoother 106

3.8 A close-up of the outlier in month 46 107

3.9 Combining Hanning with a median smoother 108

3.10 Comparing span-3 and span-9 smoothers 109

3.11 Cycle (unobserved) and cycle + residual (observed) 110

3.12 Three smoothers and cyclical data 111

3.13 The span-9 median smoother performs poorly on cyclical data . . . 112

3.14 Odd-span median smoothers tend to produce flat spots 113

3.15 The TED spread, 1/2/87–12/30/11 117

3.16 The TED spread and a smoothed version, 8/1/09–12/31/09 119

3.17 The difference between the smooth and the TED spread 120

3.18 The relative performance of a complex and a simple smoother 121

3.19 Nonparametric and EWMA smoothers applied to cyclical data 129

3.20 Nonparametric and double-exponential smoothers applied to the TED spread 131

3.21 U.S. GDP with Holt–Winters and EWMA smoothers 134

3.22 U.S. GDP with Holt–Winters and EWMA smoothers 135

3.23 U.S. GDP with Holt–Winters and EWMA smoothers 136

4.1 U.S. civilian unemployment rate, January 1948 through March 2012 148

4.2 EWMA forecasts with different projection dates 150

4.3 DEWMA forecasts with different projection dates 151

4.4 EWMA and DEWMA forecast errors 152

4.5 One-step-ahead forecast errors for three methods 154

4.6 Distribution of one-step-ahead forecast errors (in tenths of a percent) 158

4.7 Comparison of DEWMA and Holt–Winters forecasts of the unemployment rate 159

4.8 Weekly currency component of M1 160

4.9 Currency component in recent years 161

4.10 Seasonal Holt–Winters forecast 163

4.11 Backtesting the seasonal Holt–Winters forecast 165

5.1 Primary and secondary mortgage rates 170

5.2 The spread between the primary and secondary mortgage rates . . . 171

5.3 Current residuals versus lagged residuals 176

5.4 In-sample fit of three estimation strategies 198

6.1 Exponential decay in a stationary AR(1) model 211

6.2 Decay when $\phi_1 < 0$ 212

6.3 Decay in an AR(2) model with complex roots 213

7.1 Log of United States real GDP 219

7.2 Trend line and Holt–Winters smooth for real GDP 222

7.3 Comparison of trend line and Holt–Winters residuals 223

7.4 Comparison of trend line and Holt–Winters forecasts 224

7.5 Converting the linear trend to a one-step-ahead forecast 225

7.6 Autocorrelations of the log of real GDP 232

7.7 Autocorrelations of real GDP growth 233

7.8 Autocorrelation and partial autocorrelation functions of white noise 235

7.9 Autocorrelation and partial autocorrelation functions of an AR(1) series where $\phi_1 = 0.9$ 236

7.10 Autocorrelation and partial autocorrelation functions of an AR(2) series 237

7.11 Autocorrelation and partial autocorrelation functions of an MA(1) series 239

7.12 Autocorrelation and partial autocorrelation functions of an MA(2) series 240

7.13 Autocorrelation and partial autocorrelation functions of an ARMA(1,1) series 241

7.14 Autocorrelation and partial autocorrelation functions of real GDP
      growth, 1947:2 to 2012:1 243

7.15 Cumulative periodogram of the residuals from the MA(2) specification 256

7.16 Comparing four different uses of the `predict` command 261

7.17 Within-sample and out-of-sample forecasts of real GDP, 2001:2–2003:1 266

8.1   The TED spread, 1/2/1987–12/30/2012 273

8.2   The distribution of one-day changes in the TED spread 274

8.3   Variation in the volatility of the TED spread 275

8.4   Variation in the volatility of real GDP growth 276

8.5   Five-year standard deviations of real GDP growth 277

8.6   The conditional variance of ϵ_t 284

8.7   Monthly and annual consumer price inflation in the United States . 288

8.8   Comparison of GARCH and ARCH estimates of conditional variance . 291

8.9   Conditional variance of monthly consumer inflation 292

9.1   U.S. inflation, unemployment, and the Federal funds rate, 1960:1–
      2012:1 304

9.2   Eigenvalues of the companion matrix 312

9.3   U.S. inflation, forecasts and actuals, 2002:2–2012:1 318

9.4   Forecast errors of U.S. inflation, 2001:2–2012:1 319

9.5   Forecasts of U.S. inflation, unemployment, and the Federal funds
      rate, 2001:3–2011:2 320

9.6   Dynamic forecasts of U.S. inflation, unemployment, and the
      Federal funds rate, 2002:2–2012:1 322

9.7   Dynamic and one-step-ahead forecasts 325

9.8   Cross correlations of inflation and unemployment rates, 1960:2–2001:2 332

9.9   Cross correlations of inflation, unemployment, and the Federal
      funds rate 334

9.10 Orthogonalized impulse–response functions for the `iuf` ordering . . 353

9.11 Comparison of OIRFs for two different orderings 355

9.12 Cumulative OIRFs for two different orderings 357

9.13 Comparison of recursive VAR and SVAR results 365

9.14 Impact of monetary policy in a recursive VAR and an SVAR 369

9.15 An example of the sensitivity of estimates to the sample period . . . 370

10.1 Construction wages in DC–MA–VA, millions of dollars 396

10.2 Construction wages in DC–MA–VA, 1990:1=100 398

10.3 Autocorrelations of quarterly construction wages per capita in VA . 400

10.4 Estimated cointegrating relationships 418

10.5 Autocorrelation functions of the cointegrating relationships 419

10.6 Orthogonalized impulse–response functions 420

10.7 Projected construction wage indices in DC, MD, and VA 421

10.8 Evolution of the cointegrating relationship 422

10.9 Comparison of two forecasts of the construction wage index in DC . 423

Preface

Welcome.

Time-series analysis is a relatively new branch of statistics. Most of the techniques described in this book did not exist prior to World War II, and many of the techniques date from just the last few decades. The novelty of these techniques is somewhat surprising, given the importance of forecasting in general and of predicting the future consequences of today's policy actions in particular. The explanation lies in the relative difficulty of the statistical theory for time series. When I was in graduate school, one of my econometrics professors admitted that he had switched his focus from time series when he realized he could produce three research papers a year on cross-section topics but only one paper per year on time-series topics.

Why another book on time series?

The explosion of research in recent decades has delivered a host of powerful and complex tools for time-series analysis. However, it can take a little while to become comfortable with applying these tools, even for experienced empirical researchers. And in industry, these tools sometimes are applied indiscriminately with little appreciation for their subtleties and limitations. There are several excellent books on time-series analysis at varying levels of difficulty and abstraction. But few of those books are linked to software tools that can immediately be applied to data analysis.

I wrote this book to provide a step-by-step guide to essential time-series techniques— from the incredibly simple to the quite complex—and, at the same time, to demonstrate how these techniques can be applied in the Stata statistical package.

Why Stata? There are, after all, a number of established, powerful statistical packages offering time-series tools. Interestingly, the conventions adopted by these programs for describing and analyzing time series vary widely, much more widely than the conventions used for cross-section techniques and classical hypothesis testing. Some of these packages focus primarily on time series and can be used on non-time-series questions only with a bit of difficulty. Others have to twist their time-series procedures into a form that fits the rest of the structure of their package.

I helped out in a small way when Stata was first introduced. At that time, the most frequent question posed by users (and potential users) was, "When will time series be available?" For a long time, we would tell users (completely sincerely) that these techniques would appear in the next release, in six to twelve months. However, we repeatedly failed to deliver on this promise. Version after version appeared with many new features, but not time series. I moved on to other endeavors, remaining a Stata user but not a participant in its production. Like other users, I kept asking for time-series features—I needed them in my own research. I finally became frustrated and, using Stata's programming capabilities, cobbled together some primitive Stata functions that helped a bit.

Why the delay? Part of the reason was other, more time-critical demands on what was, at the beginning, a small company. However, I think the primary reason was StataCorp's commitment to what they call the "human-machine interface". There are lots of packages that reliably calculate estimates of time-series models. Many of them are difficult to use. They present a series of obstacles that must be overcome before you can test your hypotheses on data. Frequently, it is challenging to thoroughly examine all aspects of your data. And they make it onerous to switch directions as the data begin to reveal their structure.

Stata makes these tasks easy—at least, easy by comparison to the alternatives. I find that the facility of Stata contributes to better analyses. I attempt more, I look more deeply, because it is easy. The teams that work for me use several different packages, not just Stata, depending on the task at hand. I find that I get better, more thorough analyses from the team members using Stata. I do not think it is a coincidence.

When Stata finally gained time-series capabilities, it incorporated a design that retains the ease of use and intuitiveness that has always been the hallmark of this package. That is why I use Stata rather than any of the other candidate packages.

Despite the good design poured into Stata, time-series analysis is still tough. That is just the nature of the time-series inference task. I tend to learn new programs by picking up the manual and playing around. I certainly have learned a lot of the newer, more complex features of Stata that way. However, I do not think it is easy to learn the time-series techniques of Stata just from reading the Stata *Time-Series Reference Manual*—and it is a very well-written manual. I know—I tried. For a long time, I stuck with my old, home-brew Stata functions to avoid the task of learning something different, even after members of my staff had adopted the new Stata tools.

Writing this book provided me with the opportunity to break out of my bad habits and make the transition to Stata's powerful time-series features. And I am glad I did. Once you come up the learning curve, I think these tools will knock your socks off. They certainly lower the barrier to many ambitious types of empirical research.

I hope you are the beneficiary of my learning process. I have attempted in these pages to link theory with tools in a way that smooths the path for you. Please let me know if I have succeeded. Contact the folks at Stata Press with your feedback—good or bad—and they will pass it along to me.

Why a revised edition?

The first edition of this book was written using Stata 12. The revised edition has been updated for Stata 16. Specifically, chapter 1 includes updated discussions of Stata's interface, datasets, and commands for importing data. Stata's default random-number generator (RNG) changed from the 32-bit KISS RNG to the 64-bit Mersenne Twister RNG in Stata 14. Therefore, simulated datasets for examples in chapters 3, 5, 7, and 10 have changed. Results of these examples, and in some cases the random-number seed used for reproducibility, have been updated. Finally, chapter 11 was updated with brief overviews of time-series features that have been added since Stata 12.

Who should read this book?

Stata users trying to figure out Stata's time-series tools. You will find detailed descriptions of the tools and how to apply them combined with detailed examples and an intuitive explanation of the theory underlying each tool.

Time-series researchers considering Stata for their work. Each commercial time-series package takes a different approach to characterizing time-series data and models. Stata's unique approach offers distinct advantages that this book highlights.

Researchers who know a bit about time series but want to know more. The gestalt of time-series analysis is not immediately intuitive, even to researchers with a deep background in other statistical techniques.

Researchers who want more extensive help than the manual can provide. It is clear and well written, but, at the end of the day, it is a manual, not a tutorial.

How is this book organized?

Like Gaul, this book is divided into three parts.

Preliminaries. Preparation for reading the rest of the book.

> **Chapter 1: Just enough Stata.** A quick and easy introduction for the complete novice. Also useful if you have not used Stata for a while.

> **Chapter 2: Just enough Statistics.** A cheat sheet for the statistical knowledge assumed in later chapters.

Filtering and Forecasting. A nontechnical introduction to the basic ways to analyze and forecast time series. Lots of practical advice.

> **Chapter 3: Filtering time-series data.** A checklist of questions to answer before your analysis. The four components of a time series. Using filters to suppress the random noise and reveal the underlying structure.

Chapter 4: A first pass at forecasting. Forecast fundamentals. Filters that forecast.

Time-series models. Modern approaches to time-series models.

Chapter 5: Autocorrelated disturbances. What is autocorrelation? Regression models with autocorrelation. Testing for autocorrelation. Estimation with first-order autocorrelated data.

Chapter 6: Univariate time-series models. The general linear process. Notation conventions. The mixed autoregressive moving-average model. Stationarity and invertibility.

Chapter 7: Modeling a real-world time series: The example of U.S. gross domestic product. Getting ready to model a time series. The Box–Jenkins approach. How to specify, estimate, and test an autoregressive moving-average model. Forecasting with autoregressive integrated moving-average models. Comparing forecasts.

Chapter 8: Time-varying volatility: Autoregressive conditional heteroskedasticity and generalized autoregressive conditional heteroskedasticity models. Examples of time-varying volatility. A model of time-varying volatility. Extensions to the autoregressive conditional heteroskedasticity model.

Chapter 9: Models of multiple time series. Vector autoregressions. A vector autoregression of the U.S. macroeconomy. Cross correlations, causality, impulse–response functions, and forecast-error decompositions. Structural vector autoregressions.

Chapter 10: Models of nonstationary time series. Trends and unit roots. Cointegration. From intuition to vector error-correction models.

Chapter 11: Closing observations. Making sense of it all. What did we miss?

Ready, set, . . .

I am a reporter. I am reporting on the work of others. Work on the statistical theory of time-series processes. Work on the Stata statistical package to apply this theory. As a reporter, I must give you an unvarnished view of these topics. However, as we are frequently reminded in this postmodern world, none of us can be completely objective, try as we will. Each of us has a perspective, a slant informed by our life experiences.

Here is my slant. I was trained as an academic economist. I became a software developer to pay my way through graduate school and found I liked the challenges of good software design as much as I liked economic research. I began my postgraduate career in academics, transitioned to the Federal Reserve System, and eventually ended up in research in the financial services industry, where I have worked for a number of leading firms (some of them still in existence). I believe I have learned something valuable at each stage along the way.

For the purposes of this book, the most important experience has been to see how statistical research, good and bad, is performed in academics, the Fed, and industry. Good academic research applies cutting-edge research to thorny problems. Bad academic research gets caught up in footnotes and trivia and loses sight of real-world phenomena. The Federal Reserve produces high-quality research, frequently published in the best academic journals. A signature characteristic of research within the Fed is a deep knowledge of the institutional details that can influence statistical relationships. However, Fed research occasionally exhibits an oversimplified perspective of the workings of the financial services industry. Industry has to make decisions in real time. Accordingly, industry research has to generate answers quickly. Good industry research makes wise tactical choices and selects reasonable shortcuts around technical obstacles. Bad industry research is "quick and dirty".

Embrace the good, avoid the bad. Perhaps because the latter half of my career has been spent in industry, my personal bent is to recognize the limitations of the tools I use without becoming distressed over them. I am more interested in intuition than in proofs.

Here are three articles that sum up the approach I try to emulate:

- Diaconis, P. 1985. Theories of data analysis: From magical thinking through classical statistics. In *Exploring Data Tables, Trends, and Shapes*, ed. D. C. Hoaglin, F. Mosteller, and J. W. Tukey, 1–36. New York: Wiley.

- Ehrenberg, A. S. C. 1977. Rudiments of numeracy. *Journal of the Royal Statistical Society, Series A* 140: 277–297. https://doi.org/10.2307/2344922.

- Wainer, H. 1984. How to display data badly. *American Statistician* 38: 137–147. https://doi.org/10.2307/2683253.

Do not say I did not warn you. Now get cracking and learn some stuff.

Acknowledgments

Just once, I would like to see an author say, "I did it all on my own, with no help from anyone else". Of course, it would not be true, but it would be fun.

Craig Hakkio of the Federal Reserve Bank of Kansas City generously provided me with essential reference material I could not lay my hands on elsewhere. He also gave me his unvarnished opinions on some early chapters. Thank you.

The editors at Stata Press were a joy to work with, although at times when I was tired, it seemed that they had too many good ideas. The tight structure of chapter 5 owes much to David Drukker's help. Brian Poi saved me from myself more times than I can count. His suggestions were practical, reasonable, and expressed without giving any sign that he recognized how slipshod I had been. Much appreciated. Also appreciated is Stata Press's patience with a very slow author. In my defense, I believe I still came in well under the time it took Stata to finally release a version with time series.

I would like to thank the Keurig company, without whose excellent coffee maker this book would never have been finished. As I frequently tell my staff, sleep is overrated.

The Internet provided musical inspiration throughout. The careful reader should be able to detect which sections were written while listening to Glenn Gould's early recording of the Goldberg Variations and which were written while listening to AC/DC's "Highway to Hell".

I dedicate this book to my extraordinary wife, Linda. She bore the brunt of my extended mental and physical absences for writing. Without her love and support, there would be no point to all this.

1 Just enough Stata

Chapter map

1.1 Getting started. Starting and stopping. Navigating the interface. Getting help. The gestalt of Stata.

1.2 All about data. Getting data in. Getting data out. Doing things to data. Seeing what you have.

1.3 Looking at data. Exploring your data by looking at simple statistics. Looking at graphs of your data.

1.4 Statistics. Basics. Estimation.

1.5 Odds and ends. Repeating yourself. Matrices. Randomness.

1.6 Making a date. How Stata thinks about time series. Tools for handling dates and times.

1.7 Typing dates and date variables. Literally typing a date. Time-series operators.

1.8 Looking ahead. Statistical background.

If you have never used Stata before, this chapter is for you. Reading this chapter will not make you a Stata expert, but it will teach you just enough Stata to read the rest of this book without any trouble. You will be able to follow the examples, rerun them (using the datasets available on the Stata Press website), and change them around however you wish.

You may find this chapter helpful even if you have used Stata already, particularly if you have never used Stata's time-series features. The treatment of time series is sufficiently different from the rest of Stata so you will benefit from a little extra explanation.

Do not be daunted by the length of this chapter. It's 99.99% simple examples that illustrate Stata features you will encounter later in the book. In other words, lots of pictures and no math. I recommend you try out the examples as you read along. Of course, you will need access to a computer with Stata already installed on it.[1]

---

1. If you do not have a copy of Stata yet, the folks at StataCorp will be glad to help you. You can order a copy and get additional information at https://www.stata.com.

1.1 Getting started

1.1.1 Action first, explanation later

Start Stata. Type `sysuse auto` in the Command window and press *Enter*.[2] In the Results window, you will see

```
. sysuse auto
(1978 Automobile Data)

.
```

The first line is an echo of the command you typed. It follows the Stata prompt, a period (`.`) at the beginning of the line that indicates Stata is ready for your command. In this example, you asked Stata to load into memory one of the example datasets (`auto.dta`, in this case) that come with Stata.

The second line is a message from Stata. A descriptive label—`(1978 Automobile Data)`—is attached to `auto.dta`, and Stata displays this label as it loads the dataset. The next line is blank, separating this command and its results from what comes next; finally, Stata displays the period prompt indicating that it is ready and waiting for your next command.

Let's find out what is in `auto.dta`. Type `describe` and press *Enter*.[3] Now you see

```
. describe
Contains data from /Applications/Stata/ado/base/a/auto.dta
  obs:            74                          1978 Automobile Data
 vars:            12                          13 Apr 2018 17:45
                                              (_dta has notes)

              storage   display    value
variable name   type    format     label      variable label

make            str18   %-18s                 Make and Model
price           int     %8.0gc                Price
mpg             int     %8.0g                 Mileage (mpg)
rep78           int     %8.0g                 Repair Record 1978
headroom        float   %6.1f                 Headroom (in.)
trunk           int     %8.0g                 Trunk space (cu. ft.)
weight          int     %8.0gc                Weight (lbs.)
length          int     %8.0g                 Length (in.)
turn            int     %8.0g                 Turn Circle (ft.)
displacement    int     %8.0g                 Displacement (cu. in.)
gear_ratio      float   %6.2f                 Gear Ratio
foreign         byte    %8.0g      origin     Car type

Sorted by:  foreign
```

---

2. If none of this makes sense yet, just read along. We will explain more in the next subsection.
3. From here on, I assume you will type the commands you see in the examples, that is, the text following the Stata prompts.

Hmmm, lots of information here. Seventy-four observations on twelve variables. That descriptive data label appears again near the top right. There is a list of each of the 12 variables with indications of how Stata is storing the information, how Stata intends to format any values it displays, and a descriptive label for each variable. Finally, it appears the data are already sorted by one of the variables (`foreign`).

Let's look at the contents of a few of these variables.

```
. list make price mpg foreign in 1/5
```

	make	price	mpg	foreign
1.	AMC Concord	4,099	22	Domestic
2.	AMC Pacer	4,749	17	Domestic
3.	AMC Spirit	3,799	22	Domestic
4.	Buick Century	4,816	20	Domestic
5.	Buick Electra	7,827	15	Domestic

Those 1978 prices look pretty good.

Now let's get some simple distributional information about these data.

```
. summarize
```

Variable	Obs	Mean	Std. Dev.	Min	Max
make	0				
price	74	6165.257	2949.496	3291	15906
mpg	74	21.2973	5.785503	12	41
rep78	69	3.405797	.9899323	1	5
headroom	74	2.993243	.8459948	1.5	5
trunk	74	13.75676	4.277404	5	23
weight	74	3019.459	777.1936	1760	4840
length	74	187.9324	22.26634	142	233
turn	74	39.64865	4.399354	31	51
displacement	74	197.2973	91.83722	79	425
gear_ratio	74	3.014865	.4562871	2.19	3.89
foreign	74	.2972973	.4601885	0	1

The average car price is $6,165 with an average mileage of 21 MPG (but at least one car gets 41 MPG). Not bad for 1978.

There are some mysteries here. How come there are no observations for the variable `make`? We saw information on `make` listed by the previous command. And it looks as if the variable `foreign` is really a number ranging between 0 and 1, even though it was displayed as text (`Domestic`) when we listed it. We will worry about this later. For now, let's push on.

I wonder how many of the cars in this dataset are domestic and how many are foreign.

```
. tabulate foreign

    Car type |      Freq.     Percent        Cum.
-------------+-----------------------------------
    Domestic |         52       70.27       70.27
     Foreign |         22       29.73      100.00
-------------+-----------------------------------
       Total |         74      100.00
```

In 1978, foreign cars were actually foreign, that is, not produced in the United States. Typically, they were smaller than domestic cars and got better mileage. Let's check.

```
. table foreign, contents(mean mpg) row

----------------------
    Car type | mean(mpg)
-------------+--------
    Domestic |  19.8269
     Foreign |  24.7727
             |
       Total |  21.2973
----------------------
```

I wonder if that difference in MPG is statistically significant.[4]

```
. test mpg
last estimates not found
r(301);
```

Well now I have done something wrong.[5] Stata has printed an error message (`last estimates not found`) and a return code (`r(301)`). Oh, I forgot to indicate the variable that defines the two populations. Let's see if this works.

```
. test mpg, by(foreign)
option by() not allowed
r(198);
```

---

4. I know—this does not exactly seem like a random sample from a large population with potentially different distributions of some characteristic across identifiable subpopulations. But we're not doing statistics here; we're learning Stata. Go with the flow.

5. Actually, I have done something right—I guessed. For a complex and powerful tool, Stata actually is reasonably intuitive. Also it's pretty difficult to break something, so it's usually a good strategy when you are learning a new feature to start by guessing and seeing what happens.

That did not work either. I guess I will have to search Stata's interactive help system or read the manual. I will save you that step for now and just show you what I should have typed.

```
. ttest mpg, by(foreign)
Two-sample t test with equal variances
```

Group	Obs	Mean	Std. Err.	Std. Dev.	[95% Conf. Interval]
Domestic	52	19.82692	.657777	4.743297	18.50638 21.14747
Foreign	22	24.77273	1.40951	6.611187	21.84149 27.70396
combined	74	21.2973	.6725511	5.785503	19.9569 22.63769
diff		-4.945804	1.362162		-7.661225 -2.230384

```
       diff = mean(Domestic) - mean(Foreign)                      t =  -3.6308
Ho: diff = 0                                     degrees of freedom =        72

     Ha: diff < 0                 Ha: diff != 0                 Ha: diff > 0
 Pr(T < t) = 0.0003       Pr(|T| > |t|) = 0.0005         Pr(T > t) = 0.9997
```

This *t* test indicates the difference is highly significant. However, this version of the *t* test assumes the variance of MPG is identical in the two groups (see the reminder just below the **ttest** command). I wonder if that assumption is appropriate.

```
. table foreign, contents(mean mpg sd mpg) row
```

Car type	mean(mpg)	sd(mpg)
Domestic	19.8269	4.743297
Foreign	24.7727	6.611187
Total	21.2973	5.785503

```
. ttest mpg, by(foreign) unequal
Two-sample t test with unequal variances
```

Group	Obs	Mean	Std. Err.	Std. Dev.	[95% Conf. Interval]
Domestic	52	19.82692	.657777	4.743297	18.50638 21.14747
Foreign	22	24.77273	1.40951	6.611187	21.84149 27.70396
combined	74	21.2973	.6725511	5.785503	19.9569 22.63769
diff		-4.945804	1.555438		-8.120053 -1.771556

```
       diff = mean(Domestic) - mean(Foreign)                      t =  -3.1797
Ho: diff = 0                     Satterthwaite's degrees of freedom =   30.5463

     Ha: diff < 0                 Ha: diff != 0                 Ha: diff > 0
 Pr(T < t) = 0.0017       Pr(|T| > |t|) = 0.0034         Pr(T > t) = 0.9983
```

The standard deviation of MPG is 40% higher among the foreign cars, so we reran the *t* test. But this did not change the answer much.

I mentioned that foreign cars tended to be smaller than domestic cars in 1978. Maybe the difference in MPG is attributable to the weight of the car rather than the country of manufacture.

```
. scatter mpg weight, by(foreign)
```

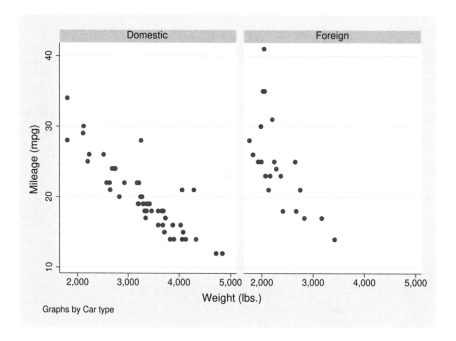

Figure 1.1. The relationship between automobile mileage and automobile weight

This could go on forever, but by now, you should have a feel for what a Stata session is like. We will stop playing with data for a bit and start explaining Stata in more detail.

1.1.2 Now some explanation

You start Stata the same way you start any other program on your computer. In other words, the way you start Stata depends on the computer you use. I use Stata on a Windows computer at work and on a Mac at home. I use Stata frequently enough that I have its icon in the program tray at the bottom of the screen on my Windows computer. Sometimes I forget and start Stata the old-fashioned way—by clicking on the **Start** button in the lower left corner of the screen and clicking through until I find Stata.

On my Mac, I start Stata by clicking on the Stata icon on the Dock. I could also start Stata by using Finder to navigate to Stata in the Applications folder, then double-clicking on the Stata icon.

1.1.3   Navigating the interface

What you see next depends on your computer and whether someone has configured the appearance of Stata in any special way. The five main windows are the History, Results, Command, Variables, and Properties windows. The largest window is the Results window, and it is stacked on top of the small Command window. The third window, labeled History, appears as a sidebar on the left. The Variables and Properties windows are stacked as a sidebar to the right. Figure 1.2 illustrates this appearance.

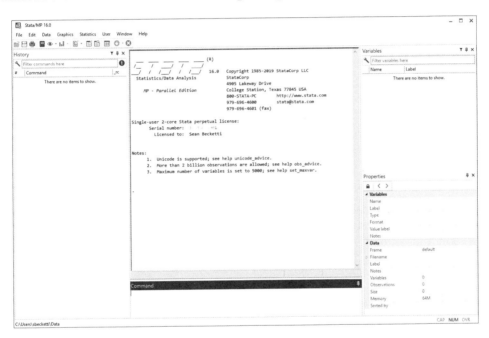

Figure 1.2. Stata for Windows opening screen

As you have probably already guessed, you type your commands in the Command window, and the results of your commands appear in the Results window. (Graphs appear in a separate Graph window. More on that later.)

Let's take a closer look at figure 1.2. As you can see, I am using Stata/MP 16. If you are using a different version, the name of your version will appear instead. The Stata menu appears below the title bar. The menu begins with some familiar choices: **File**, **Edit**, **Data**, **Graphics**, **Statistics**, **User**, **Window**, and the ever-popular **Help**. The Stata menu provides an easy way to learn Stata. Virtually everything that Stata can do can be done through the menu system. We will provide some examples in a bit, but let's finish looking at figure 1.2 first.

The toolbar at the top of the Stata window displays several icons. At the left of the toolbar, there are three icons: open a file (which can contain a Stata dataset, a script of Stata commands, and a few other types of things), save the dataset you are using currently, and print your results.

The next seven icons are the following:

Log opens a submenu for starting, stopping, and viewing log files, that is, records of your Stata session.

Viewer opens a new window that can display log files or the results of a `help` command.

Graph opens the Graph window if you have already created one or more graphs. Otherwise, the icon is grayed out.

Do-file Editor opens a window for editing a script of Stata commands. These scripts end in the file extension `.do` (for example, `myscript.do`) and are called do-files. To run the script `myscript.do`, you would type `do myscript` in the Command window.

Data Editor opens a window that looks a bit like a spreadsheet. The Data Editor provides a convenient tool for editing the dataset currently in use. You can add variables, drop variables, add observations, change the values of individual observations, and change the names or other properties of variables.

Data Browser is a "safe" version of the Data Editor. You can view, but not change, your data. As a convenience, you can switch between edit and browse modes at any time in both the Data Editor and the Data Browser.

Variables Manager opens a window for managing the properties of variables.

There are two remaining icons—a circle with a down arrow labeled "Show more results" and an "X" icon labeled "Break". By default, Stata will not pause in its display of results. You can tell Stata to pause once the screen is full by typing `set more on` in the Command window. Clicking on the **More** icon tells Stata you are done reading the current screen and you are ready for more. You can also tell Stata to display another screenful of output by pressing (almost) any key. If you press the *Return* key, Stata will display one additional line of output rather than a screenful. Clicking on the **Break** icon will interrupt Stata and return control to you—very handy when you have accidentally generated a lot more output than you expected. And, yes, there are keyboard shortcuts for **Break** as well, but they vary by computer type. Try what usually works for you. On my Mac, I use Command-. (that is, I hold down the *Command* key while pressing the *period* key).

From the **Help** menu, Stata provides a search feature to easily access Stata's vast documentation system. To access this search feature, click on **Help** from the menu

and then select **Search....** For instance, click on the *Search all* radio button and type
`regress` in the search window and then press the *Enter* or *Return* key (or whatever it's
called on your keyboard). A Viewer window that looks something like figure 1.3 will
pop up.

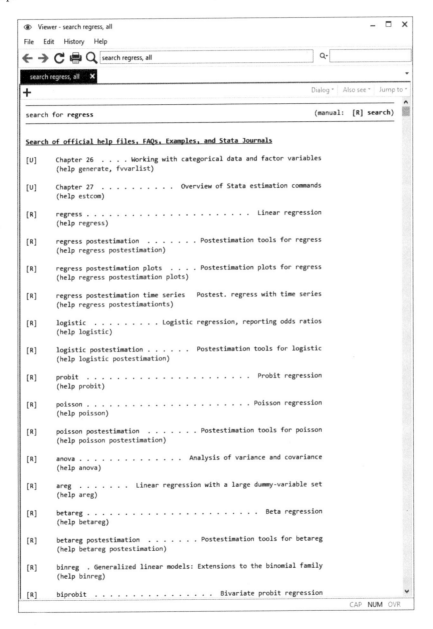

Figure 1.3. Viewer window displaying `search` results

In this example, Stata searched for the keyword "regress" in 1) the official help files, videos, blog post, frequently asked questions, examples, and the *Stata Journal* and 2) web resources from Stata and other users. The results returned by the `search` command are quite extensive, and the screenshot displays only a fraction of them. The third item in the results list is an entry from the *Stata Base Reference Manual* that describes the `regress` command. If you click on the highlighted word "regress" in that listing, the viewer will display the help file for the `regress` command. You can search for things more complicated than specific commands. For example, type `tests of normality` and see what you get.

If you wish, you can search Stata's documentation by typing the `search` command in the Command window rather than typing a keyword in the search window. The result will be identical either way. (Try it. Type `search regress` in the Command window.) Stata also offers a `help` command. I generally use the `help` command to get information about a specific command. To see the difference between `search` and `help`, type `help regress`. Then type `help help`. Then type `help search`.[6] There is a lot of useful information there, so make a mental note to come back to these screens when you get stuck. Because this chapter is intended to provide you just enough Stata to get started, I frequently will suggest you type `help` *this* or `search` *that* for more detailed information.

---

6. Type `help keyboard` to learn about useful keyboard shortcuts.

As I mentioned above, almost every Stata command can be launched from the menu system. Let's look at one example. On the menu bar, click on the word **Data**. The **Data** menu will drop down. The first item is **Describe data**. Click on it and a submenu will appear. Your screen should look something like figure 1.4.

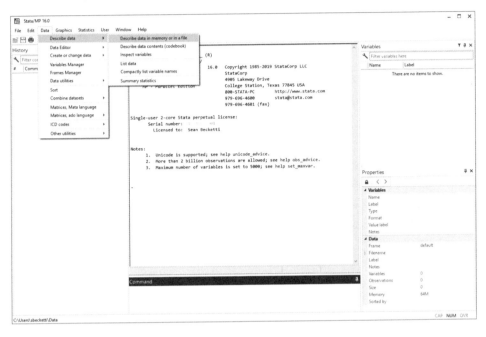

Figure 1.4. **Data** menu

Click on the item **Describe data in memory or in a file**. You should see a dialog box that looks something like figure 1.5.

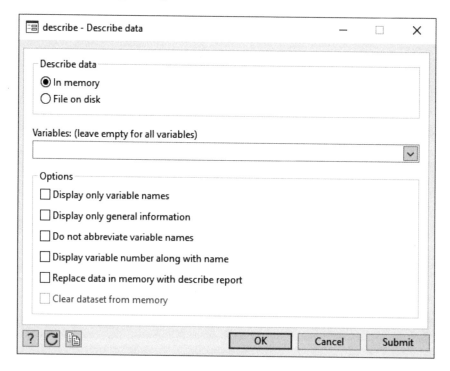

Figure 1.5. The `describe` dialog box

Press *Return* or click on **OK**, and you will see the following in the Results window:

```
. describe
Contains data
   obs:              0
  vars:              0
Sorted by:
```

Accurate, but not very interesting.[7] Stata echoes the command you could have typed in the Command window (`describe`), executes the command, and displays the results of the command. (The period [.] before the `describe` is the Stata prompt, the indication that Stata is ready for you to type another command.) In this case, Stata tells you that you have no observations and no variables in your dataset yet.

---

7. I assume you are following along in a fresh Stata session. If you still have the 1978 Automobile Data in memory, I applaud your stick-to-itiveness. However, the results of these menu actions will look different—but familiar.

From the screenshots, you can see that there are many commands to manage your data and many options within those commands to give you minute control over your data. (Click on some other commands to see what the dialog boxes offer.) Some users rely on the menu system for all their work in Stata. Others use the menu system as a set of "training wheels" for Stata: they use the menu system to familiarize themselves with the commands they need, then start typing the commands directly in the Command window, relying on the menu system only for unfamiliar commands as they gain experience. Both approaches work fine. You should use the approach that suits you best. I am old fashioned; I type all my commands in the Command window. When I need to learn about an unfamiliar feature of Stata, I type the `help` command to get the information I need. But that is just my style.

Before we go any further, you should learn how to end a Stata session. Your computer probably has methods for exiting programs, and you can use those. Stata also provides the `exit` command to end your session. If you have changed the dataset in Stata and you have not saved a copy yet, Stata will refuse to exit. The Results window will display the following information:

```
. exit
no; dataset in memory has changed since last saved
    Save the data or specify option clear to exit anyway.
r(4);
```

Stata gave the error message `no; dataset in memory has changed since last saved` and a return code (`r(4);`). For now, the only thing you need to learn from the return code is that Stata would not carry out your command. If you do not care to save the changes you have made, just type `exit, clear`, and Stata will exit without a complaint.

Now you can start and stop a Stata session, and you can find your way around the user interface. We are just about ready to dig into the details of Stata, but before we do, I would like to give you some tips that make it easier to understand Stata.

1.1.4   The gestalt of Stata

All programs have a certain logic to them, a preferred way of operating, and Stata is no exception. Once you get the hang of Stata's view of the world, using Stata and learning new features is easy. Here are some simple things to remember:

Stata is designed as an interactive program. As you saw above, you type a command (or use the menu system to type it for you), then Stata executes it and displays the results. You can collect long sequences of commands in do-files and execute all of them at once—particularly useful if you want to do the same sequence of operations to more than one dataset—but Stata still processes the commands as though you are typing them one at a time.[8]

Stata commands act on the current dataset, the data in memory. This convention contrasts with some other programs you may have used that operate directly on data stored in files on disk. Stata can import data from files and export data to files, but for the most part, the focus is on whatever data happen to be in memory at the time.[9] This convention may seem limiting—the size of your dataset is limited to what will fit in Stata's memory—but with modern computers, this limit is rarely binding.[10]

Stata commands act on the data rectangle. I think of the current dataset as the data rectangle (my own term)—essentially a spreadsheet where each column contains one variable and each row contains one observation or case. (And that is how it looks in the Data Editor.) You can add clauses to Stata commands to restrict the scope of the command to an arbitrary subset of the data rectangle, but by default, commands typically apply to the entire data rectangle.

You can limit the scope of a command without trimming your dataset. In some other statistical programs, you create subsets of your data, then execute your statistical command (regression, analysis of variance and covariance, etc.) on the subset in a subsequent step. In Stata, you can easily limit the scope of a command to a subset. This ability cuts out a lot of housekeeping steps in your analysis.

All Stata commands have the same structure (except when they do not). By design, Stata commands obey a uniform syntax. This uniformity makes it easy to learn unfamiliar Stata commands because their syntax is the same as the commands you already know. If you know (or can guess) the name of the command you need, you pretty much know how to specify the command. This feature of Stata turns out to be one of the most useful and powerful characteristics

---

8. You can also turn sequences of Stata commands into Stata programs, which are stored in ado-files, for instance, `myprog.ado`. ("ado" stands for automatically loaded do-file.) To the user, these Stata programs look just like built-in Stata commands. In fact, a lot of the Stata commands covered in this book are actually Stata programs.

9. In fact, Stata can hold multiple datasets in memory at once, storing each one in a separate data frame. One frame is selected as the current frame, and Stata commands act on the data in this frame. Throughout this book, we load only one dataset in memory at a time. Thus, our data are always stored in the current frame. Type `help frames intro` for information on working with multiple datasets in memory.

10. By default, the maximum number of variables is 2,048 in Stata/IC and 5,000 in Stata/SE and Stata/MP, but these maximums can be increased in Stata/SE and Stata/MP. In principle, Stata/IC and Stata/SE can hold a maximum of 2,147,483,647 observations, and Stata/MP can hold a maximum of 1,099,511,627,775 observations. The practical limit, which depends on your computer's capacity, is lower. Type `help memory` for details.

of the system. There are exceptions (hence, the warning in parentheses above), but they do not lessen the importance of the standard Stata syntax.

These points may not seem earth-shatteringly important to you, but they are the secret to Stata's usefulness as a tool for data exploration. A lot of thought and work went into the design of Stata's human-machine interface with the goal of making the computer "disappear" as you are doing data analysis. As you become more familiar with Stata, you will forget about the computer and the commands and will come to feel as if there is no barrier between you and the data you are examining. I know this is a bold claim, but I speak from experience.

1.1.5   The parts of Stata speech

Because the Stata command syntax is an essential element—perhaps *the* essential element—of Stata's ease of use, let's spend a little time looking at the details.

The essential "parts" of Stata "speech" are arranged as follows:

$\big[\, prefix\, :\, \big]$ *command* $\big[\, varlist\, \big]$ $\big[\, =exp \big]$ $\big[\, if\, \big]$ $\big[\, in\, \big]$ $\big[\, weight\, \big]$ $\big[\, \text{using } filename \big]$ $\big[\, ,$
$options\, \big]$

You replace the items in italics above (*command*, for instance) with whatever makes sense (for example, `describe` or `exit`). The square brackets ([ ]) enclose items that are optional. (Do not type the square brackets.) Type the rest—in this case, the ":" after the *prefix*, the "=" before the *exp*, the word "`using`" before the *filename*, and the "," before the *options*—exactly as they appear.

As the syntax diagram above indicates, sometimes you need to type only a command name (`exit`), while at other times you may need to add something (`exit, clear`). And, of course, commands typically do not incorporate all of these elements, just the ones that make sense for that command.

Let's take the building blocks of a Stata command one at a time.

- A *prefix*—usually, but not always, followed by a colon (`:`)—is a "minicommand" that operates on the Stata command that follows it. In this book, we use the `quietly` prefix command, which suppresses the display of any results, a handy feature when you want Stata to do something to your data but you want to avoid cluttering the screen with an excessive amount of output. For instance,

    `. quietly regress y x1 x2 x3 x4 x5 x6 x7 x8 x9 x10`

  estimates a linear regression of the variable `y` on all of those `x`'s but does not display the regression output.[11] The `regress` command still stores information about the estimated regression for later use, such as calculating predicted values.

---

11. `quietly` is one of the prefixes that is not followed by a colon.

- *command* is the easiest one. It's the name of the command you want Stata to execute. In the beginning, Stata tried to make every command name a simple English language verb that described the action you wanted Stata to take: `describe`, `exit`, `summarize`, `regress`, etc. As Stata added functionality—and as some commands became more sophisticated and subtle—it became challenging to find verbs to describe the desired actions; but Stata still attempts to make the command names as readable as possible.

- *varlist* is a list of variable names. In many cases, omitting the *varlist* indicates you would like the command to apply to all the variables in the current dataset. In these cases, the *varlist* provides a way to restrict the scope of the command to a subset of the variables in the current dataset.

  The rules for naming variables in Stata are straightforward. Variable names consist of 1 to 32 letters (`A-Z` and `a-z`), digits (`0-9`), and underscores (`_`). The first character must be either a letter or an underscore, but you should avoid starting your variable names with an underscore. Stata uses the underscore a lot, and you could run into trouble. Stata is case sensitive, so `myvar`, `Myvar`, and `MYVAR` are three distinct names.

  In Stata commands that include *varlists*, the variables in the list must all be either existing variables or new variables—no mixing existing and new variables in the same list.

- The `=`*exp* clause is used by commands that need to assign the value of an expression (*exp*) to a variable. For instance, the `=`*exp* clause appears in the `generate` command, which creates new Stata variables.

      . generate sinofx = sin(x)

  In this example, there is an existing variable called `x`. The `generate` command calculates the sine of `x` (`= sin(x)`) and stores it in a new variable called `sinofx`.

  Stata can handle complex expressions, and many useful mathematical, statistical, string, and other functions are built-in features of Stata. Type `search exp` for more information about expressions and the list of functions provided by Stata.

- The `if` and `in` qualifiers allow you to restrict the scope of a command to a subset of the observations in the current dataset.

  The `in` qualifier is the simpler of the two. It restricts the observations (the rows in the data rectangle) that are affected by your command to a contiguous range (which can be as short as a single observation). For instance,

      . generate sinofx = sin(x) in 12/25

calculates the sine of x in observations 12 through 25 only and stores them in the new variable sinofx. All other observations of sinofx—observations 1 through 11 and 26 through the last observation in the current dataset—are set to a *missing value*, a special value that is displayed as a period (.).[12]

You can use the in qualifier to specify a single observation

```
. replace sinofx = sin(y) in 5
```

You can type the letters f and l to indicate the first and last observations, respectively, in the current dataset.

```
. list x sinofx in f/12
. summarize x in 23/l
```

And you can use negative integers to specify the distance from the end of the current dataset. For instance,

```
. list in -5/l
```

displays the last five observations of all the variables in the current dataset.

The if qualifier allows you to restrict the scope of a command to *any* subset of observations; they do not need to be contiguous in the data rectangle. You specify any allowable Stata expression after the if. The expression is evaluated separately for each observation in the current dataset, and the Stata command is applied to the observations for which the expression is true.

Stata does not include a separate Boolean (that is, logical) variable type. In Stata, zero is interpreted as false and any nonzero value is interpreted as true, a convention that turns out to be handy when manipulating data. For instance, instead of typing if x != 0 (which is true whenever x is not equal to 0), you can type if x. This statement evaluates to "true" whenever x is any value except 0.[13]

Here are some examples:

```
. generate psinofx = sin(x) if x>0
. generate nonzerox = x if x /* x==0 is interpreted as false */
. regress wage education experience expsq if state=="CA" & gender=="Male"
> & education>8
```

I snuck in a few new features in these examples. First, the text between the characters "/*" and "*/" is a comment. Once Stata detects the begin comment characters (/*), it ignores everything that is typed until it detects the end comment characters (*/). These comments can span multiple lines, making it easy to add detailed documentation to your do-files. You can also type single-line comments. If you begin a line with an asterisk (*), the entire line is treated as a comment. And you can begin a partial-line comment with the characters "//", so our example above could have been typed as

---

12. Missing values are extremely important. They prevent all sorts of inadvertent mathematical errors. You can type a period (.) in any expression to represent a missing value—more useful than you might think. Type help missing for more information.

13. I come from the school of lazy typists, so I tend to type if x. However, if others need to read and understand my code, I usually type if x != 0 for clarity.

```
    * x==0 is interpreted as false
    . generate nonzerox = x if x
```

or

```
    . generate nonzerox = x if x // x==0 is interpreted as false
```

The second novelty was the introduction of two string variables, state and gender in the regress command. Stata provides five numeric variable types: three integer types—from the smallest to largest range covered: byte, int, and long—and two floating-point types (that is, numbers with fractional parts)—float and double. Stata also provides string variables, which can hold between 1 and 2,045 characters, and strLs, which can hold strings of up to 2,000,000,000 characters. Type help data types for details.

Finally, you are getting your first look at some of the operators used in Stata expressions. Here is a cheat sheet, but you can type help operators anytime you want to refresh your memory.

Table 1.1. Stata operators

+	addition	&	and	>	greater than
-	subtraction			or	<	less than
*	multiplication	!	not	>=	greater than or equal
/	division	~	not	<=	less than or equal
^	power			==	equal
-	negation			!=	not equal
+	string concatenation			~ =	not equal

- The *weight* clause specifies the weights you want to apply. We do not use weighted data in this book; it does not crop up a lot in the types of time-series analyses we cover.

To perform a weighted analysis, add the clause [*weight_type* = *exp*] to your command. In this case, you must type the square brackets; they are part of the command. Stata recognizes four types of weights (*weight_type*):

1. *Frequency weights* (typed as fweight or abbreviated as fw) indicate the number of duplicated observations represented by a single observation.
2. *Sampling weights* (pweight or pw) denote the inverse of the probability of inclusion for each observation.
3. *Analytic weights* (aweight or aw) are inversely proportional to the variance of an observation.
4. *Importance weights* (iweight or iw) are a Stata placeholder for Stata programmers who need to apply some sort of custom weighting scheme not handled by the other three types of weights. You are unlikely to ever come across iweights.

Your weights need not be stored in a variable. The *weight* clause allows you to define your weights with a general Stata expression. For example,

```
. regress y x1 x2 x3 [aweight = log(income)]
```

- I know I told you that Stata commands act on the dataset in memory, but every so often, you need to refer to a file on disk. The `using` *filename* clause provides the means to do so. Depending on the command, you may or may not need to include the file extension—the characters after the final period in the filename. If your filename contains spaces (`my data file.dta`), you will need to enclose the filename in double quotes (`"`).

- Most commands accept one or more options. Though I am oversimplifying just a bit, there are two types of options in Stata commands: logical options (they turn some feature on or off) and options that take a list of one or more values of some sort. To specify one or more options, place a comma (`,`) at the end of the command, then type your options.

You have already seen a logical option. The `clear` option in the command

```
. exit, clear
```

tells Stata that it can clear (that is, erase) the current dataset if necessary before Stata exits. Similarly,

```
. regress y x1 x2 x3 x4 x5 x6 x7 x8 x9 x10, noconstant
```

tells Stata to omit the constant term in the regression of `y` on `x1` through `x10`.[14]

Options that take lists are typed as the option name followed by the list of items in parentheses. For instance,

```
. arima gdp, ar(4)
```

fits an autoregressive model. The option `ar(4)` tells Stata to include four lags of `gdp` in the model. (We cover this command in detail in chapter 7.)

We saw a more complicated example of options in the opening section of this chapter,

```
. table foreign, contents(mean mpg sd mpg) row
```

which produced a table with rows for each of the two levels of the variable `foreign` (`Domestic` and `Foreign`) and a final `Total` row (specified by the `row` option). The option `contents(mean mpg sd mpg)` specified two columns, one for the mean and the other for the standard deviation of `mpg`.

I mentioned earlier that the syntax of Stata makes it easy to guess at how to use a new command. Options represent the one big exception to that rule. Because they are tailored to each command, you will just have to type `help` *command* to see which options are used by the command you are learning.

---

14. Preceding a logical option by the characters "no" often reverses its meaning.

One final point. Some options are not optional! For instance, in the `newey` command, Stata requires you to specify an option specifying the maximum lag order of autocorrelation.

You will encounter Stata commands that break this mold—that do not adhere to this syntax—but it will not happen often, and there will always be a good reason.

1.2   All about data

We are going to shift gears at this point. In the sections above, we took a little time to get you oriented, help you navigate Stata, and enable you to find additional help for those times you get stuck. But now we have to cover a lot of ground in a short amount of time.[15] Explanations will be brief and focus primarily on the Stata features you need to know to follow along in this book. As always, you can use the `help` and `search` commands to fill in the blanks. And do not underestimate the educational value of just trying things to see what happens.

Okay, let's get going.

A statistics, graphics, and data management program like Stata is not much use unless you can get data into and out of the program easily. You also need ways to manipulate data efficiently. There is no shortage of such tools in Stata. Let's start by looking at the ways to get data into and out of Stata.

For the most part, the examples in this book use data that have been saved in Stata's internal format. You can download the do-files and datasets for this book from within Stata. At the command prompt, type

```
. net from https://www.stata-press.com/data/itsus1r/
. net get itsus1r_files
. net get itsus1r_data
```

Those commands will access the Stata Press website for this book and download all the files into your current directory. The filenames of Stata datasets always end in the file extension `.dta` (for example, `mydata.dta`).

In chapter 9, we also make use of a command I wrote called `varbench` to implement some goodness-of-fit statistics after fitting vector autoregressions. To install it, after typing the previous three commands, type

```
. net install itsus1r_files
```

That will install the `varbench` command in your PLUS directory, where Stata installs user-written commands. To see where your PLUS directory is located, you can type

```
. sysdir
```

---

15. Stata is a big program. The Stata 16 documentation set includes 31 manuals.

Remember that Stata commands act on the current dataset, that is, the data in memory. Thus these data files have to be imported (loaded into memory) before they can be analyzed. To import the data file `mydata.dta`, you would type the command

```
. use mydata
```

Notice that you do not have to type the `.dta` extension, although Stata will not complain if you do.

The syntax of the `use` command is[16]

<u>u</u>se *filename* [, `clear`]

The `clear` option tells Stata to import the data file even if there is already a current dataset that has not been saved to disk yet.

The line under the letter `u` in `use` indicates that you can abbreviate the command to just the first letter. Typing

```
. u mydata
```

has the same effect as typing

```
. use mydata
```

You can abbreviate a lot of things in Stata,[17] and the allowed abbreviations are always underlined in the syntax diagrams. Note that none of the letters in `clear` are abbreviated; you have to type the full option name. As a rule, any dangerous or destructive options (`clear`, `replace`, etc.) must be typed in full.

The inverse of `use` is <u>s</u>ave. In other words, typing

```
. save newdata
```

saves the current dataset to a file in Stata's internal format named `newdata.dta`. If the file `newdata.dta` already exists, Stata will not replace it unless you add the `replace` option.

---

16. Actually, the full syntax includes a `nolabel` option and a form of the command for loading a subset of a data file. Many options are used infrequently or only by advanced users. In the interest of clarity, I will omit them freely. You can always type `help` *command* to see the full list of options for a command.

17. It took a few versions—and some improvements in computer-user interfaces—before Stata included a menu system and command-line editing (type `help keyboard`), and the folks at Stata wanted to economize on typing.

Because `save` is the only way to create a Stata dataset, there has to be some way to populate the current dataset besides the `use` command. I will touch on the basics quickly—you will not need any of this information to read this book—but you will want to read the Stata documentation carefully when you start working with your own data.[18]

You can use the `input` command to type data directly into the current dataset.

```
. input odd even
              odd        even
  1. 1 2
  2. 3 4
  3. 5 6
  4. 7 8
  5. 9 10
  6. end

. describe
Contains data
    obs:              5
   vars:              2

              storage  display   value
variable name   type   format    label      variable label

odd            float   %9.0g
even           float   %9.0g

Sorted by:
      Note: Dataset has changed since last saved.
. list
```

	odd	even
1.	1	2
2.	3	4
3.	5	6
4.	7	8
5.	9	10

```
. outfile using oddeven, replace
```

You can do essentially the same thing by opening the Data Editor (icon at the top of the Results window) and typing your data. I use the `input` command more than you might expect. It's often the fastest way to enter a small amount of data to make a graph.

For larger amounts of data, you will typically want to import data that are already stored in a file, but not in a Stata `.dta` file. Stata provides a dizzying array of methods for importing data. I will demonstrate a couple, then point you to information about the others.

---

18. In addition to the Stata documentation, several useful books on data management are on the Stata Press website.

I have a text file called `oddeven.raw` I would like to import. Let's use the `type` command to list the contents of the file.[19]

```
. type oddeven.raw
        1           2
        3           4
        5           6
        7           8
        9          10
```

Now let's import the data.

```
. clear

. describe
Contains data
  obs:            0
 vars:            0
Sorted by:
. infile even odd using oddeven.raw
(5 observations read)
. list

        even    odd

 1.        1      2
 2.        3      4
 3.        5      6
 4.        7      8
 5.        9     10
```

The file `oddeven.raw` contains no information about its contents, so I was able to inadvertently apply misleading names to the variables. Stata also reads dictionary files, which document the data in the file.

```
. type oddeven.dct
dictionary {
int     odd     "Odd"
int     even    "Even"
}
        1           2
        3           4
        5           6
        7           8
        9          10
```

---

19. In a bit, we will talk about how to point to specific directories or folders when operating on files. In this example, the file `oddeven.raw` happens to be in the current directory, so I do not need to add any information about its location.

```
. infile using oddeven, clear
dictionary {
int     odd      "Odd"
int     even     "Even"
}
(5 observations read)
. list
```

	odd	even
1.	1	2
2.	3	4
3.	5	6
4.	7	8
5.	9	10

We have just scratched the surface. Table 1.2 summarizes the Stata commands for importing and exporting data. As always, type `help` *command* for more information.

Table 1.2. Importing and exporting data

Import	Export	Type of data
input	—	Type data directly into the current dataset
infile	outfile	Free-form or dictionary files
infix	—	Fixed-format files
odbc	odbc	ODBC format data
import delimited	export delimited	Comma-separated or tab-delimited data
import dbase	export dbase	dBase files
import excel	export excel	Microsoft® Excel worksheets
import fred	—	Federal Reserve Economic Data (FRED)
import haver	—	Haver Analytics® database files
import sasxport5	export sasxport5	SAS® XPORT Version 5 Transport files
import sasxport8	export sasxport8	SAS® XPORT Version 8 Transport files
import spss	—	IBM® SPSS Statistics files
import sas	—	SAS® files

I should mention one more approach to importing data. In this book, I frequently create simulated data to illustrate a point. For instance, I might start an example by creating a 100-observation dataset with a variable containing normally distributed pseudorandom numbers.

```
. clear
. set obs 100
number of observations (_N) was 0, now 100
. generate epsilon = rnormal()
```

I mentioned above that you can limit the scope of a command without trimming the current dataset. Nonetheless, sometimes you will want to trim the dataset. For those times, you will need the **drop** and **keep** commands. **drop** specifies either variables or observations to eliminate. **keep** works the opposite way; you specify which variables or observations to keep, and the rest are eliminated. The syntax is identical for both commands, and there are three variations. To eliminate variables, type

drop *varlist*

If you type **drop** _all, Stata will erase the current dataset completely. To eliminate observations, type

drop if *exp*

or

drop in *range* [**if** *exp*]

Once we have imported data into Stata, we need ways to manipulate it. We have already demonstrated the **generate** command, which creates new variables. Stata guesses at the storage type (integer, floating point, string) by the type of expression you assign to the new variable. For instance, typing

```
. generate x = 17.3
```

creates a floating-point variable, while typing

```
generate name = "Sean Becketti"
```

creates a string variable.[20]

Use the **replace** command to modify the contents of an existing variable. The syntax is identical to that of the **generate** command.

Stata provides a rich set of operators and functions you can use in generating new variables. Sometimes, though, you need to create something a little more complicated. For example, in our **auto.dta**, imagine that you want to calculate the average mileages of foreign and domestic cars and store those averages in variables. In other words, for every domestic car, this new variable will contain the mean MPG for domestic cars, and for every foreign car, it will contain the mean MPG for foreign cars. There are lots of ways to do this, but here is the easiest way:

---

20. You can override Stata's default choice of storage type. Type **help generate** for details.

```
. sysuse auto, clear
(1978 Automobile Data)
. egen avgmpg = mean(mpg), by(foreign)
. by foreign: summarize mpg avgmpg
```

| -> foreign = Domestic | | | | | |
Variable	Obs	Mean	Std. Dev.	Min	Max
mpg	52	19.82692	4.743297	12	34
avgmpg	52	19.82692	0	19.82692	19.82692

| -> foreign = Foreign | | | | | |
Variable	Obs	Mean	Std. Dev.	Min	Max
mpg	22	24.77273	6.611187	14	41
avgmpg	22	24.77273	0	24.77273	24.77273

The egen (extended generate) command offers over three dozen specialized functions. Type help egen to see the list.[21]

You can control the format of the values displayed by Stata. We will use avgmpg as an example.

```
. list avgmpg in f
```

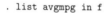

```
. describe avgmpg
```

variable name	storage type	display format	value label	variable label
avgmpg	float	%9.0g		

Just a reminder: you can abbreviate most things in Stata. In this book, we spell out commands and variable names to make the examples more readable, but you will want to take advantage of the abbreviations when you are working with Stata interactively. For instance, that last command can be abbreviated as

```
. d a
```

variable name	storage type	display format	value label	variable label
avgmpg	float	%9.0g		

If more than one variable in this dataset started with the letter a, we would have to type until the abbreviation uniquely identified our variable.

---

21. In the example above, I snuck in by, a prefix command that repeats a command for one or more subpopulations defined by the variables that appear between the by and the colon (:).

Back to our formatting example. I do not think those decimal places are helping us. Let's tell Stata to display values in six-character-wide spaces with no decimal point or fractional parts.

```
. format avgmpg %6.0f
. l a in f
```

Type `help format` to see all the different ways you can format values—and there are many alternatives. We will come back to the `format` command when we talk about date and time variables.

A note that may rub you database types the wrong way. As a rule, you want to keep variable names short but readable. It's a lot easier to type

```
summarize avgmpg
```

than

```
summarize float_average_miles_per_gallon
```

However, you may want to label your data more fully, both to document the data and to make your output more readable to others. The folks at Stata stuck a label on `auto.dta` by typing

```
. label data "1978 Automobile Data"
```

We can label individual variables as well:

```
. label variable avgmpg "Average MPG by origin"
. describe avgmpg
```

variable name	storage type	display format	value label	variable label
avgmpg	float	%6.0f		Average MPG by origin

We can also attach labels to individual values of variables. For example, we could create labels to identify domestic and foreign cars and attach the list of labels to our indicator variable by typing

```
. label define origin 0 Domestic 1 Foreign
. label values foreign origin
```

In fact, that is what the folks at Stata did:

```
. label list
origin:
           0 Domestic
           1 Foreign
. tabulate foreign
```

Car type	Freq.	Percent	Cum.
Domestic	52	70.27	70.27
Foreign	22	29.73	100.00
Total	74	100.00	

```
. tabulate foreign, nolabel
```

Car type	Freq.	Percent	Cum.
0	52	70.27	70.27
1	22	29.73	100.00
Total	74	100.00	

Frequently, you want to review features of the current dataset, especially when the data are new to you. We have already seen a few tools Stata provides for this purpose. The describe command displays an overview of the contents of the current dataset. The list command displays the contents of one or more variables. Some commands do double duty: they are for both inspecting data and calculating simple statistics. For instance, the summarize command calculates ranges, means, variances, and other statistics (type summarize avgmpg, detail to see more), but often is used to get a rough sense of the contents of a variable. The tabulate command used in the example above details the frequency of individual values of a variable—again a handy tool when you are trying to puzzle out the contents of an unfamiliar dataset. Another handy tool is the compare command, which compares the contents of two variables. I often use compare to verify that a calculation is producing the expected results.

The display command provides a general purpose tool for poking around in data. display has too many capabilities to list here, but a few examples should give you the idea.

```
. display "Hello, world"
Hello, world
. display 5/2
2.5
. display cos(2*_pi)
1

. decode foreign, generate(flabel)
. replace flabel = lower(flabel)
(74 real changes made)
. display _new "The mileage of the 1978 " make[5] " was " mpg[5] " MPG,"
> %3.0f 100 * mpg[5]/avgmpg[5] " percent of the average for " flabel[5] " cars
> in 1978."

The mileage of the 1978 Buick Electra was 15 MPG, 76 percent of the average for
> domestic cars in 1978.
```

You should learn several things from these examples. First, `display` can intermingle text and numbers. Second, `display` can function as a powerful "pocket calculator". (You may never use the calculator app that comes with your computer again.) Third, Stata has some useful built-in constants (`_pi`). Finally, I like to show off from time to time. Just tuck away that last example for the future, when you want one of your scripts to report complicated results in a customized display. Stata can help.

1.3 Looking at data

We are going to leave `auto.dta` behind for now and explore a messier dataset. The 1978 Automobile dataset demonstrates a lot of different aspects of Stata, but the data have been prepared so carefully that you may overlook the tools Stata provides for investigating unfamiliar data. To follow along in this section, you need to download an example dataset from the Stata Press website. I showed you how to do that in section 1.2.

When you import data, Stata has to know where to look for the data file. If you do not provide any information besides the filename, Stata will look in the current directory or folder, which may be a default location or may be the last folder you reached as you navigated around your computer. So what is your current folder? You can find out by typing `pwd`.[22]

```
. pwd
/Users/sbecketti/Data/itsus1r/juststata
```

As I write this chapter, I am preparing my examples in a folder called `/Users/sbecketti/Data/itsus1r/juststata`.[23] But what if I want to import a file from another directory? In that case, I type the *fully qualified filename*, that is, the full location of the file.

Let's import a file containing data I downloaded from the World Bank website. The file is not in the current directory, so I type

```
. use ${ITSUS_DATA}/wbdata
```

---

22. No, Stata does not have something against vowels. This command is an acronym for print working directory. As it happens, the Unix command for displaying the name of the current directory also is `pwd`. (Used in Unix first.) Several commands for rooting around in the file system reuse the names of Unix or DOS commands. For instance, you can type either `ls` (Unix) or `dir` (DOS) to list the names of the files in the current directory.

23. I keep a wide variety of data files in folders inside `/Users/sbecketti/Data`. `itsus1r` is shorthand for *Introduction to Time Series Using Stata, Revised Edition*, and `juststata` is this chapter.

What the heck is that? It does not look like a filename. The string ITSUS_DATA is a Stata macro, that is, a name that stands in place of some other text. I have stored all the data files used in this book in a separate directory, and I have stored the name of that directory in the macro ITSUS_DATA. To tell Stata that ITSUS_DATA is a macro (and to replace the macro with the text it stands for), I preceded the macro name with a dollar sign ($).[24,25]

```
. display "${ITSUS_DATA}"
/Users/sbecketti/Data/itsus1r/data
```

In this instance, the curly braces ({ and }) are not necessary, but they can be used to make it clear where the macro begins and ends in cases when Stata might get confused.

I do not know how your computer is organized or where you are likely to store the downloaded data files for this book, so most of the examples start with

```
. use ${ITSUS_DATA}/example-file-name
```

To follow along, all you need to do is type

```
. global macro ITSUS_DATA "your-data-directory"
```

before you start the example.

Enough preliminaries. The World Bank provides an enormous number of intriguing datasets for download on its website. An Excel workbook titled "Climate Change Data" piqued my curiosity, so I downloaded it and converted it to a Stata dataset.[26] Let's look at what we have.

---

24. Replacing a macro with the text it contains is called expanding the macro.

25. Stata provides two kinds of macros: global macros and local macros. You create a global macro by typing `global MY_GLOBAL_MACRO "Some text"` and expand it by preceding it with a dollar sign (`$MY_GLOBAL_MACRO`). You create a local macro by typing `local my_local_macro "Some other text"` and expand it by enclosing it in single quotes (`'my_local_macro'`). Note that the single quotes we use are a grave accent (`` ` ``) on the left and an apostrophe (`'`) on the right. As you might expect from the names, global macros can be referenced anywhere in Stata, in your interactive session or deep in a do-file or Stata program. Global macros are handy but risky. Because they can be referenced anywhere, they can be changed anywhere. For instance, if you inadvertently create a macro with a name that is already in use, you will destroy the current macro and replace it with your new version. Difficult-to-diagnose chaos will ensue. Local macros are safer. They can be seen only within the current "level". For instance, if you create a local macro in your interactive session, you can safely call a Stata program that creates a local macro with the same name. Stata will keep these local macros separate and will use the correct one at the correct time.

26. You can find the original spreadsheet at https://datacatalog.worldbank.org/dataset/climate-change-data. The World Bank provides a host of related information at its Climate Change Knowledge Portal (https://climateknowledgeportal.worldbank.org/). It does take a bit of Stata legerdemain to extract the data and convert them to a convenient format for use in Stata.

```
. use ${ITSUS_DATA}/wbdata

. describe
Contains data from /Users/sbecketti/Data/itsus1r/data/wbdata.dta
  obs:         1,936
  vars:           56                          27 Aug 2012 17:01
```

variable name	storage type	display format	value label	variable label
country	str30	%30s		Country
code	long	%8.0g	code	Country code
year	int	%8.0g		Year
busease	int	%12.0fc		Ease of doing business (ranking 1-183; 1=best)
cereal	double	%12.0fc		Cereal yield (kg per hectare)
cleanprojects	str5	%9s		Hosted Clean Development Mechanism (CDM) projects
co2	double	%12.1fc		CO2 emissions, total (KtCO2)
(output omitted)				
urbpop	double	%12.0fc		Urban population
urbpopgrow	double	%12.1fc		Urban population growth (annual %)

```
Sorted by:  country  year
```

There is a lot of information in this dataset, and not all of it is directly related to climate change, as far as I can tell. These data constitute a panel dataset, that is, a time series of observations on a panel of individuals (countries in this example). We will not work with panel datasets in the remainder of this book, but that is no reason we cannot poke around in this dataset. Let's see what the country coverage looks like.

```
. tabulate country
```

Country	Freq.	Percent	Cum.
Argentina	22	1.14	1.14
Australia	22	1.14	2.27
Bolivia	22	1.14	3.41
Brazil	22	1.14	4.55
Cambodia	22	1.14	5.68
Canada	22	1.14	6.82
Chad	22	1.14	7.95
(output omitted)			
United Kingdom	22	1.14	90.91
United States	22	1.14	92.05
Upper middle income	22	1.14	93.18
Venezuela, RB	22	1.14	94.32
Vietnam	22	1.14	95.45
Virgin Islands (U.S.)	22	1.14	96.59
World	22	1.14	97.73
Zambia	22	1.14	98.86
Zimbabwe	22	1.14	100.00
Total	1,936	100.00	

In addition to individual countries, the data contain observations on collections of countries (for example, Upper middle income and World).

What years are included?

```
. summarize year
    Variable |        Obs        Mean    Std. Dev.        Min        Max
-------------+--------------------------------------------------------
        year |      1,936      2000.5     6.345928       1990       2011
. count if year==2011
      88
```

The data cover the 22 years from 1990 through 2011, and it looks as if there are observations on 88 countries or aggregates of countries ($88 \times 22 = 1936$, the number of observations in the dataset).

Look back at the output from the describe command. There is a mysterious variable named code (which has a value label named code as well). Let's see what is in it.

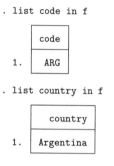

```
. list code in f

      code

1.    ARG
```

```
. list country in f

      country

1.    Argentina
```

Ah, it's a country code, a number that stands for a country and is labeled with an abbreviation for the country name. That can be handy.

It's time to see what is in the data in more detail.

```
. summarize
```

Variable	Obs	Mean	Std. Dev.	Min	Max
country	0				
code	1,936	44.5	25.40833	1	88
year	1,936	2000.5	6.345928	1990	2011
busease	204	75.69608	55.3507	1	183
cereal	1,638	3004.549	1788.97	309.4	9032.1
cleanproje~s	0				
co2	1,606	1277018	3628367	3.667	3.06e+07
co2gdp	1,504	464.8336	310.4788	.0269741	2675.182
co2pcap	1,606	4.928568	4.58914	.0005569	20.17687
degrees	0				
disaster	34	3.558824	.5541128	2.25	4.5
(output omitted)					
under5mort	1,764	51.42871	52.01535	2.4	226.6
urbgtmil	1,639	23.97409	17.45635	4.264831	100
urbpop	1,848	1.64e+08	4.51e+08	44313.2	3.47e+09
urbpopgrow	1,827	2.084769	1.515676	-3.10399	12.82905

Some variables (`code`, `year`) are fully populated—all 1,936 observations have nonmissing values. A lot of variables are only partially populated. Perhaps their information was not reported every year or for every country.

That `degrees` variable looks interesting.

```
. describe degrees
```

variable name	storage type	display format	value label	variable label
degrees	str12	%12s		Average daily min/max temperature (1961-1990, Celsius)

```
. count if degrees!=""
  88
```

```
. tabulate degrees
```

Average daily min/max temperature (1961-1990, Celsius)	Freq.	Percent	Cum.
-1.1 / 4.6	1	1.14	1.14
-10.1 / -0.1	1	1.14	2.27
-10.1 / -0.6	1	1.14	3.41
-2.0 / 5.0	1	1.14	4.55
-2.1 / 6.3	1	1.14	5.68
-2.4 / 5.8	1	1.14	6.82
-22.6 / -9.5	1	1.14	7.95
..	1	1.14	9.09
0.3 / 11.1	1	1.14	10.23
(output omitted)			
9.5 / 17.4	1	1.14	82.95
n/a	15	17.05	100.00
Total	88	100.00	

Well, that is inconvenient: the minimum and maximum temperatures are combined in a string variable. If we wanted to do some statistical analyses with these data, we would need to extract these values from the string and store them in separate values. Like this:

```
. generate degmin = real(substr(degrees,1,strpos(degrees,"/")-1))
(1,864 missing values generated)
. generate degmax = real(substr(degrees,strpos(degrees,"/")+1,.))
(1,864 missing values generated)
. list deg* in 22
```

	degrees	degmin	degmax
22.	8.2 / 21.4	8.2	21.4

The strpos() function locates the position of one string in another string. We use the location of the "/" character in the substr() function to extract the relevant portion of the string, then we apply the real() function to convert the string of digits to a number. Not difficult, but you have to know how.[27]

---

27. Stata Press offers several books on data management that can save you a lot of head scratching.

This dataset contains some income data. Let's take a look.

```
. describe g*

              storage  display    value
variable name type     format     label    variable label
```
gdp	double	%12.0fc		GDP ($)
ghg	double	%12.1fc		GHG net emissions/removals by LUCF (MtCO2e)
girlstoboys	double	%12.0fc		Ratio of girls to boys in primary & secondary school (%)
gnppcap	double	%12.0fc		GNI per capita (Atlas $)

Gross national income (GNI) is the new gross national product, and Atlas $ is a World Bank construct that attempts to convert local currencies to U.S. dollars in a way that minimizes the impact of exchange rate fluctuations. (See the World Bank website for details of the conversion formula.) Let's see what sort of coverage we have for gnppcap.

```
. table year, contents(count gnppcap)
```

Year	N(gnppcap)
1990	76
1991	76
1992	78
1993	79
1994	79
1995	81
1996	81
1997	82
1998	82
1999	83
2000	83
2001	83
2002	82
2003	82
2004	83
2005	83
2006	84
2007	84
2008	84
2009	83
2010	76
2011	0

We do not have any data for 2011, and we do not have all 88 countries represented in any single year. Let's look at the data in 2007, a year with good coverage and prior to the global recession.

```
. summarize gnppcap if year==2007
```

Variable	Obs	Mean	Std. Dev.	Min	Max
gnppcap	84	14332.44	18068.87	150	76190

The average annual per capita national income is \$14,332.44, and the range is from \$150 to \$76,190. This is the sort of variable that is likely to have a positively skewed distribution. We can use the **detail** option of the **summarize** command to get more information.

```
. summarize gnppcap if year==2007, detail
                      GNI per capita (Atlas $)

              Percentiles      Smallest
      1%          150            150
      5%          400            150
     10%          590            380       Obs                  84
     25%         1480         395.527      Sum of Wgt.          84

     50%         4980                      Mean           14332.44
                             Largest       Std. Dev.      18068.87
     75%        28435         48900
     90%        44050         54700        Variance       3.26e+08
     95%        48590         58430        Skewness       1.293259
     99%        76190         76190        Kurtosis        3.62269
```

The skewness is positive. It's still a little difficult to visualize the distribution from examining these statistics. A histogram would be helpful.

```
. histogram gnppcap if year==2007, percent color(gray) blcolor(black)
(bin=9, start=150, width=8448.8889)
```

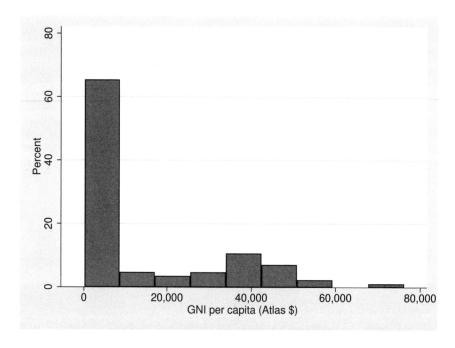

Figure 1.6. A histogram of GNI per capita, Atlas \$

We have specified the `percent` option to the `histogram` command to direct Stata to display the share of the data represented by each bar.[28] Over 60% of the countries (and aggregates of countries) in our data had a per capita GNI less than $10,000 in 2007. However, it might be more meaningful to display this histogram in terms of the share of the world's population in each income category. Let's display a histogram that is weighted by the population in each country. This type of weight is a frequency weight, abbreviated as `fw`.

```
. histogram gnppcap [fw=pop], percent
may not use noninteger frequency weights
r(401);
```

Oops. We specified a frequency weight—that is, a count of occurrences—but the population variable, `pop`, is a floating-point variable. Easily fixed. We will add 0.5 to `pop` and take the integer part of the result.[29]

```
. histogram gnppcap [fw=int(pop + 0.5)], percent
(bin=118, start=80, width=722.54237)
```

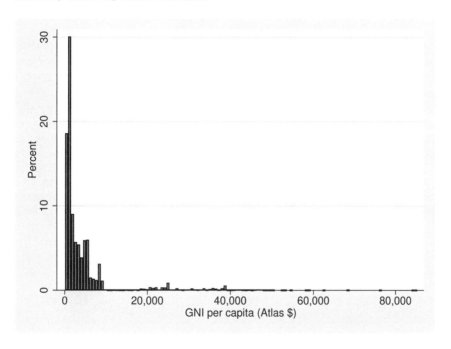

Figure 1.7. A histogram of GNI per capita, weighted by population

28. We specified the fill color `color(gray)` and the outline color `blcolor(black)` for the bars to make them easier to read on the printed page.
29. I know—nothing important is preserved by adding 0.5, but persnickety habits are hard to break.

The `histogram` command chose to break the data into smaller bins this time. We could specify additional options to control the number of bins, their widths, and their starting points if we liked, but that is not important right now. It's clear from this histogram that annual income is very low for the vast majority of the world's population.

The `histogram` command is the second Stata graph command we have seen. (We displayed a scatterplot of `auto.dta` in the first section of this chapter.) Stata offers an enormous variety of graphs via the `graph` command and its many variations and options. Stata also provides some "convenience" commands—`histogram` is an example—that produce graph commands that would be onerous to construct from scratch.

Graph commands have the form[30]

`graph` *graph_subcommand a_lot_of_other_stuff*

where the subcommand indicates the general type of graph to display. For example, a wide variety of bar charts can be produced by commands that start with the words `graph bar` .... Type `help graph` to see the full array of choices and `help graph intro` for a tutorial.[31]

The workhorse of the `graph` family is `graph twoway`, which displays scatterplots, line plots, and a host of variations on these themes. For instance, `twoway scatter` displays scatterplots, `twoway line` displays line plots, `twoway connected` displays line plots with marker symbols indicating each plotted value, and so on. For convenience, you can omit the `graph twoway` for some of the most common versions. For instance, in the first section of this chapter, we typed `scatter` to indicate the `graph twoway scatter` command. Most of the figures in this book are produced by some version of a `twoway` graph.

We cannot possibly cover even a sliver of the things you can do with Stata's `graph` command, but a few more examples may serve to give you a general idea of how to use this tool.

---

30. We warned you there would be some commands that violated the standard Stata syntax. The `graph` command is a prime offender, but these variations are needed to specify the enormous variety of graphs that can be produced.
31. My most dog-eared Stata Press book is *A Visual Guide to Stata Graphics, Third Edition* by Michael N. Mitchell (2012), an indispensable cookbook for Stata graphs.

When we `described` the variables that start with the letter "g" (see above), my eye was caught by `girlstoboys`, the ratio of girls to boys in primary and secondary school. Let's poke around a bit.

```
. summarize girlstoboys if year==2007
    Variable |       Obs        Mean    Std. Dev.       Min        Max
-------------+--------------------------------------------------------
 girlstoboys |        74    96.40255    8.689697     53.454    104.589
. histogram girlstoboys [fw=int(pop + 0.5)], percent
(bin=116, start=40.916, width=.6409569)
```

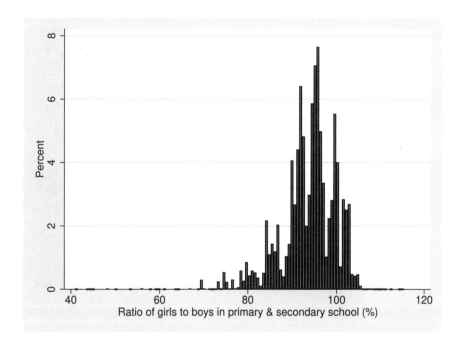

Figure 1.8. The ratio of girls to boys in primary and secondary school

This distribution is skewed left. Let's see if this variable is associated with income per capita.

```
. scatter gnppcap girlstoboys if year==2007
```

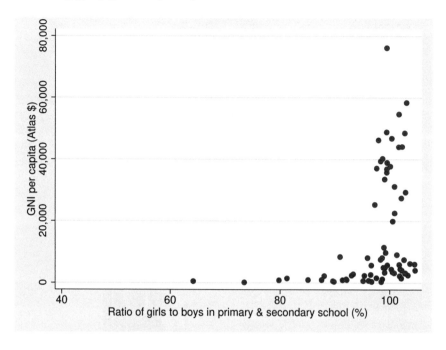

Figure 1.9. Income per capita and the ratio of girls to boys in school

Countries with a low ratio of girls to boys in primary and secondary education are invariably poor (at least in 2007), but there is a vertical stack of observations in the neighborhood of a 1-to-1 ratio. Maybe this graph would be easier to read if we used the log of income per capita rather than the level. Actually, the `graph` command will rescale the axis without requiring us to transform the variable.

```
. scatter gnppcap girlstoboys if year==2007, yscale(log)
```

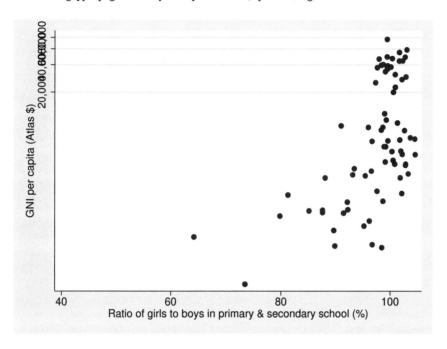

Figure 1.10. Log income per capita and the ratio of girls to boys in school

The log transformation spreads the data out a bit, but now the graph is hard to read, especially the overprinted labels on the y axis. Let's clean things up a bit.

```
. scatter gnppcap girlstoboys if year==2007, yscale(log)
> xlabel(60 70 80 90 100 110) ylabel(100 1000 10000 100000)
```

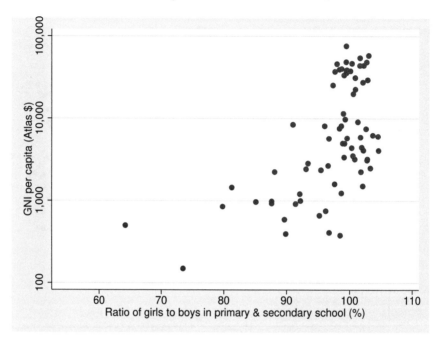

Figure 1.11. Log income per capita and the ratio of girls to boys in school

What happens if we ask Stata to weight this graph by population?

```
. scatter gnppcap girlstoboys if year==2007 [fw=int(pop + 0.5)], yscale(log)
> xlabel(60 70 80 90 100 110) ylabel(100 1000 10000 100000)
```

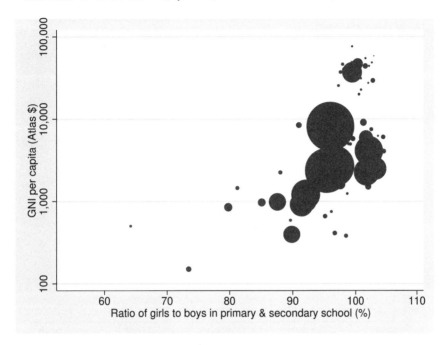

Figure 1.12. Population-weighted relationship between income and female education

Now the markers for each observation are scaled in proportion to that country's population.

Time series are most naturally displayed in line graphs. We will drop all the non-U.S. observations and demonstrate some variations.[32]

```
. keep if code=="USA":code
(1,914 observations deleted)
. twoway line gdp year
```

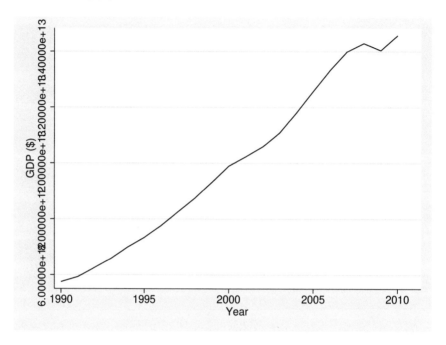

Figure 1.13. U.S. GDP, 1990–2010

We abbreviated `graph twoway line ...` as `twoway line ....` We could have shortened it further to just `line ....` Once again, the graph looks fine, but the axis labels are overcrowded.

---

32. We used a trick to keep just the observations for the United States. Remember that `code` is an integer variable with attached value labels. The `keep` command tells Stata to keep observations whose integer code maps to a label of "USA" according to the value labels stored under the name `code`. Alternatively, we could have typed `keep if country=="United States"`.

Let's look at the quantity of CO_2 emissions per year in the United States.

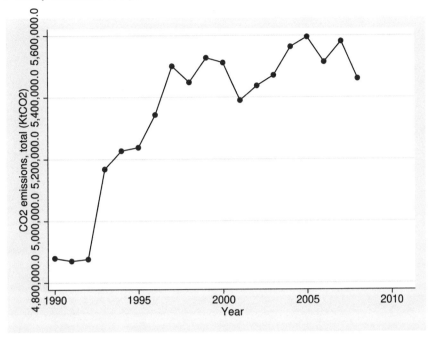

Figure 1.14. CO_2 emissions in the United States, 1990–2008

This line chart adds markers at each observed value. We also can overlay the CO_2 and gross domestic product (GDP) plots.

```
. twoway (connected co2 year) (line gdp year)
```

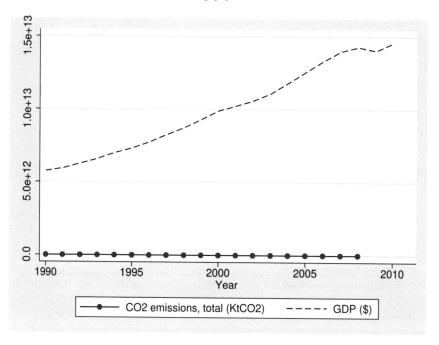

Figure 1.15. CO_2 emissions and GDP in the United States

That did not work too well. The large values of GDP squash the CO_2 line to the bottom of the graph. We need to display these plots with independent vertical axes.

```
. twoway (connected co2 year) (line gdp year, yaxis(2))
```

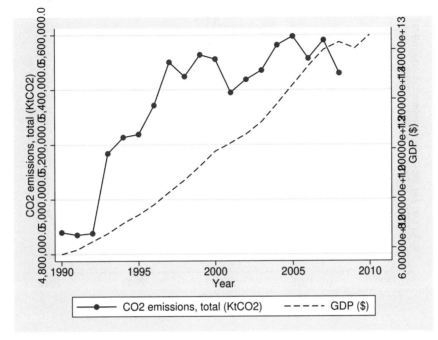

Figure 1.16. CO_2 emissions and GDP in the United States

It can sometimes be useful to modify the observation markers. Here is one example:

```
. twoway connected kgpcap year, mlabel(year)
```

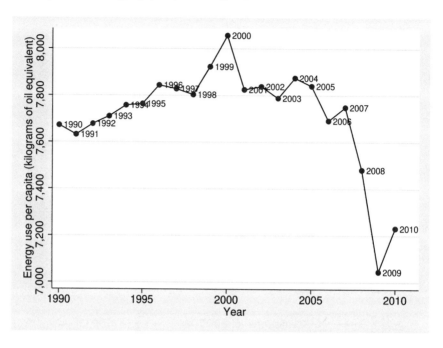

Figure 1.17. Energy use per capita in the United States (kg of oil equivalent)

Because line graphs are so common in time-series analysis, Stata has provided another abbreviation, `tsline`, for them. To use `tsline`, though, Stata has to know your data are time-series data and which variable represents your time or date index. The `tsset` command handles that chore.

```
. tsset year, yearly
        time variable:  year, 1990 to 2011
               delta:  1 year

. tsline gdp
```

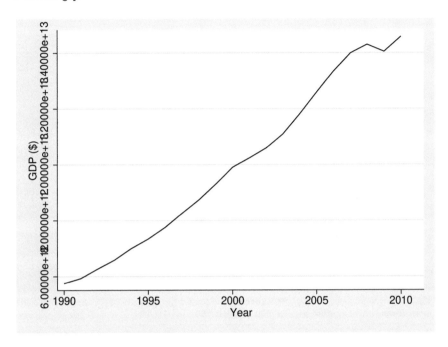

Figure 1.18. U.S. GDP, 1990–2010

While we have barely scratched the surface of Stata's graph capabilities, this introduction should get you through the rest of the book.

1.4 Statistics

1.4.1 Basics

We have already covered some of Stata's basic statistical commands, but let's recap and fill in some gaps.

The simplest of all is the `count` command. Let's load a fresh copy of our data and count the number of observations. The syntax of the `count` command is

count [*if*] [*in*]

```
. use ${ITSUS_DATA}/wbdata, clear
. count
  1,936
. count if country=="Iceland"
    22
```

Simple but extremely useful.

A more sophisticated type of counting is tabulation.[33],[34]

```
. generate int inccat =
> recode(gnppcap, 0, 100, 1000, 10000, 25000, 50000, 75000)
(232 missing values generated)
. format inccat %6.0fc
. tabulate inccat
```

inccat	Freq.	Percent	Cum.
100	5	0.35	0.35
1,000	436	30.47	30.82
10,000	729	50.94	81.76
25,000	261	18.24	100.00
Total	1,431	100.00	

```
. generate int girlcat = 10 * int(girlstoboys/10)
(822 missing values generated)
. tabulate girlcat inccat
```

girlcat	100	inccat 1,000	10,000	25,000	Total
40	0	6	0	0	6
50	0	8	0	0	8
60	0	10	1	0	11
70	2	29	8	0	39
80	0	72	38	0	110
90	0	111	211	77	399
100	0	2	190	105	297
110	0	0	1	1	2
Total	2	238	449	183	872

---

33. We are slipping in some new functions and methods for manipulating data without much explanation. I think you can follow along at this point. If you get puzzled, use the **search** and **help** commands to clear things up.
34. As a convenience, the command **tab1** *var1 var2* ... will execute a sequence of **tabulate** commands, one for each of the variables listed.

```
. tabulate girlcat inccat, nofreq cell chi2
```

			inccat		
girlcat	100	1,000	10,000	25,000	Total
40	0.00	0.69	0.00	0.00	0.69
50	0.00	0.92	0.00	0.00	0.92
60	0.00	1.15	0.11	0.00	1.26
70	0.23	3.33	0.92	0.00	4.47
80	0.00	8.26	4.36	0.00	12.61
90	0.00	12.73	24.20	8.83	45.76
100	0.00	0.23	21.79	12.04	34.06
110	0.00	0.00	0.11	0.11	0.23
Total	0.23	27.29	51.49	20.99	100.00

```
          Pearson chi2(21) = 354.8469   Pr = 0.000
```

Another workhorse is the **summarize** command, which displays the most commonly used summary statistics. We have seen both the basic and the detailed versions of this command.

```
. summarize girlstoboys if year==2007
```

Variable	Obs	Mean	Std. Dev.	Min	Max
girlstoboys	74	96.40255	8.689697	53.454	104.589

```
. summarize girlstoboys if year==2007, detail
```

```
              Ratio of girls to boys in primary & secondary
                            school (%)
```

	Percentiles	Smallest		
1%	53.454	53.454		
5%	79.791	64.174		
10%	87.597	73.445	Obs	74
25%	95.407	79.791	Sum of Wgt.	74
50%	99.0775		Mean	96.40255
		Largest	Std. Dev.	8.689697
75%	101.6	103.284		
90%	102.751	103.659	Variance	75.51084
95%	103.284	104.501	Skewness	-2.762963
99%	104.589	104.589	Kurtosis	12.0873

The `correlate` command displays a matrix of correlation coefficients.[35]

```
. describe co2pcap

              storage  display    value
variable name  type    format     label      variable label
```
```
co2pcap       double  %12.1fc                CO2 emissions per capita (metric
                                             tons)
```
```
. correlate gnppcap girlstoboys co2pcap
(obs=1,032)
```

	gnppcap	girlst~s	co2pcap
gnppcap	1.0000		
girlstoboys	0.3774	1.0000	
co2pcap	0.7054	0.4522	1.0000

There are many options to the `correlate` command. For instance, the `covariance` option displays covariances in place of correlations. Type `help correlate` for more details.

The `table` command is a powerful tool for displaying a wide variety of univariate statistics.

```
. table year, contents(count gnppcap mean gnppcap max co2pcap sd girlstoboys)
> row format(%6.1f)
```

Year	N(gnppcap)	mean(gnppcap)	max(co2pcap)	sd(girlst~s)
1990	76	6872.0	19.5	14.9
1991	76	7222.2	19.3	9.8
1992	78	7668.5	19.0	9.9
1993	79	7632.0	19.9	9.7
1994	79	7930.1	19.9	12.6
1995	81	8234.5	19.7	12.7
1996	81	8778.3	19.8	13.4
1997	82	8934.1	20.2	12.1
1998	82	8632.6	19.8	10.1
1999	83	8555.4	19.8	9.9
2000	83	8749.8	19.5	9.3
2001	83	8595.8	18.9	8.5
2002	82	8473.7	18.9	8.5
2003	82	9313.5	18.8	7.8
2004	83	10895.2	19.0	7.6
2005	83	12447.8	18.9	6.8
2006	84	13259.4	18.5	6.7
2007	84	14332.4	18.5	8.7
2008	84	15229.8	18.6	6.7
2009	83	15121.1		7.0
2010	76	14577.3		
2011	0			
Total	1,704	10107.4	20.2	9.3

---

35. `correlate` reports the Pearson product-moment correlations. The `spearman` and `ktau` commands report the Spearman rank correlation coefficient and Kendall's τ rank correlation coefficient.

1.4.2   Estimation

In the 1930s and 1940s, Metro–Goldwyn–Mayer Pictures claimed to have more stars than there are in heaven. Stata may be able to make a similar claim about estimation commands. Type **help estimation commands** if you think I am exaggerating. Fortunately, we only need to concern ourselves with a manageable subset in this book. In addition, we will defer discussion of the commands for fitting time-series models to later chapters, where we will dig into the details.

Because we are deferring time-series models (and we are not even considering models for panel data), let's extract a cross-section dataset from the World Bank data by restricting our attention to the year 2008.

```
. keep if year==2008
(1,848 observations deleted)
```

We will start with **regress**, the command that fits linear regression models. The syntax is

regress *depvar* $\big[\,indepvars\,\big]$ $\big[\,if\,\big]$ $\big[\,in\,\big]$ $\big[\,weight\,\big]$ $\big[$, *lots_of_options* $\big]$

Let's see if greenhouse gas emissions per capita are related (linearly) to income per capita, that is, if the rich countries are accounting for more than their fair share of greenhouse gases.

```
. regress co2pcap gnppcap
```

Source	SS	df	MS		Number of obs	=	83
					F(1, 81)	=	71.16
Model	774.533113	1	774.533113		Prob > F	=	0.0000
Residual	881.580495	81	10.8837098		R-squared	=	0.4677
					Adj R-squared	=	0.4611
Total	1656.11361	82	20.1965074		Root MSE	=	3.299

co2pcap	Coef.	Std. Err.	t	P>|t|	[95% Conf.	Interval]
gnppcap	.0001629	.0000193	8.44	0.000	.0001245	.0002013
_cons	2.642862	.4674281	5.65	0.000	1.712827	3.572898

regress automatically drops any observations in which any of the variables have missing values. In this example, $88 - 83 = 5$ observations were dropped.

Is affluence really the culprit? Let's condition on energy use and see if income remains significant.

```
. regress co2pcap gnppcap kgpcap
```

Source	SS	df	MS		Number of obs	=	78
					F(2, 75)	=	50.95
Model	893.326407	2	446.663203		Prob > F	=	0.0000
Residual	657.521929	75	8.76695905		R-squared	=	0.5760
					Adj R-squared	=	0.5647
Total	1550.84834	77	20.1408875		Root MSE	=	2.9609

co2pcap	Coef.	Std. Err.	t	P>|t|	[95% Conf.	Interval]
gnppcap	.0000695	.0000256	2.72	0.008	.0000185	.0001205
kgpcap	.0009344	.000198	4.72	0.000	.00054	.0013288
_cons	1.901774	.4754076	4.00	0.000	.9547135	2.848835

Apparently, income matters even when energy use is accounted for.

Some single-equation models cannot be transformed to a linear regression. In those cases, the nl command can provide nonlinear least-squares estimates.[36] While it offers no advantage, we will demonstrate nl by replicating the ordinary least-squares estimates above. We do have to eliminate missing values first—nl does not do the housekeeping for us.

```
. drop if missing(co2pcap,gnppcap,kgpcap)
(10 observations deleted)
. nl (co2pcap = {b0} + {b1}*gnppcap + {b2}*kgpcap)
(obs = 78)
Iteration 0:  residual SS =  657.5219
Iteration 1:  residual SS =  657.5219
```

Source	SS	df	MS		Number of obs	=	78
Model	893.326407	2	446.663203		R-squared	=	0.5760
Residual	657.521929	75	8.76695905		Adj R-squared	=	0.5647
					Root MSE	=	2.960905
Total	1550.84834	77	20.1408875		Res. dev.	=	387.6324

co2pcap	Coef.	Std. Err.	t	P>|t|	[95% Conf.	Interval]
/b0	1.901774	.4754076	4.00	0.000	.9547135	2.848835
/b1	.0000695	.0000256	2.72	0.008	.0000185	.0001205
/b2	.0009344	.000198	4.72	0.000	.00054	.0013288

```
Parameter b0 taken as constant term in model & ANOVA table
```

These results match the **regress** results. However, the point of the nl command is its ability to impose nonlinear constraints. To demonstrate, we will impose a nonsensical constraint here:

---

36. Stata also provides a powerful suite of commands for maximum likelihood estimation. We will not need those commands in this book because the Stata time-series estimation commands already provide maximum likelihood estimates of the models we want to use.

```
. nl (co2pcap = {b0} + {b1}*gnppcap + {b1}*kgpcap)
(obs = 78)
Iteration 0:  residual SS =  796.4424
Iteration 1:  residual SS =  796.4424
```

Source	SS	df	MS		
				Number of obs =	78
Model	754.405956	1	754.405956	R-squared =	0.4864
Residual	796.44238	76	10.479505	Adj R-squared =	0.4797
				Root MSE =	3.237206
Total	1550.84834	77	20.1408875	Res. dev. =	402.5832

co2pcap	Coef.	Std. Err.	t	P>|t|	[95% Conf. Interval]	
/b0	2.567251	.4865744	5.28	0.000	1.598154	3.536348
/b1	.0001489	.0000176	8.48	0.000	.000114	.0001839

```
Parameter b0 taken as constant term in model & ANOVA table
```

To impose constraints in some of the more complex models we consider in this book (especially the models with multiple equations), we will use the `constraint` command. While we cannot use this approach with the `regress` or `nl` commands, we could define the same nonsensical constraint we used above as follows:

```
. constraint define 1 gnppcap = kgpcap
. constraint list
    1:  gnppcap = kgpcap
```

We can define an arbitrary number of constraints in this manner, and they can incorporate arbitrarily complex expressions. In estimation commands that accept constraints, there is an option (often called something such as `constraints()`) that takes a list of numbers that identify the constraints we want to impose.

Aside from the many time-series estimation commands we will discuss later, we will use three additional estimation commands, and their syntax is similar to that of the `regress` command:

- The `newey` command estimates a linear regression with Newey–West standard errors that account for heteroskedasticity and autocorrelation of the error term.

- The `prais` command is specialized to linear regressions with first-order autocorrelated errors.

- The `ivregress` command fits single-equation linear models via instrumental-variables regression. The syntax is

  `ivregress` *estimator depvar* $\big[$ *varlist1* $\big]$ (*varlist2* = *varlist_iv*) $\big[$ *if* $\big]$ $\big[$ *in* $\big]$ $\big[$ *weight* $\big]$ $\big[$, *options* $\big]$

  The estimator is two-stage least squares (`2sls`), limited-information maximum-likelihood (`liml`), or generalized method of moments (`gmm`). The variables in *varlist2* are correlated with the error term, and the variables in *varlist_iv* are the instruments.

Often the most interesting part of an analysis takes place after a model is estimated. Stata provides a rich set of postestimation commands for calculating fitted values and residuals and for testing hypotheses about a model. The `predict` command may be the most ubiquitous of these. It can be used after most estimation commands, although its syntax and options vary with the estimation command it follows. For instance, to review fitted values and residuals from our greenhouse gas model, we could type

```
. quietly regress co2pcap gnppcap kgpcap
. predict fit
(option xb assumed; fitted values)
. twoway (line fit kgpcap, sort) (scatter co2pcap kgpcap, mlabel(code))
```

Iceland is a notable outlier. It combines the highest energy use per capita by far with just an average level of CO_2 emissions per capita. Iceland is the beneficiary of ample geothermal power thanks to its location on a major fault line with a large amount of volcanic activity. All of Iceland's electricity and building heating is supplied by geothermal power.

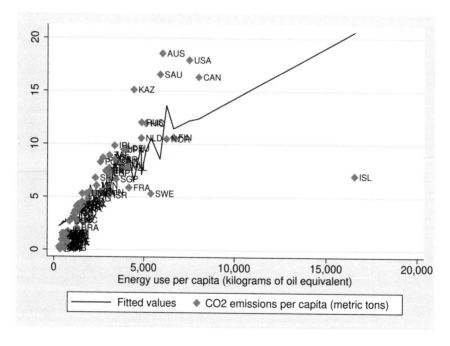

Figure 1.19. Fitted and actual values of CO_2 emissions

```
. predict resid, residual
(496 missing values generated)

. twoway scatter resid kgpcap, yline(0)
```

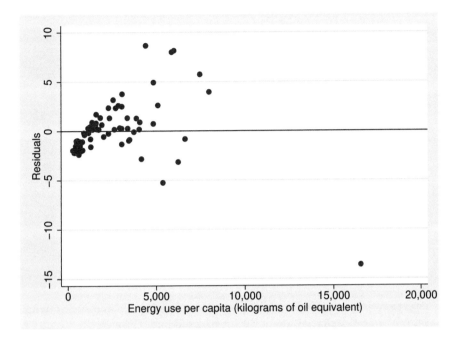

Figure 1.20. Residuals from the CO_2 regression

The large negative residual on the far right corresponds to Iceland.

There are several ways to test hypotheses in Stata. The most commonly used tool is the **test** command.

```
. regress
```

Source	SS	df	MS		
Model	893.326407	2	446.663203		
Residual	657.521929	75	8.76695905		
Total	1550.84834	77	20.1408875		

	Number of obs	=	78
	F(2, 75)	=	50.95
	Prob > F	=	0.0000
	R-squared	=	0.5760
	Adj R-squared	=	0.5647
	Root MSE	=	2.9609

co2pcap	Coef.	Std. Err.	t	P>|t|	[95% Conf. Interval]	
gnppcap	.0000695	.0000256	2.72	0.008	.0000185	.0001205
kgpcap	.0009344	.000198	4.72	0.000	.00054	.0013288
_cons	1.901774	.4754076	4.00	0.000	.9547135	2.848835

```
. test kgpcap

 ( 1)  kgpcap = 0

       F(  1,    75) =    22.28
            Prob > F =     0.0000

. test gnppcap, accumulate

 ( 1)  kgpcap = 0
 ( 2)  gnppcap = 0

       F(  2,    75) =    50.95
            Prob > F =     0.0000

. testparm kgpcap gnppcap

 ( 1)  gnppcap = 0
 ( 2)  kgpcap = 0

       F(  2,    75) =    50.95
            Prob > F =     0.0000

. test kgpcap = 0.00165

 ( 1)  kgpcap = .00165

       F(  1,    75) =    13.06
            Prob > F =     0.0005

. test kgpcap = gnppcap

 ( 1)   - gnppcap + kgpcap = 0

       F(  1,    75) =    15.85
            Prob > F =     0.0002
```

Typing **regress** by itself redisplays the previous regression. It is a handy trick if you want to refresh your memory but do not want to scroll back through the Results window or rerun the regression. To test joint hypotheses, you can use the **accumulate** option. However, if you just want to test the simple hypothesis that two or more variables are jointly 0, you can use the **testparm** command.

You can use the **nlcom** command to perform an approximate test of nonlinear hypotheses. The command is a little more cumbersome to type. You cannot refer to a parameter by typing the name of the variable it multiplies. Instead you indicate the coefficient on, say, kgpcap by typing _b[kgpcap]. For instance,

```
. display _b[kgpcap]
.00093444

. nlcom _b[kgpcap]*_b[gnppcap]

      _nl_1:  _b[kgpcap]*_b[gnppcap]
```

co2pcap	Coef.	Std. Err.	t	P>|t|	[95% Conf. Interval]	
_nl_1	6.50e-08	1.69e-08	3.86	0.000	3.14e-08	9.86e-08

Recall that the **egen** (extended generate) command handles some calculations that are just too complicated for the **generate** command by itself. Similarly, the **estat** command provides some specialized postestimation statistics and tests that do not fit neatly into one of our other commands. For example, we can use **estat** to summarize the data subset used in our regression.

```
. estat summarize
    Estimation sample regress          Number of obs =      78
```

Variable	Mean	Std. Dev.	Min	Max
co2pcap	5.282355	4.48786	.045078	18.5696
gnppcap	15787.55	19150.7	160	84850
kgpcap	2442.751	2476.527	284.843	16541.8

We can also calculate information criteria for the regression.

```
. estat ic
Akaike's information criterion and Bayesian information criterion
```

Model	Obs	ll(null)	ll(model)	df	AIC	BIC
.	78	-227.2813	-193.8162	3	393.6324	400.7025

Note: BIC uses N = number of observations. See [R] BIC note.

Or we can display the variance–covariance matrix of the estimated coefficients.

```
. estat vce
Covariance matrix of coefficients of regress model
```

e(V)	gnppcap	kgpcap	_cons
gnppcap	6.555e-10		
kgpcap	-3.678e-09	3.920e-08	
_cons	-1.365e-06	-.00003769	.2260124

estat also provides estimation-command-specific tests. We will see some examples of this use of estat in later chapters.

Stata can store detailed information about the results of an estimation command for later use.

```
. quietly regress co2pcap gnppcap kgpcap

. estimates store big

. quietly regress co2pcap gnppcap

. estimates store little

. estimates dir
```

name	command	depvar	npar	title
big	regress	co2pcap	3	
little	regress	co2pcap	2	

```
. estimates describe big
  Estimation results produced by

      . regress co2pcap gnppcap kgpcap
. estimates table big
```

Variable	big
gnppcap	.00006955
kgpcap	.00093444
_cons	1.9017741

This is interesting, but more useful is the ability to use the `lrtest` command to perform likelihood-ratio tests on a pair of previously stored estimates. One of the models must be nested within the other, but this approach provides a convenient way to test to complex hypotheses.

```
. lrtest big little
Likelihood-ratio test                                   LR chi2(1)  =      20.29
(Assumption: little nested in big)                      Prob > chi2 =     0.0000
```

Overkill for this test, but you get the idea.

1.5 Odds and ends

Here are a few things you will encounter later in the book, but they do not fit anywhere else in this chapter.

Sometimes you want to repeat one or more commands while changing one thing on each repetition (for instance, the variable you are working on). In computerese, what you need is a loop, something that runs through a sequence of commands a certain number of times. Stata provides the `foreach` and `forvalues` commands for that purpose, and we will use the `foreach` command in a later example. `foreach` is a very flexible command, and it has several forms. The syntax we will use is

`foreach` *local_macro_name* `in` *list_of_stuff* `{` *bunch_of_commands* `}`

Here is a trivial example:

```
. foreach name in co2pcap gnppcap kgppcap {
  2. display "Current variable is `name'"
  3. }
Current variable is co2pcap
Current variable is gnppcap
Current variable is kgppcap
```

`name` is a local macro (we explained global and local macros in section 1.3). In each iteration of the loop, `name` contains one of the variable names in the list. There can be an arbitrary number of commands between the opening and the closing braces.

Near the end of the book, we will need to define some matrices of constraints to be used in an estimation command. It will look something like this:

```
. matrix define A = [1,2,3\4,5,6]
. matrix list A
A[2,3]
    c1  c2  c3
r1   1   2   3
r2   4   5   6
```

Stata can do a lot more with matrices—much, much more—than we need. In fact, there are two complementary systems for handling matrices within Stata. The `matrix` suite of commands provides a host of matrix calculations. Stata also provides Mata, a complete matrix programming language. Type `help matrix` and `help mata` for more information.

As I mentioned earlier, for some examples, we generate simulated data with equation errors represented by pseudorandom numbers. Of course, computers cannot generate truly random numbers, just sequences of numbers that resemble random numbers. Although difficult to anticipate, the sequence is completely determined by the initial seed.

To guarantee that you can get the same results on your computer if you try to replicate these examples, we explicitly set the initial seed of the sequence by typing something like

```
set seed 53
```

at the beginning of the example.

1.6   Making a date

The rest of this book is devoted to introducing a wealth of Stata features for analyzing time-series data and fitting time-series models. However, Stata also provides some useful tools that make the "housekeeping" features of time-series analysis easier. This section covers those housekeeping features, so we can concentrate on the more substantive features later.

Most of the time, Stata can figure out what you are trying to do from context without a lot of extra help. For example, when we wanted to display scatterplots of mileage against automobile weight separately for domestic and foreign cars, all we had to do was type

```
. scatter mpg weight, by(foreign)
```

With time series, though, it helps to give Stata a little more information. By identifying in advance the frequency of the data and the variable that contains the time or date index, Stata can handle your analysis requests more conveniently.

As we saw above, the `tsset` command handles this task. The syntax of `tsset` is[37]

`tsset` *timevar* [, *options*]

The most commonly used options indicate the frequency of your data. For timed data, use the `clocktime` option. For dates, the alternatives are `daily`, `weekly`, `monthly`, `quarterly`, `halfyearly`, and `yearly`. There is also a `generic` option, which is the default if you do not specify another choice. The `generic` option is useful when you want to use Stata's time-series features but your data do not fit into one of the predefined categories. We use the `generic` designation later in this book in examples that use simulated data.[38]

How does Stata handle time and dates? What is Stata expecting to see in our time and date variables? As usual, it's easier to understand from an example than from a long explanation. We will begin by creating a daily date variable for the first few days in 1492, a memorable year for exploration.

```
. clear
. input month day year

          month         day        year
  1. 1 1 1492
  2. 1 2 1492
  3. 1 3 1492
  4. end
```

---

37. You also use `tsset` to inform Stata about the structure of your panel datasets. In that case, type
    `tsset` *panelvar timevar, options*.
38. The `delta()` option lets you override Stata's assumption about the period between observations.
    For instance, if you have daily observations spaced two days apart, you would specify `delta(2)` or
    `delta(2 days)`. Type `help tsset` for more details.

```
. generate date = mdy(month,day,year)
. list
```

	month	day	year	date
1.	1	1	1492	-170933
2.	1	2	1492	-170932
3.	1	3	1492	-170931

We have used `mdy()`, one of Stata's date functions, to convert individual month, day, and year values to Stata's internal coding for days. Stata anchors its coding for daily data to 1 January 1960.

```
. display mdy(1,1,1960)
0
```

For other dates, Stata counts forward (or backward) one unit for every day.

```
. display mdy(1,2,1960)
1
. display mdy(12,31,1959)
-1
```

Apparently, 1 January 1492 was 170,933 days prior to 1 January 1960.

1.6.1   How to look good

While this convention solves a lot of problems for Stata, it's not very easy to read. For instance, it's not very useful to label the x axis of a time-series graph with values like $-170,933$. It would be nice if our date variable were readable by humans, not just the computer. Not to worry. Stata has a solution.

```
. tsset date, daily
        time variable:  date, 01jan1492 to 03jan1492
                delta:  1 day
. list
```

	month	day	year	date
1.	1	1	1492	01jan1492
2.	1	2	1492	02jan1492
3.	1	3	1492	03jan1492

In addition to noting that we have daily data, tsset reformatted date to make it readable.

```
. describe date

                  storage   display     value
variable name      type     format      label        variable label

date                float    %td
```

The display format %td tells Stata to display $-170,933$ as 01jan1492. We could have done this ourselves by typing format date %td. Conveniently, tsset takes care of the reformatting for us. tsset also sorts the data in order of the *timevar*.

Note that Stata changed only the display format. The values stored in date are unchanged.

```
. format date %9.0g
. list

        month    day    year      date

  1.       1      1     1492    -170933
  2.       1      2     1492    -170932
  3.       1      3     1492    -170931
```

The table below displays examples of the default formats for each data frequency. The examples show how the value 0 is displayed for each frequency.[39]

Table 1.3. Date and time formats

Frequency	Format	Example
Datetime	%tc	01jan1960 00:00:00
Daily	%td	01jan1960
Weekly	%tw	1960w1
Monthly	%tm	1960m1
Quarterly	%tq	1960q1
Half-yearly	%th	1960h1
Yearly	%ty	0

Stata threw us a curve ball with the yearly format. Zero is displayed as 0, not 1960. And 1960 is displayed as 1960.[40]

What if you do not like these default formats? Personally, I am not crazy about 1960m1 for January 1960. Fortunately, Stata provides lots of ways to gussy up date

---

39. We do not use clocktime data in this book, so we will focus only on the date-related features of Stata in what follows.
40. But, as Emerson said, "a foolish consistency is the hobgoblin of little minds". We certainly do not want to be accused of having little minds.

formats—so many, in fact, that it can be tough to choose.[41] You can use the `display` command to experiment with different formats. For instance, you can see how Stata would choose to display a monthly value of January 1960 (an internal value of 0) by typing

```
. display %tm 0
  1960m1
```

I prefer to use a different format for monthly data.

```
. display %tmm_Y 0
Jan 60
```

The first three characters (`%tm`) indicate we are formatting a monthly value. The next `m` specifies a three-letter abbreviation for the month with the first letter capitalized. The underscore (`_`) inserts a space, and the `Y` specifies a two-digit year.

For quarterly data, I use

```
. display %tqCCYY:q 0
  1960:1
```

And for daily data, I use

```
. display %tdn/D/Y 0
  1/01/60
```

You might prefer something different. Experiment until you find what you like. And remember, you can change format as often as you want in a session or a script.

1.6.2 Transformers

Time-series data rarely arrive in Stata internal coding. Typically, dates appear in some sort of readable format: `Mar 27, 2009`, 3/27/09, 03-27-2009, etc. To work with these types of data, you need a way to convert this text to `17983`, Stata's internal code for 27 March 2009.

We have seen one way of transforming data—the `mdy()` function that converts individual month, day, and year values into a daily date in Stata's internal coding. When confronted with dates in a human-readable format, we could use Stata's string functions and our ingenuity to extract the month, day, and year into separate variables, but that is a lot of work. Stata's `date()` function saves us the trouble.

---

41. Type `help datetime display formats` for a complete list.

`date()` takes two strings and returns a Stata date. The first string is the input, the human-readable stuff you are trying to convert. The second string describes the input string to help Stata decipher it. Again an example is worth a thousand words.

```
. display mdy(3,27,2009)
17983
. display date("3/27/2009","MDY")
17983
```

`date()` is flexible and tolerant of different formats.

```
. display date("03/27/2009","MDY")
17983
. display date("2009-03-27","YMD")
17983
```

`date()` also can navigate the ambiguity presented by two-digit years.

```
. display date("3/27/09","MD20Y")
17983
. display %td date("3/27/09","MD19Y")
27mar1909
```

Once your data are in Stata's internal coding, you may want to extract components of the date. Stata provides a full complement of functions for that purpose.

```
. drop month day year
. format date %tdn/D/Y
. list
```

	date
1.	1/01/92
2.	1/02/92
3.	1/03/92

```
. generate day = day(date)
. generate week = week(date)
. generate month = month(date)
. generate quarter = quarter(date)
. generate half = halfyear(date)
. generate year = year(date)
. list
```

	date	day	week	month	quarter	half	year
1.	1/01/92	1	1	1	1	1	1492
2.	1/02/92	2	1	1	1	1	1492
3.	1/03/92	3	1	1	1	1	1492

There are also functions for extracting the day of the week (`dow()`) and the Julian date (`doy()`), that is, the day (by count) of the year.

```
. generate dow = dow(date)
. generate doy = doy(date)
. format date %tdDayname,_Month_dd,_CCYY
. list date dow doy
```

	date	dow	doy
1.	Friday, January 1, 1492	5	1
2.	Saturday, January 2, 1492	6	2
3.	Sunday, January 3, 1492	0	3

You will probably want to make the day of the week more readable.

```
. label define dow 0 Sunday 1 Monday 2 Tuesday 3 Wednesday 4 Thursday 5 Friday
> 6 Saturday
. label values dow dow
. list dow
```

	dow
1.	Friday
2.	Saturday
3.	Sunday

The functions displayed above work only on daily data—the `date` variable is stored in Stata's internal coding for daily data. However, Stata also provides functions for translating one encoding to another, say, from weekly to daily. The names of these functions all have the same structure, $xofy()$, where y indicates the frequency of the argument of the function and x indicates the frequency of the output. For example,

```
. drop day-doy
. format date %tdn/D/Y
. generate month = mofd(date)
. list
```

	date	month
1.	1/01/92	-5616
2.	1/02/92	-5616
3.	1/03/92	-5616

Of course, you have to format the converted data.

```
. format month %tmm_Y
. list
```

	date	month
1.	1/01/92	Jan 92
2.	1/02/92	Jan 92
3.	1/03/92	Jan 92

1.7   Typing dates and date variables

Frequently, you will need to type a date literally. It would be inconvenient to figure out what strange number Stata has assigned to represent your specific date. Stata provides functions to convert a single, human-readable date to its internal format. The names of these functions look like date format specifications without the initial percent sign. In other words, you use `td()` to specify a daily date, `tm()` to specify a monthly date, and so on.

With these functions, you easily can type things like

```
. summarize unemployment if date < tm(2008m9)
```

Time-series variables rarely come singly. For instance, consider the fourth-order autoregression

$$y_t = \mu + \beta_1 y_{t-1} + \beta_2 y_{t-2} + \beta_3 y_{t-3} + \beta_4 y_{t-4} + \epsilon_t$$

To estimate this regression in Stata, we could create four new variables to hold the lagged values of y_t. For instance, we could create y_{t-1} by typing

```
generate y1 = y[_n-1]
```

We have not seen this syntax before. The square brackets ([and]) allow us to specify observation numbers, and _n is a built-in Stata variable that stands for the current observation number. So this statement creates a new variable (y1) that contains the one-period lag of y. We could use this same approach to create the other lags, then estimate the regression

```
regress y y1 y2 y3 y4
```

There is a better way. Once you have `tsset` your data, Stata allows you to precede your variable names with time-series operators. You type the operator you want, a period (.), and the name of the variable you want to operate on. For instance, L is the time-series operator for a lag. L.y specifies y_{t-1}, L2.y specifies y_{t-2}, and so on. With this operator, you could specify the regression above as

```
regress y L.y L2.y L3.y L4.y
```

We no longer need to create superfluous variables to hold the lags, but this is still clumsy. We can run this regression more easily by typing

```
regress L(0/4).y
```

Stata interprets `L(0/4).y` as the five variables `y`, `L.y`, ..., `L4.y`.

Stata provides three additional time-series operators. The lead operator, `F`, is the inverse of the lag operator. `F.y` specifies y_{t+1}. The difference operator, `D`, indicates the arithmetic difference of adjacent observations. `D.y` specifies $\Delta y_t \equiv y_t - y_{t-1}$. Note that `D2.y` indicates

$$\Delta^2 y_t \equiv \Delta(\Delta y_t) \equiv y_t - 2y_{t-1} + y_{t-2}$$

not $y_t - y_{t-2}$.

Finally, the seasonal difference operator, `S`, takes differences at seasonal frequencies rather than differences of adjacent observations. For instance, if `y` is a quarterly time series, the first seasonal difference, `S4.y`, is the same as `y - L4.y`.

1.8 Looking ahead

You now know all the Stata you need to know to read the rest of this book. There will be no big surprises, and for the small surprises, you know how to use the `help` and `search` commands to get more information. The next chapter summarizes the statistical background you will need to follow along.

2 Just enough statistics

Chapter map

2.1 Random variables and their moments. Random variables, probability distribution and density functions, theoretical and sample moments.

2.2 Hypothesis tests. Null and alternative hypotheses. Type I and type II errors. Test statistics, critical values, and p-values.

2.3 Linear regression. Ordinary least squares. Instrumental variables. Feasible generalized least squares (FGLS).

2.4 Multiple-equation models. Seemingly unrelated regression.

2.5 Time series. White noise, autocorrelation, stationarity, and unit roots. Autoregressive moving-average (ARMA) models.

The previous chapter was long; this one is terse. The mathematical and statistical knowledge you need to read this book is highlighted below. Many readers will be familiar with most, if not all, of this information, but it never hurts to refresh your memory.

Invariably, some readers will find this chapter elementary and tedious. If you are among them, you may prefer to consult *Time Series Analysis* by James D. Hamilton (Princeton University Press, 1994), the definitive summary of the theory underlying the topics in this book. The classic *Time Series Analysis: Forecasting and Control* by George E. P. Box, Gwilym M. Jenkins, and Gregory C. Reinsel (Wiley, 4th ed., 2008) provides an alternative slant that I often found helpful, particularly in chapters 6 and 7.

Other readers will find this chapter opaque or touching on topics they have not thought about in years. If you are in this group (as I often am), *Introduction to Econometrics* by James H. Stock and Mark W. Watson (Pearson, 4th ed., 2019) is a well-written and approachable text. A book that attempts to cover the whole spectrum (and largely succeeds) is *Introduction to the Theory and Practice of Econometrics* by George G. Judge, R. Carter Hill, William E. Griffiths, Helmut Lütkepohl, and Tsoung-Chao Lee (Wiley, 2nd ed., 1988). Do not confuse this book with the also-excellent *The Theory and Practice of Econometrics* (Wiley, 2nd ed., 1985) by the same authors. The word "Introduction" in the title is important.

2.1   Random variables and their moments

Statistics is concerned with the properties of random variables, variables that are not perfectly predictable but whose repeated realizations are described by a probability distribution function. Let y be a random variable. We denote its distribution function by $F(y)$. The random variables in this book all are assumed to follow distributions with well-defined probability density functions, $f(y)$, where $f(y) \equiv dF(y)$. In this book, we are concerned exclusively with continuous random variables.

The distributions of random variables are characterized by their moments. These measures provide information about the shape of a distribution—its most likely values and the dispersion of values likely to be encountered in any given sample.

The kth moment of the random variable y about a value c is defined as

$$E\left\{(y - c)^k\right\} \equiv \int_{-\infty}^{\infty} (y - c)^k df(y) dy$$

where E is described as the expectation of the expression. The expected value of y, also called the mean of y, is defined as the first moment about 0.

$$E(y) \equiv \mu_y$$

The mean of a random variable provides a measure of its location. For symmetric distributions, the mean is identical to the median, the central value of the distribution. Moments about the mean are called central moments. Our analyses of higher moments focus exclusively on central moments. The variance of a random variable is defined as the second central moment and provides a measure of the dispersion of a random variable.

$$\sigma_y^2 \equiv E\left\{(y - \mu)^2\right\}$$

The square root of the variance, σ_y, is called the standard deviation of y. Most of our attention will be focused on the mean and variance (and standard deviation). We rarely will refer to higher moments.

Moments of random variables are not guaranteed to exist. A frequently encountered example in time-series analysis is the random walk

$$y_t = y_{t-1} + \epsilon_t$$

where ϵ_t is identically and independently distributed with mean 0 and finite variance. It is straightforward to show that $E(y_t) = 0$, but $E(y_t^2) = \infty$, that is, the second moment of y_t does not exist.

Given a sample, $y_1, y_2, \ldots, y_T$, from a random distribution, we can calculate sample estimates of these moments. The sample mean is defined as the sum of the observed values divided by the number of observations:

$$\bar{y} = \frac{\sum_{t=1}^{T} y_t}{T}$$

The sample variance is

$$\text{Var } y \equiv \widehat{\sigma}_y^2 = \frac{\sum_{t=1}^{T}(y_t - \overline{y})^2}{T}$$

and the sample standard deviation is

$$\widehat{\sigma}_y \equiv \sqrt{\widehat{\sigma}_y^2}$$

2.2   Hypothesis tests

Calculating estimates of statistical quantities is a first step. Interpreting the estimates—deciding if they are meaningful—is the next step. In an experimental setting, a researcher may estimate the impact of a treatment—a new drug for preventing a disease, a new fertilizer for increasing crop yields, a new fuel additive for increasing gas mileage—by comparing the outcomes between a treated group and a control (that is, untreated) group. The estimated impact, say, the mean difference in outcomes between the two groups, is a random variable. Any observed reduction in the incidence of the disease, increase in crop yields, or improvement in gas mileage may occur by chance and thus may not be representative of the outcome that would be achieved in the population.

Hypothesis tests quantify the likelihood that the observed results occurred by chance. If this likelihood is very low, we may conclude the observed results represent a meaningful impact of the treatment. Otherwise, we may conclude the treatment is ineffective. In classical hypothesis testing, we articulate two competing hypotheses. The null hypothesis, denoted by H_0, usually asserts the "no effect" outcome—the treatment has no impact. The alternative hypothesis (H_a) asserts that the treatment has an impact.

Frequently, a specific statistic, call it γ, quantifies the impact of the treatment.[1] For example, γ may be the mean difference in crop yields between the treatment and the control groups, or γ may be the regression coefficient that multiplies the quantity of the chemical that is added to the gas tank of the treated cars. The estimated value of γ is denoted by $\widehat{\gamma}$. In this case, the null hypothesis is

$$H_0 : \gamma = 0$$

If we are interested only in the possibility that, say, the treatment increases crop yields, then we will test the one-sided alternative[2]

$$H_a : \gamma > 0$$

If we are interested in any impact of the treatment, good or bad, we will test the two-sided alternative

$$H_a : \gamma \neq 0$$

---

1. Statistical hypothesis tests take many forms. To aid the intuition, this discussion focuses on a specific and commonly encountered approach.
2. If we were interested in testing a treatment that reduced an outcome, we would reverse the inequality in this one-sided alternative.

There may be multiple ways to test these hypotheses, and for any choice of test, we are exposed to two potential errors. A type I error occurs if we incorrectly reject the null hypothesis and accept the alternative. For a particular hypothesis test, the probability of a type I error is denoted by α and is called the size of the test. A type II error occurs if we incorrectly accept the null hypothesis and reject the alternative. The probability of a type II error is denoted by β, and the probability $(1 - \beta)$ is called the power of the test.

Because the estimate $\widehat{\gamma}$ is a random variable, it has some dispersion. If we can estimate the disperson of $\widehat{\gamma}$ under the null hypothesis (that is, when H_0 is true), we can calculate the probability of a type I error for any value of $\widehat{\gamma}$. For any given distribution of $\widehat{\gamma}$, this probability increases with the difference between the estimated value of γ and the value under the null (in this case, with $|\widehat{\gamma} - 0| = |\widehat{\gamma}|$). If $\widehat{\sigma}_\gamma$ is an estimate of the standard deviation of $\widehat{\gamma}$, the ratio $\widehat{\kappa} \equiv \widehat{\gamma}/\widehat{\sigma}_\gamma$ provides a measure of the distance between the estimated value $\widehat{\gamma}$ and the value under the null ($\gamma = 0$) in standard deviation units.

If we can determine the distribution of $\widehat{\kappa}$—or at a minimum, the asymptotic distribution of $\widehat{\kappa}$—we can define the following decision rule: choose H_a (reject H_0) if $\widehat{\kappa} > c$, where the critical value, c, is chosen so that α, the size of the test, is set to a predetermined level, often 5% (0.05). If, in fact, we reject the null, we say that $\widehat{\kappa}$ is different from 0 at the $100\alpha\%$ level of significance. For a two-sided alternative, we need two critical values: choose H_a (reject H_0) if $\widehat{\kappa}$ lies outside the interval $[c_L, c_H]$, where[3]

$$\text{Prob}\left(\widehat{\kappa} < c_L\right) = \text{Prob}\left(c_H < \widehat{\kappa}\right) = \alpha/2$$

Historically, statisticians relied on conventional significance levels as benchmarks. The 5% level was most common, but 1% and 10% were used as well. In part, this practice reflected the availability of tables of the 1%, 5%, and 10% critical values for the t, F, and χ^2 distributions, which characterize many test statistics. Modern statistical programs report the marginal significance or p-values, that is, the size of the test if the critical value happened to be set equal to the test statistic. However, the conventional significance levels are still used as benchmarks of importance—a $\widehat{\kappa}$ with a p-value of 0.03 is described as statistically significant, where a $\widehat{\kappa}$ with a p-value of 0.17 is not.

2.3 Linear regression

2.3.1 Ordinary least squares

Let the variable y_t be a linear function of a variable x_t plus an unobservable random disturbance ϵ_t

$$y_t = \mu + \beta x_t + \epsilon_t \tag{2.1}$$

where

$$E(\epsilon_t) = 0, \ E(\epsilon_t^2) = \sigma^2$$

---

3. If the distribution of $\widehat{\kappa}$ is symmetric about 0, then $c_L = -c_H$.

and

$$E(x_t \epsilon_t) = 0, \forall t$$

The coefficients μ and β are unknown parameters. We wish to use a sample of T pairs of y_t and x_t to "fit" this equation; that is, we want to calculate values of $\widehat{\mu}$ and $\widehat{\beta}$ that provide a good approximation to the unobservable (2.1). Because $E(\epsilon_t) = 0$, an intuitive solution is to choose values of $\widehat{\mu}$ and $\widehat{\beta}$ that make the residuals $y_t - \widehat{\mu} - \widehat{\beta} x_t$ small. A practical solution is to fit the equation by ordinary least squares (OLS),[4] that is, choose $\widehat{\mu}$ and $\widehat{\beta}$ to minimize the sum of squared residuals:

$$\sum_{i=1}^{T} \left(y_t - \widehat{\mu} - \widehat{\beta} x_t \right)^2$$

It is straightforward to show that

$$\widehat{\beta} = \frac{\sum_{i=1}^{T} (y_t - \overline{y})(x_t - \overline{x})}{\sum_{i=1}^{T} (x_t - \overline{x})^2}$$

and

$$\widehat{\mu} = \overline{y} - \widehat{\beta}\overline{x}$$

It follows immediately from this last equation that the line defined by this equation passes through the point of means: $\overline{y} = \widehat{\mu} + \widehat{\beta}\overline{x}$.

More generally, y_t may be a function of multiple variables.

$$y_t = \beta_0 + \beta_1 x_{1,t} + \beta_2 x_{2,t} + \cdots + \beta_K x_{K,t} + \epsilon_t$$

For a given sample of T observations, we can write this model as

$$\mathbf{y} = \mathbf{X}\boldsymbol{\beta} + \boldsymbol{\epsilon} \tag{2.2}$$

where

$$\mathbf{y} = \begin{bmatrix} y_1 \\ y_2 \\ \vdots \\ y_T \end{bmatrix}, \quad \mathbf{X} = \begin{bmatrix} 1 & x_{1,1} & x_{2,1} & \cdots & x_{K,1} \\ 1 & x_{1,2} & x_{2,2} & \cdots & x_{K,2} \\ \vdots & \vdots & \vdots & \ddots & \vdots \\ 1 & x_{1,T} & x_{2,T} & \cdots & x_{K,T} \end{bmatrix}, \quad \boldsymbol{\beta} = \begin{bmatrix} \beta_0 \\ \beta_1 \\ \vdots \\ \beta_K \end{bmatrix}$$

and

$$\boldsymbol{\epsilon} = \begin{bmatrix} \epsilon_1 \\ \epsilon_2 \\ \vdots \\ \epsilon_T \end{bmatrix}$$

---

4. There is nothing ordinary about ordinary least squares. OLS is a powerful estimation technique with broad applicability. Accordingly, some authors prefer to call it classical least squares to emphasize its foundation in classical assumptions about the distributions of x_t and ϵ_t.

We add the classical assumptions

$$E\left(\epsilon\epsilon'\right) = \sigma^2\mathbf{I}$$

and

$$E\left(\mathbf{X}'\epsilon\right) = 0$$

The first of these two assumptions, the assumption of homoskedasticity, states that the random disturbances are uncorrelated across observations and that all the disturbances have identical variances, σ^2. The second assumption—that the explanatory variables are uncorrelated with the random disturbances—can be weakened to an assumption that the probability limit (denoted by "plim") of these correlations is 0:

$$\text{plim}\frac{1}{n}\mathbf{X}'\epsilon = 0$$

As before, we solve for the $\widehat{\boldsymbol{\beta}}$ that minimizes the sum of squared residuals:

$$(\mathbf{y} - \mathbf{X}\boldsymbol{\beta})'\,(\mathbf{y} - \mathbf{X}\boldsymbol{\beta})$$

The least-squares solution is

$$\widehat{\boldsymbol{\beta}} = (\mathbf{X}'\mathbf{X})^{-1}\mathbf{X}'\mathbf{y}$$

The random vector $\widehat{\boldsymbol{\beta}}$ is an unbiased estimator of $\boldsymbol{\beta}$

$$E\left(\widehat{\boldsymbol{\beta}}\right) = E\left\{(\mathbf{X}'\mathbf{X})^{-1}\mathbf{X}'\mathbf{y}\right\} = E\left\{(\mathbf{X}'\mathbf{X})^{-1}(\mathbf{X}'\mathbf{X}\boldsymbol{\beta} + \mathbf{X}'\epsilon)\right\} = \boldsymbol{\beta}$$

with covariance matrix

$$\sigma^2(\mathbf{X}'\mathbf{X})^{-1}$$

Because σ^2 is an unknown parameter, we replace it with its unbiased estimator

$$\widehat{\sigma}^2 = \frac{\widehat{\epsilon}'\widehat{\epsilon}}{T - K - 1}$$

where

$$\widehat{\epsilon} \equiv \mathbf{y} - \mathbf{X}\boldsymbol{\beta}$$

If ϵ is normally distributed, $\widehat{\boldsymbol{\beta}}$ is normally distributed independently of $\widehat{\epsilon}$ and functions of $\widehat{\epsilon}$. Let $(\mathbf{X}'\mathbf{X})^{jj}$ be the (j,j)th element (the jth element of the main diagonal) of $(\mathbf{X}'\mathbf{X})^{-1}$. It can be shown that the ratio

$$t_j = \frac{\widehat{\beta}_j - \beta_j}{\sqrt{\widehat{\sigma}^2(\mathbf{X}'\mathbf{X})^{jj}}}$$

follows the t distribution with $T - K - 1$ degrees of freedom. This ratio can be used as a test statistic for hypothesis tests about an individual coefficient, β_j.

The approach for hypothesis tests involving multiple coefficients is slightly different. Because $\widehat{\beta}$ minimizes the sum of squared residuals, an alternative hypothesis that produces a different estimate, $\widetilde{\beta}$, will incur a loss of fit, that is, an increase in the sum of squared residuals. If the alternative hypothesis comprises J linearly independent restrictions on β, then

$$F = \frac{\left(\widetilde{\epsilon}'\widetilde{\epsilon} - \widehat{\epsilon}'\widehat{\epsilon}\right)/J}{\widehat{\epsilon}'\widehat{\epsilon}/(T - K - 1)}$$

follows an F distribution with $(J, T - K - 1)$ degrees of freedom, where $\widetilde{\epsilon}$ is the residual vector under the alternative hypothesis. Because the unrestricted OLS estimator $\widehat{\beta}$ minimizes the sum of squared residuals, we are guaranteed that $F > 0$.

2.3.2   Instrumental variables

The derivation of the OLS estimator above depends crucially on the assumption that the explanatory variables are uncorrelated in the limit with the random disturbance, that is, $\operatorname{plim}\frac{1}{n}\mathbf{X}'\epsilon = 0$. That assumption is not always tenable. For instance, we may be estimating a single equation from a larger system of equations for jointly endogenous variables. In this case, the application of OLS will produce inconsistent estimates of β.

We can overcome this obstacle if we can find suitable instrumental variables, that is, variables that are correlated in the limit with $\mathbf{X}$ but not ϵ. Imagine that we have identified a suitable set of $(K + 1)$ instruments, $\mathbf{Z}$, such that

$$\operatorname{plim}\frac{1}{n}\mathbf{Z}'\epsilon = 0, \qquad \operatorname{plim}\frac{1}{n}\mathbf{Z}'\mathbf{X} \neq 0$$

Then the simple instrumental variables estimator of β is

$$\widetilde{\beta} \equiv \left(\mathbf{Z}'\mathbf{X}\right)^{-1}\mathbf{Z}'\mathbf{y}$$

It can be shown that $\widetilde{\beta}$ is a consistent estimator of β, that is,

$$\operatorname{plim}\widetilde{\beta} = \beta$$

2.3.3   FGLS

The assumption of homoskedasticity may be too strong. In time-series regressions, random disturbances may be autocorrelated, that is, today's disturbance may be correlated with yesterday's disturbance. In cross-section regressions, disturbances may have unequal variances if, for instance, observations represent average outcomes in groups of varying size. In either case, we replace the assumption of homoskedasticity with the more general assumption

$$E\left(\epsilon\epsilon'\right) = \sigma^2\mathbf{\Omega}$$

In this case, the OLS estimator $\widehat{\beta}$ is still unbiased, but $\widehat{\sigma}^2(\mathbf{X}'\mathbf{X})^{-1}$, the OLS estimator of the covariance matrix of $\widehat{\beta}$, is incorrect.

If $\mathbf{\Omega}$ were known, we could calculate a matrix $\mathbf{P}$ such that[5]

$$\mathbf{\Omega}^{-1} = \mathbf{P}\mathbf{P}'$$

If we premultiply (2.2) by $\mathbf{P}$, we obtain

$$\mathbf{P}\mathbf{y} = \mathbf{P}\mathbf{X}\boldsymbol{\beta} + \mathbf{P}\boldsymbol{\epsilon} \tag{2.3}$$

With this transformation, we have

$$E\left(\mathbf{P}\epsilon\epsilon\mathbf{P}'\right) = \mathbf{P}\sigma^2\mathbf{\Omega}\mathbf{P}' = \sigma^2 I$$

Thus the transformed equation incorporating $\mathbf{P}\mathbf{y}$ and $\mathbf{P}\mathbf{X}$ can be estimated by OLS.

In practice, $\mathbf{\Omega}$ is an unknown matrix. However, in many cases, we can find $\widetilde{\mathbf{\Omega}}$, a consistent estimator of $\mathbf{\Omega}$. The FGLS estimator uses $\widetilde{\mathbf{\Omega}}$ in place of $\mathbf{\Omega}$ to calculate the transformation in (2.3).[6]

2.4   Multiple-equation models

Consider the case where we have M linear equations of the form of (2.2). We can write the mth equation as[7]

$$\mathbf{y}_m = \mathbf{X}_m\boldsymbol{\beta}_m + \boldsymbol{\epsilon}_m$$

and the entire system of M equations as

$$\begin{bmatrix} \mathbf{y}_1 \\ \mathbf{y}_2 \\ \vdots \\ \mathbf{y}_M \end{bmatrix} = \begin{bmatrix} \mathbf{X}_1 & \mathbf{0} & \cdots & \mathbf{0} \\ \mathbf{0} & \mathbf{X}_2 & \cdots & \mathbf{0} \\ \vdots & \vdots & \ddots & \vdots \\ \mathbf{0} & \mathbf{0} & \cdots & \mathbf{X}_M \end{bmatrix} \begin{bmatrix} \boldsymbol{\beta}_1 \\ \boldsymbol{\beta}_2 \\ \vdots \\ \boldsymbol{\beta}_M \end{bmatrix} + \begin{bmatrix} \boldsymbol{\epsilon}_1 \\ \boldsymbol{\epsilon}_2 \\ \vdots \\ \boldsymbol{\epsilon}_M \end{bmatrix}$$

which we rewrite more compactly as

$$\mathbf{y} = \mathbf{X}\boldsymbol{\beta} + \boldsymbol{\epsilon}$$

where now, for example,

$$\mathbf{y} \equiv \begin{bmatrix} \mathbf{y}_1 \\ \mathbf{y}_2 \\ \vdots \\ \mathbf{y}_M \end{bmatrix}$$

We assume

$$E\left(\epsilon_{ms}\epsilon_{nt}\right) = \sigma_{mn} \quad \text{if } t = s \text{ and } 0 \text{ otherwise}$$

---

5. Because $\mathbf{\Omega}$ is symmetric and positive definite, it is always possible to solve for $\mathbf{P}$.
6. As long as $\widetilde{\mathbf{\Omega}}$ is a consistent estimator of $\mathbf{\Omega}$, FGLS is asymptotically efficient.
7. This discussion follows the treatment in Greene (2018, chap. 10).

Thus

$$E\left(\boldsymbol{\epsilon}_m \boldsymbol{\epsilon}'_n\right) = \sigma_{mn} I_T$$

and

$$E\left(\boldsymbol{\epsilon}\boldsymbol{\epsilon}'\right) \equiv \mathbf{Q} = \begin{bmatrix} \sigma_{11}\mathbf{I} & \sigma_{12}\mathbf{I} & \cdots & \sigma_{1M}\mathbf{I} \\ \sigma_{21}\mathbf{I} & \sigma_{22}\mathbf{I} & \cdots & \sigma_{2M}\mathbf{I} \\ \vdots & \vdots & \ddots & \vdots \\ \sigma_{M1}\mathbf{I} & \sigma_{M2}\mathbf{I} & \cdots & \sigma_{MM}\mathbf{I} \end{bmatrix}$$

Define

$$\boldsymbol{\Sigma} = \begin{bmatrix} \sigma_{11} & \sigma_{12} & \cdots & \sigma_{1M} \\ \sigma_{21} & \sigma_{22} & \cdots & \sigma_{2M} \\ \vdots & \vdots & \ddots & \vdots \\ \sigma_{M1} & \sigma_{M2} & \cdots & \sigma_{MM} \end{bmatrix}$$

So we can write

$$\mathbf{Q} = \boldsymbol{\Sigma} \otimes \mathbf{I}$$

Each of the M equations obeys the classical assumptions and can be estimated appropriately by OLS. However, $\mathbf{Q}$, the covariance matrix of the random disturbances for the system of equations as a whole, is not diagonal; hence, the generalized least-squares (GLS) estimator is the asymptotically efficient estimator for the system. This type of system of equations is called a seemingly unrelated regressions model because there are no observable links between the equations. The only connections are the correlations of the unobservable random disturbances across equations.

There are some special cases where GLS offers no advantage over OLS for this model. First, if the random disturbances are, in fact, uncorrelated across equations ($\sigma_{mn} = 0, \forall m \neq n$), then $\mathbf{Q}$ is diagonal, this is a system of actually unrelated equations, and the GLS and OLS estimators are identical. Second, if all the equations have identical explanatory variables ($\mathbf{X}_m = \mathbf{X}_n, \forall m, n$), it can be shown that OLS and GLS are identical.[8] Third, if the explanatory variables in one block of equations are a subset of the regressors in the rest of the system, GLS provides no improvement for the "smaller" subblock.

The residuals from the OLS estimates, $\widehat{\boldsymbol{\epsilon}}_m$, can be used to form consistent estimates of the elements of $\boldsymbol{\Sigma}$:

$$\widehat{\sigma}_{mn} = \frac{\boldsymbol{\epsilon}'_m \boldsymbol{\epsilon}_n}{T}$$

These estimates can be used to construct the FGLS.

2.5 Time series

We will cover the fundamentals of time series in models in detail in chapter 6 and subsequent chapters, but it may be helpful to highlight a few essentials here.

---

8. This result also holds if all the $\mathbf{X}_m$ are nonsingular and span the same linear subspace.

2.5.1   White noise, autocorrelation, and stationarity

Our focus is discrete time series. We assume the variables in our Stata datasets are vectors of random variables observed at regular intervals and indexed by some measure of time, t.

$$\mathbf{y} = \begin{bmatrix} y_1 \\ y_2 \\ \vdots \\ y_T \end{bmatrix}$$

The essential feature of time-series analysis is the characterization of the dependence of the current value of y_t on its own past. In this book, we consider time series that can be represented by a linear stochastic difference equation of the form

$$y_t = \phi_1 y_{t-1} + \phi_2 y_{t-2} + \cdots + \phi_p y_{t-p} + \epsilon_t \tag{2.4}$$

We call (2.4) an autoregression in y_t. In the simplest case, ϵ_t has no dynamic properties of its own. Its only role in this case is to contribute random disturbances to the stochastic difference equation in (2.4). We can (and will) add lots of complications to this structure, but a stochastic difference equation in y_t is at the heart of all our models.

An essential building block of time-series models is white noise. A random time series, ϵ_t, is white noise if $E(\epsilon_t) = 0$, $E(\epsilon_t^2) = \sigma_\epsilon^2$, and $E(\epsilon_t \epsilon_{t'}) = 0, \forall t \neq t'$. Independent white noise obeys the stronger condition that ϵ_t is independent of $\epsilon_{t'}, \forall t \neq t'$. ϵ_t may obey a wide variety of probability distributions, but we frequently assume that ϵ_t is Gaussian white noise; that is, ϵ_t follows a normal distribution.

A defining property of a nonwhite time series, y_t, is autocorrelation, that is, nonzero covariances $E(y_t y_{t-j})$. For instance, consider the first-order stochastic difference equation

$$y_t = \phi_1 y_{t-1} + \epsilon_t$$

Thus

$$y_t = \sum_{i=0}^{\infty} \phi_1^i \epsilon_{t-i}$$

and

$$E(y_t) = \sum_{i=0}^{\infty} \phi_1^i E(\epsilon_{t-i}) = 0$$

and

$$E\left(y_t^2\right) = \sum_{i=0}^{\infty} \phi_1^{2i} \sigma_\epsilon^2 = \frac{\sigma_\epsilon^2}{1 - \phi_1^2}$$

if $|\phi_1| < 0$. The covariance between y_t and y_{t-j} is

$$E\left(y_t y_{t-j}\right) = E\left\{(\epsilon_t + \phi_1 \epsilon_{t-1} + \phi_1^2 \epsilon_{t-2} + \cdots)(\epsilon_{t-j} + \phi_1 \epsilon_{t-j-1} + \phi_1^2 \epsilon_{t-j-2} + \cdots)\right\}$$

$$= \frac{\phi_1^j \sigma_\epsilon^2}{1 - \phi_1^2}$$

Thus the autocorrelation between y_t and y_{t-j} is ϕ_1^j.

We say that a time series is strictly stationary if the joint distribution of the vector $(y_{t_1}, y_{t_2}, y_{t_3}, \ldots, y_{t_K})$ depends only on the intervals between the y's and not on t. A time series is said to be weakly or covariance stationary if the second moments exist and depend only on the interval between the y's. Throughout this book, we use the term "stationary" to mean covariance stationary.

Because the covariances of the elements of a stationary time series depend only on their separation in time (and not on their location on the time line), we can write the autocovariances of a stationary time series as a function of the interval between the observations, $E(y_t y_{t-j}) \equiv \gamma_j$. We denote the autocorrelations by ρ_j.

Because the time series we consider in this book follow linear stochastic difference equations of the form in (2.4), the dynamic properties of these difference equations determine whether a series is stationary or not. If the difference equation implies a damped response to random shocks, all the autocorrelations will be finite and the series is stationary. If the difference equation implies an unbounded or explosive response to random shocks, the series is nonstationary.

We can use the lag operator $(Ly_t \equiv y_{t-1})$ to rewrite (2.4) in terms of y_t alone.

$$y_t = \phi_1 L y_t + \phi_2 L^2 y_t + \cdots + \phi_p L^p y_t + \epsilon_t$$

Collecting terms, we can write

$$\phi(L) y_t = \epsilon_t$$

where

$$\phi(L) \equiv 1 - \phi_1 L - \phi_2 L^2 - \cdots - \phi_p L^p$$

a polynomial in the lag operator, L.

The deterministic difference equation

$$\phi(L) y_t = 0 \tag{2.5}$$

defines the dynamic behavior of y_t, that is, its dynamic response to the random disturbances introduced by ϵ_t. Replacing the lag operator, L, with an ordinary algebraic variable, z, we can infer the dynamic behavior implied by (2.5) by calculating the roots of

$$\phi(z) = 0$$

We can always rewrite this polynomial equation as the product of p terms

$$\phi(z) = \Pi_{i=1}^{p}(1 - \lambda_i z) = 0$$

where $1/\lambda_i$ is the ith root of $\phi(z)$.

The response of y_t to random shocks is damped (the impacts eventually die out) when all the roots of $\phi(z)$ lie outside the unit circle ($|\lambda_i| < 1, \forall i$). The response of y_t to random shocks is explosive when any of the $1/\lambda_i$ lies within the unit circle. If all the roots are real numbers, shocks either die out or explode exponentially. Complex roots come in pairs of complex conjugates. Complex roots induce cyclical responses (either damped or explosive) to random shocks.

Roots that lie on the unit circle, called unit roots ($|\lambda_i| = 1$), produce permanent responses—neither damped nor explosive—to random shocks. An important example of a process with a unit root is the random walk:

$$y_t = y_{t-1} + \epsilon_t = \sum_{i=0}^{\infty} \epsilon_{t-i}$$

This process has mean 0 but infinite variance:

$$E\left(y_t^2\right) = \Sigma_{i=0}^{\infty}\sigma_\epsilon^2$$

2.5.2   ARMA models

Perhaps more than in other branches of statistics, there is an explicit recognition that time-series models are intended as approximations that characterize the dynamic behavior of the underlying series. Only in rare circumstances would we talk about a "true model" for a time series. Instead we focus on determining whether a time-series model provides an adequate approximation to observed behavior.

In many instances, it may take a large number of lagged terms to provide an adequate fit with the pure autoregression specification in (2.4). Often we can fit an observed time series more parsimoniously by including a moving-average component, that is, a sum of weighted lags of the random shock, ϵ_t. The combined ARMA model is written as

$$y_t = \phi_1 y_{t-1} + \phi_2 y_{t-2} + \cdots + \phi_p y_{t-p} + \epsilon_t - \theta_1 \epsilon_{t-1} - \cdots - \theta_q \epsilon_{t-q} \qquad (2.6)$$

Equation 2.6 is called an ARMA(p,q) model, that is, a model with a pth-order autoregressive component (difference equation in the observable time series, y_t) and a qth-order moving-average component (weighted sum of lags of the unobservable random shock, ϵ_t). Using the lag operator, we write this model more compactly as

$$\phi(L)y_t = \theta(L)\epsilon_t$$

While the roots of $\phi(z) = 0$ are fundamental to the behavior of the observable y_t, the properties of $\theta(z) = 0$ are somewhat arbitrary because ϵ_t is unobservable and are set in part by normalization.

Chapters 6 and 7 analyze the properties of ARMA models in detail. Chapter 8 extends the ARMA model to allow for time-varying σ_ϵ^2. Chapter 9 considers multiple equation time-series models, and chapter 10 characterizes relationships among nonstationary time series.

3 Filtering time-series data

Chapter map

3.1 Preparing to analyze a time series. Know your data. A checklist of questions to answer before your analysis. Extra questions that must be answered for time-series data.

3.2 The four components of a time series. Trend, cycle, seasonality, and random noise.

3.3 Some simple filters. The assumption underlying all smoothers. The strengths and weaknesses of different elementary smoothers. Some practical advice.

3.4 Additional filters. Adding an explicit connection between time periods. Exponentially weighted moving averages (EWMAs) and Holt–Winters smoothers.

3.5 Points to remember. Before analyzing data. The systematic portion of a time series. Filter performance. Decomposing the signal.

The preliminaries are finally out of the way, and we can start looking at some time series. But how exactly do we go about doing this? The initial assessment of a time series is a challenging but essential part of a successful analysis. This chapter provides some tools and advice to get you started.

When presented with fresh data, a researcher finds himself or herself in one of two situations:

- *The data are familiar.* They comprise measurements on phenomena previously studied by the researcher or exhaustively analyzed in the literature. In this case, the researcher has a good idea of the processes that generated the data, the parametric forms likely to provide a good fit, and the appropriate estimation techniques to use. The project proceeds rapidly from routine data preparation (checking for outliers that may indicate typos or transmission errors, transforming the data to stabilize variance, etc.) to confirmatory analysis—model estimation, prediction, and hypothesis testing. We will begin discussing confirmatory techniques—a lot of them—in a while, but we have a bit of preparation to do first.

- *The data are unfamiliar*, at least to the researcher. In this case, the researcher must look at the data—ask questions about the data definition and collection methods, produce graphs, calculate descriptive statistics—prior to any model specification and estimation. This exploratory analysis is much more challenging, particularly with time series, where model specification is notoriously difficult.[1] If this describes your situation, pay particular attention to this chapter and the next one.

One of the most counterproductive cliches is "Let the data speak for themselves". Data do not speak for themselves. Raw data are noisy, peppered with distracting and misleading artifacts. To make sense of the data, we have to find some method of filtering out the noise and revealing the underlying signal.

There is no perfect filter, one that cleanly separates all the random disturbances from the systematic component. Especially with nonexperimental data such as economic or financial time series, the random component of the data may be so substantial that filters struggle to distinguish signal from noise. Filters that make few assumptions about the underlying data-generation process are particularly challenged. The ability to discriminate between signal and noise can be improved by using filters that incorporate more (and more specific) assumptions about the data-generation process. However, if these assumptions are misguided, the filter is likely to produce misleading results.

This chapter begins with a checklist of "things you should do before you do anything", that is, things you should do prior to firing up Stata and analyzing your data. These preparatory steps are essential but too often overlooked. Next we present an intuitive structural decomposition of a time series into trend, cycle, and seasonal components plus random disturbances. Filtering is designed to highlight and quantify these commonsense features. Even though they are presented informally here, these concepts translate fairly directly to features of the formal models covered in later chapters. The remainder of the chapter discusses various filters that can be applied to time-series data, beginning with filters that make very few assumptions about the data-generation process and continuing with filters that incorporate incrementally more specific assumptions about the underlying structure of the data. All of these filters fall into the category of smoothers; that is, they assume at a minimum that the systematic component of a series evolves smoothly over time. This assumption allows the smoothers to identify rapid variation in the series as random noise, which can be separated from the underlying smooth systematic component of the series.

---

1. Tukey's (1977) introduction to exploratory techniques provides an excellent discussion of the relationship between exploratory and confirmatory analysis.

3.1   Preparing to analyze a time series

The first and most important step in analyzing unfamiliar data is to ask lots of questions about the definition of the data and the circumstances of their collection. In short, know your data. The availability of powerful and easy-to-use tools such as Stata makes it almost irresistible to plunge immediately into graphing and measuring the data. Resist. A firm understanding of the data reduces the risk of mistaking an artifact for a relationship and improves your chances of identifying an appropriate model for the data. In a book such as this, where data are used primarily to illustrate the concepts and tools, this essential step necessarily receives short shrift. However, the importance of thinking carefully about the phenomena that the data are intended to reflect and the strengths and weaknesses of the data collection methods cannot be overemphasized, especially for time-series analysis, where model choice is particularly challenging.

3.1.1   Questions for all types of data

How are the variables defined?

Table 3.1 lists questions you should ask and answer prior to your data analysis. Some of the questions you should ask are straightforward. How are the variables defined? What are the units of measurement? How was the sample selected?

Table 3.1. Questions to answer prior to data analysis

	General questions	Follow-up questions
All types of data	How are the variables defined?	What are the units of measurement? Do the data comprise a sample? If so, how was the sample drawn?
	What is the relationship between the data and the phenomenon of interest?	Are the variables direct measurements of the phenomenon of interest, proxies, correlates, etc.?
	Who compiled the data?	Is the data provider unbiased? Does the provider possess the skills and resources to ensure data quality and integrity?
	What processes generated the data?	What theory or theories can account for the relationships between the variables in the data?
Time-series data	What is the frequency of measurement?	Are the variables measured hourly, daily, monthly, etc.? How are gaps in the data (for example, weekends and holidays) handled?
	What is the type of measurement?	Are the data a snapshot at a point in time, an average over time, a cumulative value over time, etc.?
	Are the data seasonally adjusted?	If so, what is the adjustment method? Does this method introduce artifacts in the reported series?
	Are the data revised?	Should your analysis be based on the final revision or on the preliminary data available at the time?

What is the relationship between the data and the phenomenon of interest?

Some of the questions in table 3.1 are subtle and pertain to the relationship between the data you have and the phenomenon you aim to study. For instance, how directly do the data measure the phenomenon of interest? In many cases, measurements of the specific phenomenon are not available, and researchers look for proxies—data that provide indirect measurements. A good example is human intelligence, which is not directly observable. If you are studying the determinants of intelligence, your data are likely to contain proxies such as IQ scores, which you hope are closely related to the unobservable intelligence.

As another example, consider two frequently cited measures of the condition of the labor market in the United States: the monthly change in nonfarm payrolls and the unemployment rate. Both statistics are compiled by the Bureau of Labor Statistics (BLS) and published on the first Friday of each month. These statistics provide timely information on the strength of the U.S. economy (in contrast, gross domestic product [GDP] is published only quarterly) and frequently move financial markets and influence government policy decisions. However, even though both statistics are measurements of employment trends, they do not always provide consistent signals.

The graph below displays a scatterplot of the change in the unemployment rate against the change in nonfarm payroll employment from January 1950 through January 2012. The expected negative correlation is evident (increases in nonfarm payrolls tend to coincide with decreases in the unemployment rate) but far from perfect. The correlation coefficient is only −0.52. Furthermore, the indicators provided contradictory information—both statistics increased or decreased simultaneously—24% of the time.[2]

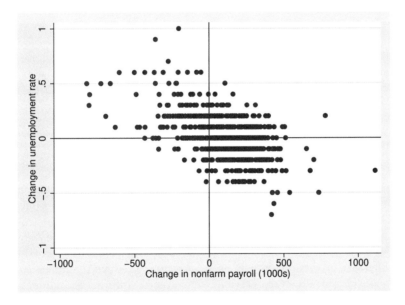

Figure 3.1. The relationship between the unemployment rate and nonfarm payroll employment, January 1950 to January 2012

_____

2. The definition of the unemployment rate—the number of unemployed persons divided by the number of persons in the labor force—muddies the water a bit. The numerator is defined as the number of persons currently out of work who have actively sought jobs in the previous four weeks, while the denominator is the sum of employed and unemployed persons. In periods of economic weakness, some workers become discouraged by their poor prospects for employment and exit the labor force (that is, stop looking for work). This phenomenon changes both the numerator and the denominator of the unemployment rate in ways that can generate misleading results. The BLS tallies alternative measures of the unemployment rate that attempt to correct for this effect.

Nonfarm payroll employment is essentially an employer-tallied head count. The unemployment rate relies on individual self-reports to determine whether someone is unemployed. To qualify as unemployed, a person must be currently out of work, willing and able to work for pay, and actively seeking employment. Clearly, the measurement of unemployment involves substantially more judgment and interpretation than the measurement of nonfarm payroll employment. As a consequence, financial markets tend to place greater weight on nonfarm payrolls than on the unemployment rate as an indicator of the strength of the economy.[3]

Who compiled the data?

Another important issue to consider is the source of your data. Many time series are published by government, academic, or private entities that adhere to the highest standards of statistical skill and integrity. Aspects of the sampling methodology, survey construction, or seasonal adjustment may inadvertently impart a bias to the data—one socioeconomic group may be underrepresented, self-reporting may be self-serving, etc.— but every effort is made to ensure data quality.[4] Some data, however, are collected by interested parties, and researchers must consider the source in those cases. For example, many aspects of accounting and valuation in financial services firms require significant management judgment. Even with the best of motives and controls, it is not surprising that the forces pressing on senior management have some influence on the pattern of accounting reports.

What processes generated the data?

Separate from any consideration of the definition or accuracy of the data, it is essential to think carefully about the real-world processes that generated the data, that is, the theory that explains the data and of which the data provide a test. Despite the many useful statistical tools available to a researcher, theoryless data mining turns out to be of limited applicability in nonexperimental time-series data—economic data, financial data, demographic data, and the like. In a later chapter, we will discuss vector autoregression, which was introduced in part as an attempt to reduce the influence of a priori theory on model choice and interpretation. While vector autoregression remains a useful tool, the experience of researchers in the last few decades has confirmed the essential contribution of theory in making sense of economic time series.

---

3. For more information on the construction of these statistics, see the *BLS Handbook of Methods* (https://www.bls.gov/opub/hom/).

4. To help pay my way through graduate school, I taught an introductory economics course in the evening at a local state college. Most of the students were far older and more worldly wise than I. In the middle of my explanation of how the BLS measures unemployment, one of my students interrupted to inform me that I was a naïve youngster and that the president simply called the BLS to tell it what unemployment rate to report. While I am older and more skeptical now, I still do not worry about the integrity of the BLS statistics.

3.1.2   Questions specifically for time-series data

What is the frequency of measurement?

For time-series data, there are a host of additional questions. What is the frequency of data collection—hourly, daily, weekly, etc.? Do the data record a snapshot at a point in time (the daily closing value of the Dow Jones Industrial Average stock index), a cumulative value (the number of new claims for unemployment insurance over the week), or an average value over time (the mean temperature in January in Seattle)? Was the data recorded at regular intervals or intermittently, perhaps triggered by an event? If the normal measurement or publication date of an economic time series falls on a business holiday, what happens? What is the impact of a holiday-shortened measurement period on the data?

Are the data seasonally adjusted?

These last two questions lead us to the issue of seasonal adjustment. Many time series have predictable seasonal (or time-of-day) variations, and many published economic time series are seasonally adjusted to remove this predictable component. In other words, the data already are filtered. When done correctly, seasonal adjustment can make it easier to analyze time-series data. However, when done incorrectly, seasonal adjustment can inject artifacts and noise into time series, making it more difficult to analyze the data. Some researchers prefer to work with unadjusted data and to apply their own filters. In many cases, though, only adjusted data are published. In any event, the researcher needs to know if any seasonal adjustment filter has been applied and, ideally, to know the details of the filtering algorithm.

Are the data revised?

Researchers encounter one additional wrinkle with economic time series. The data typically are revised, sometimes significantly, over time. There are several reasons for revisions. In survey data like the employment data discussed above, some responses arrive too late to be included in the current month's calculation. In addition, some respondents initially may submit incorrect or incomplete information that is corrected later. The agency conducting the survey will revise statistics from the previous month or two to incorporate this fuller, more accurate information. Some series also receive more thorough baseline revisions every few years that affect multiple years of data. Agencies also may change data collection methodology occasionally. These changes can reflect advances in methodology or simply the agency's adaptation to changes up or down in its data collection budget. When methodologies change, older data may be recalculated to ensure consistency in measurement over time. Or the data may be published using both the old and the new methodology for a time to allow users to determine their own splicing algorithm. More often, the break in methodology simply is noted.[5]

An important assumption in much of finance and economics is the efficient use of information. For example, investors are assumed to forecast the future price of a stock by conditioning on all the information available to them at present. In other words, they form their expectations in period t of the stock price s periods from now according to

$$P^*_{t+s} \equiv E(P_{t+s}|I_t)$$

where P is the stock price and I_t is all the relevant information available at time t. To test versions of this theory, researchers need to use the data actually available to investors in period t, not revised data available only at a later date.

3.2 The four components of a time series

One of the first steps in the exploratory analysis of cross-section data is the examination of the empirical distributions of the variables considered one at a time. Are the data unimodal? Are they distributed symmetrically? Do they appear to follow a known distribution like the normal distribution? Are there outliers? The goal is to find transformations of the data (called re-expressions in the exploratory data analysis literature) that produce well-behaved data distributions. In subsequent stages of the analysis, relationships between pairs (or sets) of variables are examined with the goal of finding transformations that produce straight-line fits between continuous variables or main effects fits (that is, fits without interactions between the explanatory variables) between continuous and categorical variables.

---

5. The Federal Reserve Bank of St. Louis provides many economic time series in their preliminary, revised, and final versions through its FRASER project (https://fraser.stlouisfed.org). I thank Craig Hakkio of the Federal Reserve Bank of Kansas City for bringing the FRASER project to my attention. The Federal Reserve Bank of St. Louis also maintains an associated public database called FRED (Federal Reserve Economic Data). Stata provides a method for downloading data from FRED into Stata. Type `help import fred` for more information.

The exploratory analysis of time-series data is significantly different. In the first place, except for trivial cases, there is no meaningful univariate analysis of a time series. So-called univariate time-series analysis actually is the analysis of the bivariate relationship between the variable of interest and time. Furthermore, the goal of univariate time-series analysis is not to find a straight-line relationship between a transformation of the variable of interest and time but rather to isolate the signal—the systematic, predictable components of the time series—from the random noise that obscures the signal and makes the series not completely predictable.

The signal in time-series data usually is divided into three components: trend, cycle, and seasonal. Each of these components describes a different mechanism by which past values of a time series may be related to the present value. A given time series may incorporate all of these components or just one or two of them. The fourth component of a time series is random noise, which muddies the water a bit and makes it hard to isolate each of the three signal components.

We discuss each of these three components of the signal in turn.

Trend

The trend component denotes a persistent, systematic tendency for a series to increase or decrease. For example, the population of the United States has trended upward for the last 200 years or so. For a series with a positive (negative) trend, the present value is likely to be higher (lower) than past values.

Trends may change over time, both in magnitude and even in direction, and it is common to talk about the current or recent or local trend, such as a recent trend in stock prices where prices trend up in a "bull" market and down in a "bear" market. This notion of changeable trends overlaps with the notion of a cycle. In common parlance, a trend is supposed to be longer lived than the upward or downward phase in a cycle, but the distinction is informal and more a matter of convention than statistics.[6]

For a real-world example of trend, let's look at U.S. GDP. The data in figure 3.2 are quarterly measurements of nominal GDP in billions of dollars, seasonally adjusted and converted to an annual rate.[7] There are important nontrend components to GDP, but the trend is the dominant visual feature.

---

6. The lack of a clear distinction between a variable trend and an aperiodic cycle foreshadows the slippery nature of time-series model specification. Components of time-series models—parameters with specific functions and interpretations—often can be "traded off" without diminishing the model fit. This fungibility introduces an element of art to time-series model identification.

7. GDP is published by the Bureau of Economic Analysis, a division of the U.S. Department of Commerce, as part of an extensive set of national income and product accounts. An explanation of the national income and product accounts and their construction can be found on the Department of Commerce website at https://www.bea.gov/resources/methodologies/nipa-handbook.

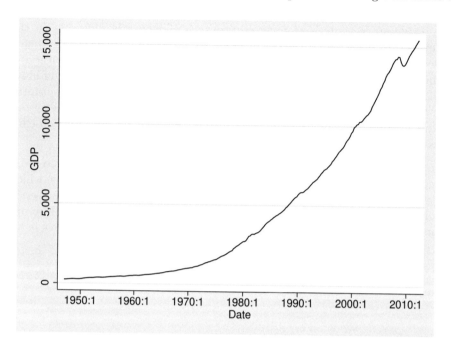

Figure 3.2. U.S. GDP, billions of dollars, 1947:1 through 2012:1

The upward trend of GDP over time is unmistakable. The downward "hook" near the end of the series is evidence of the recession that began in 2008.

In the special case of trend, it often is useful to find a transformation that induces a straight-line relationship between the variable of interest and time. From the graph, it appears that GDP is characterized by exponential growth, so a log transformation may be useful. To confirm this idea, we change the vertical axis to a log scale in figure 3.3.[8]

---

8. We also converted the units to trillions of dollars so that the vertical axis labels remained readable.

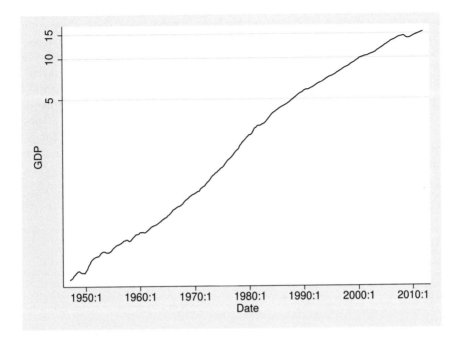

Figure 3.3. Log of U.S. GDP, trillions of dollars, 1947:1 through 2012:1

Cycle

Many time series oscillate around a trend if the series incorporates a trend, or around the mean if the series has no trend. In most real-world cases, a complete cycle covers multiple time periods. In other words, above-trend values in the recent past indicate a likelihood that the current value also will be above trend, and vice versa.

The use of the term "cycle" is a bit misleading. In most cases, the cyclical component of a time series does not resemble the regular, periodic oscillation of a trigonometric function such as the sine or cosine. Instead the cycles in most time-series data are aperiodic—the series oscillates around the trend, but the timing and duration of the excursions above and below the trend are irregular. The business cycle in the United States provides a good example of an aperiodic cycle. Expansions—periods of trend or greater economic growth—are long lasting and of varied lengths. In a time-series graph, expansions resemble hills. On the other hand, recessions—periods of negative or below-trend growth—tend to be short and sharp, usually lasting no more than a few quarters.[9] Graphs of recessions look like Vs.[10]

---

9. The recession that began in 2008 is a painful counterexample.
10. The terms "recession" and "expansion" often are used loosely to indicate periods of slower or faster growth. In the financial press, two successive quarters of negative GDP growth are termed a recession. Informally, it is often said that a recession is when your neighbor is out of work, while a depression is when you are out of work. Despite this confusion, there are official definitions of these terms. The language we use to describe business cycles was introduced and formalized by Wesley Mitchell, A. F. Burns, and other researchers at the National Bureau of Economic Research (NBER), a private, nonprofit economic research organization founded in 1920. The NBER's Business Cycle Dating Committee reviews a wide variety of data about the macroeconomy and decides when to date the peaks and troughs in the macroeconomy. According to the NBER, the periods between peaks and troughs are recessions, and the periods between troughs and peaks are expansions. These business cycle dating decisions are made retrospectively to avoid overreacting to events in real time and to allow all the relevant data to settle. In this book, we adhere to the NBER definition of recessions and expansions.

For an example of cycle, let's look at the civilian unemployment rate we discussed above.

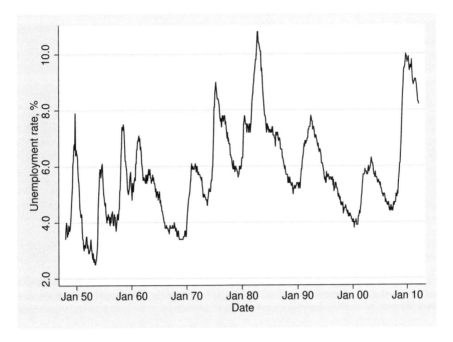

Figure 3.4. U.S. civilian unemployment rate, January 1948 through March 2012

The sharp increase in the unemployment rate at the right-hand side of the graph is another indicator of the post-2007 recession. This series does not appear to exhibit a trend, nor did we expect to find one. The unemployment rate represents the sum of frictional unemployment, a base rate of unemployment that reflects the typical time required to match the workers entering the labor force to the jobs becoming available; cyclical unemployment, the increase or decrease in the unemployment rate, relative to the frictional rate, in response to swings in the strength of the economy; and structural unemployment, a long-lasting base of unemployment attributable to mismatches between the assortment of skills in the labor force and the skills required for available jobs. The lack of an obvious trend in this graph suggests (but does not prove) that the rate of unemployment accounted for by the sum of frictional and structural unemployment has been relatively stable during the post–World War II era. Instead the shape of the time line is dominated by the cyclical component of the unemployment rate.

As we noted above, the oscillations in the unemployment rate are not smooth and symmetric; that is, they do not resemble a periodic function such as a simple sine wave. Episodes of increases in the unemployment rate tend to be explosive but short lived. Episodes of decreases in the unemployment rate—aside from the snapback from episodes of sudden increases—tend to be gradual and last many months.[11]

Seasonal

Another type of cycle is seasonality, the tendency of some series to increase or decrease in predictable ways at the same time of the day, the same day of the week, the same month of the year, etc. The average daily high temperature follows a seasonal pattern, as does the average temperature in a 24-hour period. This latter example of a seasonal component happens to have a smoothly oscillating character. Temperatures tend to rise during the morning, peak in the afternoon, and decline in the evening and overnight. Other seasonal variations are irregular, not resembling a smooth cycle at all. For instance, bank deposits rise and fall sharply around tax payment dates.

For another example of seasonality, we will look at prepayments on residential mortgages in the United States. We will restrict our attention to discount mortgages, that is, mortgages with interest rates lower than the current market rate. When a homeowner's mortgage rate is significantly lower than the rate available on new mortgages, there is no incentive for the borrower to refinance.[12] In this case, a mortgage prepayment indicates that the borrower has sold the existing home and, most likely, bought another. This type of prepayment is evidence of a life event—new job, new child, retirement, divorce, and the like—and these prepayments tend to be greater in some months than in others. Families with children try to avoid moving during the school year. Moreover, it is easier to move during the summer than during the winter, when snow and low temperatures make moving more challenging.

Figure 3.5 displays the average prepayment rate for a selection of 30-year fixed rate mortgages securitized by Fannie Mae in each of the 12 months of the year. The sample is restricted to mortgages that are at least three years old and that bear interest rates at least 0.5 percentage points lower than the prevailing rate on a new, 30-year fixed rate mortgage.[13] In other words, these prepayments represent home sales rather than refinances.

---

11. Because higher unemployment rates are evidence of weaker economic growth, the short, sharp Vs of recessions are upside down in figure 3.4.
12. For mortgages with rates close to the prevailing rate, the homeowner may refinance nonetheless to tap the equity in the home. This type of refinance is called a cash-out refinance. Cash-outs have become less common since the onset of the housing crisis and the ensuing tightening of underwriting standards.
13. Mortgages tend to have very low rates of prepayment in their early years. Homeowners tend not to move every year.

The seasonal component of prepayment rates in this sample is very strong. Prepayments are lowest after the start of the school year and over the Thanksgiving and Christmas holidays. Prepayments pick up in early spring and peak during the summer months, as expected.

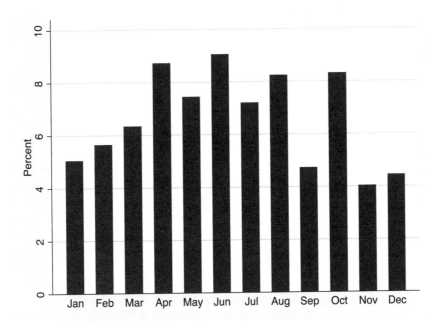

Figure 3.5. Average monthly prepayment rates (annualized) on seasoned Federal National Mortgage Association 30-year discount mortgages, 2000–2007

We have selected examples that provide clear instances of trend, cycle, and seasonal components. As we noted, it is not always this easy to distinguish these components, especially in series that incorporate more than one component. For example, look more closely at figure 3.4. It appears that the mean unemployment rate has shifted over time. There appears to have been an upward trend in the unemployment rate between January 1948 and December 1982, followed, perhaps, by a downward trend thereafter. Disentangling these components of the signal from each other and from the random noise in the series is challenging.

Smoothers provide tools for filtering out the noise component in a series, making it easier to isolate the systematic components. The next section provides a brief introduction to a simple class of smoothers useful for many types of data. These simple smoothers are motivated by the assumption that the signal in a series evolves smoothly over time, while the noise component is erratic. However, these smoothers do not attempt to decompose the signal into trend, cycle, or seasonal components. The subsequent section looks more closely at smoothers developed specifically for time-series data. These time-series smoothers add incrementally more assumptions about the structure of the signal in the series, ultimately separating out the trend, cycle, and seasonal components.

3.3   Some simple filters

In the section above, we described time-series variables as a combination of signal—a collection of systematic, predictable components—and noise, an unpredictable, random component. In other words,

$$\text{series} = \text{signal} + \text{noise}$$

A statistician might express this same decomposition as

$$\text{series} = \text{fit} + \text{residual}$$

At this stage, we assume only that the fit is tolerably well behaved, the product of a gradually evolving trend or a smoothly varying cycle.[14] The residual, on the other hand, is erratic and unsystematic. As a consequence, rapid, "jumpy" variation in the series can be attributed to the residual and subtracted from the observed series, producing a smooth estimate of the fit.

At a later stage of the analysis, we will want to specify a mathematical function for the fit and a probability distribution for the residual. At that point, we will describe the decomposition as

$$\text{series} = \text{model} + \text{error}$$

and apply an appropriate technique to recover estimates of the parameters of the function and of the probability distribution. At this exploratory stage, though, our aims are more modest. Right now, we are focusing only on filtering out as much of the noise (residual) as possible so that we can get a clearer look at the signal (fit).[15]

Although we are not proposing a model or functional form for the signal at this point, the signal is some sort of function of time. That is why graphs of the series versus time are the focus. They serve to give us clues about the type of function of time that makes the most sense.

---

14. The smoothers discussed in this section have difficulty handling a nonsmooth seasonal component, like the surge and fall in bank deposits around tax dates.
15. To add yet another bit of terminology, Tukey (1977) describes the result of smoothing as dividing the series into the smooth and the rough. Because no smoother perfectly discriminates between signal and noise, the smooth and the rough will of necessity differ somewhat from the signal and noise.

The smoothers discussed in this section can be applied to any scatterplot where there is no more than one observation of the series of interest, the y value in the scatterplot, for each value of the series on the x axis. In other words, the points in the scatterplot can be regarded as a graphic rendition of a table of function values. For time series, some measure of time defines the x axis, but for other applications, the variable on the x axis may be any measure that embodies a concept of "closeness"—location in space, temperature, age, etc. These smoothers are most appropriate if the distance between x values is uniform. Other more complex scatterplot smoothers are designed for general scatterplots with multiple y measurements for each x value and irregular spacing of x values, but they are not particularly useful in time-series analysis.[16]

Even in the restricted class of simple smoothers discussed in this section, the great number of choices can make it difficult to select one. However, the intuition behind most smoothers is similar. We can write our decomposition of a time series, y_t, into a fit and a residual as

$$y_t = f(t) + e_t$$

where $f(t)$ is the fit and e_t is the residual. For well-behaved functions—and we are assuming that $f(t)$ is tolerably well behaved—observations sufficiently close in time to t should exhibit values close to $f(t)$. In contrast, the values of the residuals are assumed to be independent, random noise, and hence, there is no reason to expect residuals that occur around the same time to bear any relation to each other.[17] Thus some measure of the typical value of the series in the neighborhood of time t should be less noisy—that is, a better estimate of $f(t)$—than y_t by itself.[18]

Consider N observations at times t_j, which are reasonably close in time to t_i. One possible smoother is the mean of the associated y values, that is,

$$y_{t_i}^* = 1/N \sum y_{t_j} = 1/N \sum f(t_j) + 1/N \sum e_{t_j} \approx f(t_i) + 1/N \sum e_{t_j}$$

If the residuals are independent with mean 0 and constant variance σ^2, the variance of the sum of the residuals is σ^2/N^2. The smoother does not completely eliminate the residual component from the smoothed series, but it does reduce the variance of the residual component, making the fit more prominent.[19] In this approach, we define a

---

16. There also are surface smoothers for y variables that are functions of several x's, but they are a little more complicated to describe. For our purposes, the bivariate scatterplot smoothers are sufficient.

17. Looking ahead to future chapters, we will point out that many time-series models incorporate serially correlated error terms. Smoothers discussed here extract the predictable, that is, serially correlated, component of the error term from the residual and add it to $f(t)$.

18. Somewhat more formally, and not necessarily restricted to the case of time series, we are assuming that if

$$y_i = f(x_i) + e_i$$

then

$$|x_i - x_j| \text{ small} \Rightarrow |f(x_i) - f(x_j)| \text{ small}$$

19. This variance-reducing property holds under more general assumptions about the residual than independence and constant variance.

smoother by just two characteristics: 1) span, the number of adjacent points included in the calculation; and 2) the type of estimator—median, mean, weighted mean, etc.—used to calculate $y_{t_i}^*$.

Stata offers a wide variety of time-series smoothers through its family of `tssmooth` commands. All the `tssmooth` commands have the form

`tssmooth` *type_of_smoother newvar* = *exp*

where the *type_of_smoother* subcommand selects the general smoothing method to apply. Depending on the command, various options are provided to control the details of the smoothing.

We will begin with the `tssmooth nl` command, which provides median smoothers and a special case of a weighted-average smoother. As the name suggests, median smoothers set the fit to the median of the observations within each span. For instance, a span-5 median smoother calculates the tth value of the fit as

$$y_t^* = \text{median of } (y_{t-2}, y_{t-1}, y_t, y_{t+1}, y_{t+2})$$

`tssmooth nl` allows you to calculate median smoothers with spans from one to nine observations. `tssmooth nl` also offers a span-3 weighted mean, called the Hanning smoother, that is defined as

$$y_t^* = (y_{t-1} + 2y_t + y_{t+1})/4$$

The smoothers in the `tssmooth nl`, while easy to calculate and understand, are fairly sophisticated mathematically, and the theory underlying them is challenging. Nonetheless, they provide a useful point of entry to the broader topic of extracting the signal from a noisy time series.[20]

The syntax of the `tssmooth nl` command is

`tssmooth nl` [*type*] *newvar* = *exp* [*if*] [*in*], <u>sm</u>oother(*smoother*[, <u>twice</u>]) [`replace`]

---

20. Stata provides another command called `smooth` that produces results that are identical to those of the `tssmooth nl` command. The only difference is `smooth` can be applied to non-time-series data, while all the `tssmooth` commands can be used only on time-series data after the data have been `tsset`.

The *smoother* is specified as a string of case-insensitive alphanumeric characters. The median smoothers are specified by a digit between 1 and 9, inclusive, where the digit indicates the span of the median. The Hanning smoother is specified by the letter H. `tssmooth nl` also allows compound smoothers. For instance, the string 35H indicates a sequence of three smoothers: first, a span-3 median is applied to the observed series; then a span-5 median is applied to the output of the span-3 median; and finally, a Hanning smoother is applied to the output of the previous two steps. Compound smoothers can provide further refinements in the estimated smooth and further reductions in the residual variance.

Three additional characters—R, E, and S—can appear in the smoother. The letter R indicates that the previous smoothing operation should be repeated until the fit converges. For example, the string 3R indicates that the span-3 median should be applied repeatedly until there is no further change in the fit. The R operation should be used only with odd-span median smoothers because the other smoothers are not guaranteed to converge. The E operator applies special treatment to the endpoints of a series and is described a little later in this section. The S operator is intended to improve the fit when the initial smoother produces repeated values—a common occurrence with the span-3 median smoother. S splits the time series into subseries at every occurrence of repeated values. Then it applies the E operator (which we have not explained yet) to the split ends and recombines the subseries. From this explanation, it's probably not clear exactly what S does or how it improves the fit, but be patient. All will be clear when we get to an example. The S operator may be used only after 3, 3R, or another S.

The smoothers provided by the `tssmooth nl` command do not guarantee perfect separation of the fit and the residual. That type of performance is too much to hope for with any statistical tool. One way of improving the performance of the smoother is to resmooth the estimated residual. The `twice` option automates that process. The specified smoother is applied twice—once to the observed series, then to the resulting residual. Any fit extracted from the residual in this second stage is added to the fit generated in the first stage.

Given all of these choices, how should we choose a smoother? In practice, researchers use complex compound smoothers, such as 4253H, `twice`. However, we are going to start with elementary smoothers applied to artificial data for a trend, a cycle, and a seasonal pattern to understand the strengths and weaknesses of each smoother by itself and in combination with other smoothers.

3.3.1 Smoothing a trend

First, let's construct an ideal unobserved trend, a straight line for 60 "months". To generate the "observed" series, we add random noise and an outlier.

```
. drop _all
. set obs 60
number of observations (_N) was 0, now 60
```

```
. generate int month = _n
. label variable month Month
. tsset month, monthly
        time variable:  month, 1960m2 to 1965m1
                delta:  1 month
. generate trend = _n
. label variable trend Trend
. set seed 53
. generate rawtrend = trend + 4*rnormal()
. replace rawtrend = 20 in 46                    /* Outlier */
(1 real change made)
. label variable rawtrend "Raw data"
. generate resid = rawtrend - trend
. label variable resid Residual
. format trend rawtrend residual %6.2f
. save trend
file trend.dta saved
. tsline trend rawtrend
```

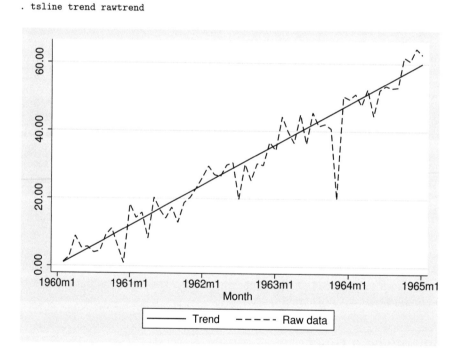

Figure 3.6. Trend (unobserved) and Trend+Residual (observed)

We saved the trend data in a Stata dataset for later use. Note that we used the `tsset` command to mark the data as time-series data with `month` as the time variable. Stata interpreted the values of `month`, which we set equal to the observation numbers, as best it could. All of Stata's time-series commands, including the `tssmooth` commands, require that the data be `tsset`.

Now let's generate some smoothers and look at the results.

```
. foreach smoother in h 3 3h h3 9 {
  2.         tssmooth nl S`smoother´ = rawtrend, smoother(`smoother´)
  3. }
```

As a convenience, we used Stata's `foreach` command to avoid typing five different `tssmooth nl` commands.[21] This command created the local macro `smoother`, then set this macro to each of the elements of the list following the keyword `in` (h 3 3h h3 9) in turn, and executed the resulting commands. In all, five separate `tssmooth nl` commands were executed. The first was

```
. tssmooth nl Sh = rawtrend, smoother(h)
```

and so on. The variable names for the smoothed series all begin with an uppercase `S` followed by the characters that define the smoother used. For instance, `S3h` is the series generated by applying the span-3 median smoother followed by the Hanning smoother to `rawtrend`.

---

21. See [P] **foreach** for a full description of the `foreach` command.

We have a lot to look at here. We will start by comparing the span-3 median smoother with the Hanning smoother.[22]

```
. tsline rawtrend S3 Sh, lpattern(solid dash shortdash) legend(cols(1))
```

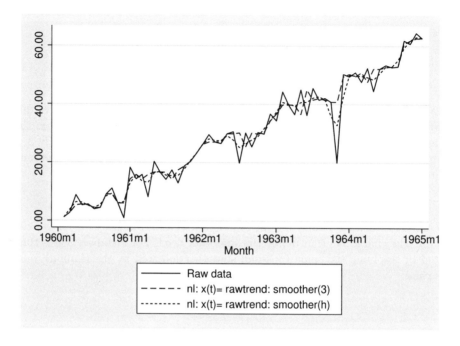

Figure 3.7. A span-3 median smoother and a Hanning smoother

We can see that both smoothed series tend to avoid the largest deviations from trend that occur in the observed series, `rawtrend`. However, it looks as if the smoothers handle the large outlier in month 46 somewhat differently. Let's zoom in on the months around this outlier to get a better look.

---

22. We specified `legend(cols(1))` to place the legend items in one column.

Figure 3.8. A close-up of the outlier in month 46

This figure highlights the vulnerability of the Hanning smoother—and any other mean-based smoother—to outliers. Median-based smoothers have greater resistance to outliers, as shown by the span-3 median smoother here, which ignores the outlier completely.

The vulnerability of Hanning to outliers can be overcome by combining it with a median-based smoother.

```
. tsline rawtrend S3 S3h Sh3 in 30/50, lpattern(solid dash shortdash longdash_dot)
> legend(cols(1))
```

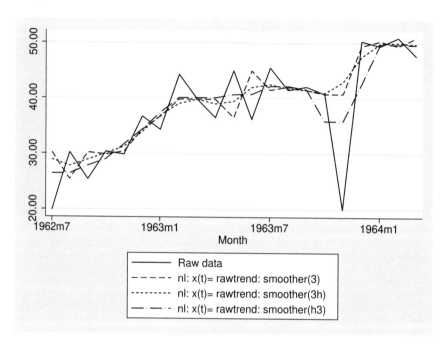

Figure 3.9. Combining Hanning with a median smoother

Both compound smoothers are less affected by the outlier than the Hanning smoother by itself, but S3h shows more resistance than Sh3. In general, compound smoothers should begin with a median-based smoother to remove outliers prior to Hanning.

Because mean-based smoothers are vulnerable to outliers, why use Hanning at all? For one thing, all the values generated by odd-span median smoothers occur in the observed series, while Hanning (and even-span median smoothers) can generate arbitrary values. It is highly unlikely that the values of the ideal fit will exactly coincide with observed values. In addition, Hanning weights the observations, allocating half the weight to the middle observation and a quarter to the observations on either side. This weighting scheme formalizes the intuition that $f(t_j)$ is less closely related to $f(t_i)$ as t_j is further from t_i.[23]

Note that while all the smoothers examined so far diminish the large deviations from trend, the smoothed series still follow the ups and downs of the observed raw series;

---

23. `tssmooth nl` does not provide weighted median smoothers. More general scatterplot smoothers rely on distance weighting in generating estimates of the fit.

that is, they fail to impose the straight-line trend that we happen to know is the true fit for this series. This behavior can be attributed to the short span of the smoothers examined so far. Compare the performance of a span-3 median smoother on these data with that of a span-9 median smoother.

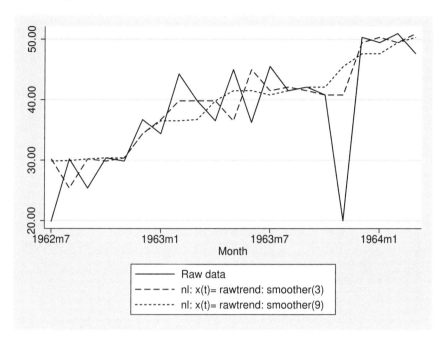

Figure 3.10. Comparing span-3 and span-9 smoothers

The span-9 smoother reduces the volatility of the smoothed series noticeably. In this special case of a linear trend with no cycle and no seasonal component, a long-span smoother performs well. In general, however, we tend to avoid long-span smoothers because they "flatten" the series whether or not flattening is appropriate.

3.3.2 Smoothing a cycle

Let's see how these elementary smoothers perform on cyclical data. We use Stata's sin() function to construct an ideal cycle and add normally distributed random noise to this cycle to create the observed series. Remember, though, that real-world data are unlikely to exhibit the regular periodicity of this sine wave.

```
. drop _all
. set obs 60
number of observations (_N) was 0, now 60
. generate int month = _n
. label variable month Month
. tsset month, monthly
        time variable:  month, 1960m2 to 1965m1
                delta:  1 month
. set seed 23
. generate cycle = 2*sin(8*_pi*_n/_N)
. label variable cycle Cycle
. generate rawcycle = cycle + rnormal()
. label variable rawcycle "Raw data"
. generate residual = rawcycle - cycle
. format cycle rawcycle residual %6.2f
. save cycle
file cycle.dta saved
. tsline cycle rawcycle
```

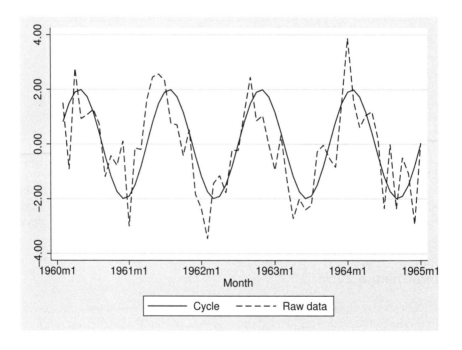

Figure 3.11. Cycle (unobserved) and cycle + residual (observed)

Our data contain four full cycles. Let's see how well the smoothers track these cycles. We can skip the Hanning smoother by itself because we highlighted its deficiencies in the previous subsection.

```
. foreach smoother in 3 3h 9 {
  2.          tssmooth nl S`smoother´ = rawcycle, smoother(`smoother´)
  3. }
. tsline rawcycle S3 S3h S9, lpattern(solid dash shortdash longdash_dot)
> legend(cols(1))
```

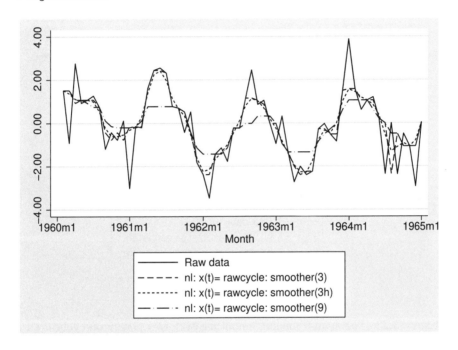

Figure 3.12. Three smoothers and cyclical data

The short-span smoothers reproduce the general cyclical pattern of the fit, but the span-9 smoother flattens the data excessively. This characteristic is easier to see if we zoom in on a subset of the data.

```
. tsline rawcycle S3 S3h S9 in 30/50, lpattern(solid dash shortdash longdash_dot)
> legend(cols(1))
```

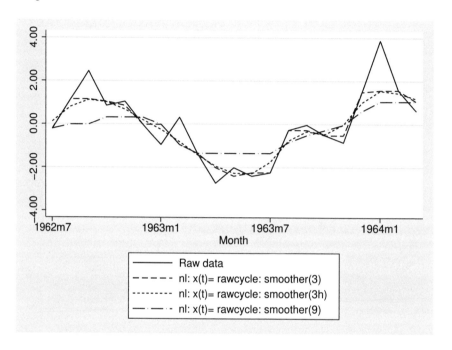

Figure 3.13. The span-9 median smoother performs poorly on cyclical data

Odd-span median smoothers tend to produce flat spots because they are restricted to values that occur in the raw data. Compare the results of the span-3 median smoother with the results of following the span-3 median smoother with Hanning.

```
. tsline S3 S3h in 30/50, lpattern(solid dash) legend(cols(1))
```

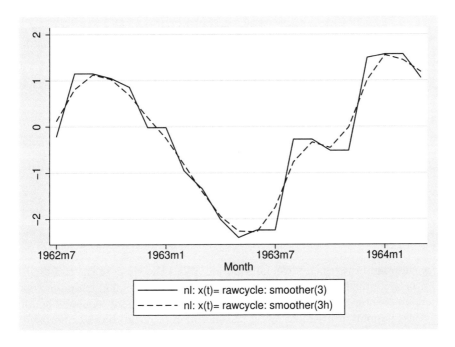

Figure 3.14. Odd-span median smoothers tend to produce flat spots

The Hanning smoother eliminates the flat spots by averaging adjacent points and generating smoothed values that do not occur in the raw data.

The `tssmooth nl` command also provides splitting, a refinement that can sometimes help with flat spots generated by a span-3 median smoother. If you type the letter S immediately after 3 in the smoother specification, Stata will split any repeated values, treat the split points as the endpoints of subsequences, apply a special treatment for endpoints (discussed below), and resmooth the series.

The endpoints of a smoothed series pose some interesting questions. Consider the first point in a series. There are no points to the "left" of this first point, so it's not clear how a smoother that spans several points can be applied. By default, Stata reduces the span of odd-span smoothers as the endpoints are approached. Each endpoint in the smoothed series is "copied down" from the original series, that is, unchanged.

Even-span smoothers are handled differently. The length of the series is extended by one observation, the smoothed points are recentered halfway between the original x values, and the span is reduced to the nearest-feasible even-span smoother as the endpoints are approached. Again the endpoints themselves are simply copied down unchanged. For you to eliminate the extra point, even-span smoothers are required to appear in pairs (although not contiguously in the sequence of smoothers), and the second even-span smoother in the pair recenters the smoothed series over the original x values and reduces the length of the series by one observation to its original length.

The `tssmooth nl` command also offers a special treatment of the endpoints specified by the letter E. In this case, the first and last points in the smoothed series are calculated as

$$y_1^{**} = \text{median}(3y_2^* - 2y_3^*, y_1^*, y_2^*)$$

and

$$y_T^{**} = \text{median}(3y_{T-2}^* - 2y_{T-1}^*, y_T^*, y_{T-1}^*)$$

where a single asterisk indicates the initial fit and a double asterisk is the (we hope) improved fit.

These endpoint adjustments use information about the slope of the two next-to-last points in the smoothed sequence to determine whether the original endpoint represents an outlier. To see how this intuition is applied, we will examine the left-hand endpoint. If $y_2^* = y_3^*$, this endpoint adjustment forces the last point to lie on the same horizontal line; that is, $y_1^{**}=y_2^* = y_3^*$ regardless of the value of y_1^*. If the smooth is increasing between the second and third points in the sequence ($y_3^* - y_2^* > 0$), the adjusted endpoint will be no greater than y_2^*—the adjustment will not allow the smooth to reverse direction at the endpoint—and y_1^{**} will lie in the interval $[y_2^* - 2(y_3^* - y_2^*), y_2^*]$—the adjustment will not allow the slope to steepen too much either. The reasoning is symmetric if the smooth is decreasing between the second and third points in the sequence.

3.3.3 Smoothing a seasonal pattern

The smoothers discussed in this section are not appropriate for smoothing a seasonal component unless your data exhibit a smoothly oscillating seasonal pattern, such as the rise and fall of the average temperature over the 24 hours of the day. A seasonal pattern that exhibits short, sharp reversals, such as the seasonal spike in bank deposits around tax dates will be misidentified as residual by these smoothers. In fact, a smoother with the same span as the seasonal cycle—say, a span-12 smoother applied to monthly data—will deseasonalize the data. The smooth will incorporate a simple form of seasonal adjustment.

The Holt–Winters smoothers discussed at the end of the next section are designed to identify seasonal variation as part of the signal rather than as part of the noise.

3.3.4   Smoothing real data

Now that we understand the strengths and weaknesses of the individual smoothers, let's apply them to a real-world data series that measures investors' estimate of the risk of default in the banking system.

The TED spread is the arithmetic difference between 3-month LIBOR and the yield on the 3-month on-the-run (OTR) Treasury bill.[24] LIBOR is an interbank lending rate, while the yield on a Treasury bill is the cost of short-term debt to the U.S. Treasury. Investors regard U.S. Treasury securities as risk free.[25] In contrast, banks—even large banks—may fail and default on their obligations, and investors typically require a higher yield on LIBOR obligations than equivalent Treasury securities. As a result, the TED spread is regarded by market participants as a measure of the perceived risk in the banking sector. When investors feel bank risk is rising, the TED spread increases, and vice versa.

---

24. Treasury securities are issued through semiregular auctions. The 1-month, 3-month, and 6-month bills are auctioned once a week, the 12-month bill is auctioned roughly once a month, and most Treasury bonds (coupon-bearing securities) are auctioned once a month. (The current auction schedule is available on the Treasury website: https://www.treasury.gov/resource-center/data-chart-center/quarterly-refunding/Documents/auctions.pdf.) As a consequence, securities with maturities exactly matching the stated maturity are available only on selected dates. For instance, the day after, say, the 3-month bill is issued, it becomes a Treasury bill with a maturity of 3-months-minus-1-day. Moreover, securities issued with a longer maturity become shorter-maturity securities as they age; the 6-month T-bill becomes a 3-month T-bill after 3 months pass. The OTR security is the security with maturity closest to, but not greater than, the stated maturity. Trading is concentrated in the OTR securities, and their greater liquidity commands a higher price (and hence, lower yield) than off-the-run securities.
25. To be clear, we are speaking here of credit risk only. Investors behave as though there is no chance that the U.S. Treasury will default on its obligation to make coupon payments on time and to redeem principal at maturity. However, Treasury securities, like all bonds, are subject to market risk, the risk that their market value will change as interest rates in general fluctuate.

While the market's estimate of bank risk varies over time, we anticipate the TED spread to be marked by sudden, dramatic increases at the onset of banking crises. Figure 3.15, which displays the daily TED spread from 1 May 1987 through 16 April 2009, confirms this intuition. The spike on the left-hand side of the graph—when the TED spread widened to 300 basis points[26]—marks the stock market crash in October 1987, a period of high investor concern about the economy as a whole and about the banking system in particular. This episode is dwarfed, however, by the spike at the right-hand side of the graph. The TED spread widened to 203 basis points in March of 2008, when the investment bank Bear Stearns collapsed over the weekend of March 15–16. In late 2008, the global credit system became gridlocked in the wake of the American International Group bailout (September 16), the Lehman Brothers bankruptcy (September 25), and the revelation of Bernard Madoff's Ponzi scheme (October 11). The TED spread peaked at 458 basis points on October 10.[27]

```
. use ${ITSUS_DATA}/daily, clear
(Daily data for ITSUS)

. generate ted = 100*(libor3m - otr3m)     /* convert rates to basis points */
(9,278 missing values generated)

. format ted %6.0f

. label variable ted "TED spread (in b.p.)"

. keep if !missing(ted)
(9,278 observations deleted)

. tsset
        time variable:  date, 1/02/87 to 12/30/11, but with gaps
                delta:  1 day

. summarize ted
    Variable |       Obs        Mean    Std. Dev.       Min        Max

         ted |     5,936    64.50632    46.57971    8.762999    457.875
. list if date==mdy(10,10,2008)

            +---------------------------------------+
            |     date    otr3m    libor3m    ted   |
            |---------------------------------------|
      5346. | 10/10/08     0.24      4.819    458   |
            +---------------------------------------+
```

---

26. A basis point is one one-hundredth of a percentage point.

27. A couple of points to note in this Stata log. Recall that ITSUS_DATA is a Stata macro, that is, a name that stands in place of some text that Stata will substitute for the macro before it executes the command. The $ is the cue to Stata to replace the macro with the text it stands for. In this case, ITSUS_DATA contains the name of the folder where example datasets for this book are stored. On my computer, the text underlying this macro is /Users/sbecketti/Data/itsus1r/data. You should define the macro to point at the folder in which you stored the example datasets.

```
. tsline ted
```

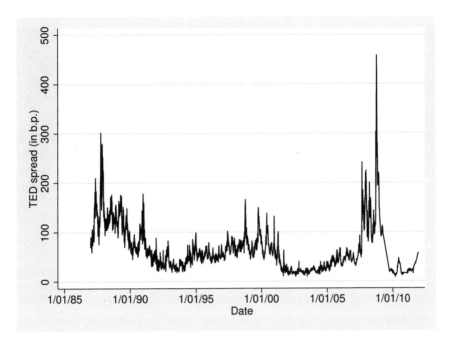

Figure 3.15. The TED spread, 1/2/87–12/30/11

Our smoothers are based on an assumption that the underlying signal in the observed time series evolves smoothly. As a result, sudden spikes such as the ones we see in the TED spread data, will be attributed by our smoothers to the residual rather than to the fit. There is nothing wrong with this characteristic of the smoothers, but we need to take it into account when we interpret our results. If we are interested in the behavior of the market under "normal" conditions, then the smoothed series is the appropriate focus of our study. If, on the other hand, our phenomenon of interest is banking crises, we will probably set the fit to one side and focus instead on the residual.[28]

---

28. A point to note is that our intuitive decomposition of time series into trend, cycle, seasonal, and residual does not seem to have any category for sudden spikes except the residual. Financial data frequently exhibit this type of unexpected jump, and this behavior greatly complicates estimation in a parametric model. Models incorporating fat-tailed distributions, mixture distributions, jump volatility, and regime changes all have been applied at various times. These types of models are active areas of research, but they are beyond the scope of this book. For now, we use our smoothers to segregate these episodes from the more tractable portion of the data.

In the subsections above, we examined elementary smoothers (3, H, and 9) and a relatively simple compound smoother (3H). In practice, researchers tend to use more elaborate compound smoothers.[29] We will apply one commonly used smoother, 3RSSH, to the TED spread.[30] Let's focus on the performance in the turbulent last five months of 2008.

```
. tssmooth nl smooth = ted, smoother(3rssh)
cannot apply nl smoothers to variables with missing values
r(198);
```

What happened? Look back at the log above. We dropped any observation where the TED spread is missing (any day prior to 1987 or after 2011). So where are the missing values? Well, they are not actually present in the data. Instead they are implied. These are daily data, but values for LIBOR and Treasury yields are recorded only on business days—no Saturdays, no Sundays, and no bank holidays. tssmooth nl regards those absent days as missing values.

---

29. See Tukey (1977) for a thorough discussion of compound smoothers.
30. Other useful smoothers are 4253H and 43RSR2H. All three of these smoothers often are combined with the twice option. For these data, all of these smoothers performed similarly.

The cure is simple. Create an artificial time variable without gaps, just as we did in the previous examples.

```
. generate n = _n
. tsset n, generic
        time variable:  n, 1 to 5936
                delta:  1 unit
. tssmooth nl smooth = ted, smoother(3rssh)
. label variable smooth "Smooth, 3RSSH"
. tsline ted smooth if date>mdy(7,31,2008) & date<mdy(1,1,2009)
```

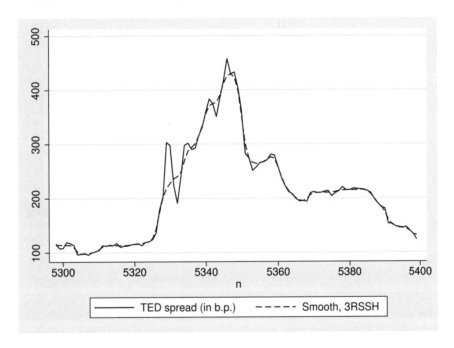

Figure 3.16. The TED spread and a smoothed version, 8/1/09–12/31/09

The smoothed series reproduces the overall shape of the TED spread, but the deviations between the smooth and the observed series are greater at the end of September and in October than elsewhere. A time line of the residuals makes this pattern unmistakable.

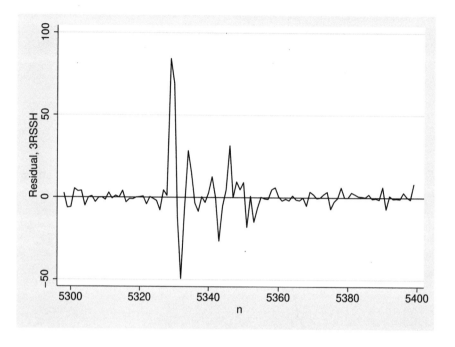

Figure 3.17. The difference between the smooth and the TED spread

In this case, the smoother is performing as it should. The events of late 2008 were unprecedented. It is not clear that any significant amount of interbank lending actually took place at the spreads reported during this episode. It makes sense that the smoother would ignore some of the volatility in the observed series during this episode.

Does this compound smoother perform any better than the simpler alternatives we analyzed above? Let's compare the performance of the 3H smoother.

```
. tssmooth nl S3h = ted, smoother(3h)

. generate R3h = S3h - ted

. label variable R3h "Residual, 3H"

. tsline residual R3h if date>mdy(7,31,2008) & date<mdy(1,1,2009), yline(0)
> lpattern(solid dash)
```

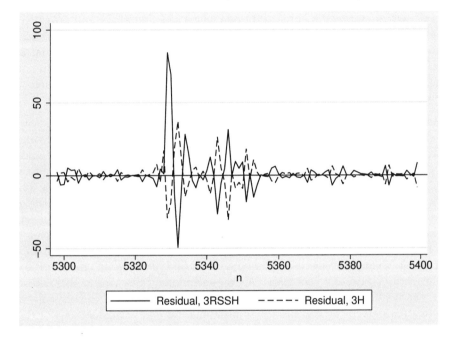

Figure 3.18. The relative performance of a complex and a simple smoother

The simpler smoother tends to follow the TED spread too closely during the period of peak turbulence.

3.4 Additional filters

The smoothers described in the previous section are classified as compound, nonlinear, resistant smoothers. "Compound" because elementary smoothers are strung together in a pipeline. "Nonlinear" and "resistant" because the compound smoothers incorporate median-based estimates of the smoothed values (not to mention the nonlinear calculations specified by the E, R, and S operators).

These smoothers make limited assumptions—the fit evolves smoothly and the residual is unsystematic—and have a limited goal: to uncover the underlying fit by filtering out, as far as possible, the random noise obscuring it. Two important tasks are left unaddressed by these smoothers. First, they do not reveal the structure of the fit; that is, they do not decompose the fit into its components: trend, cycle, and seasonal. Second, and more important, they do not specify the mechanism that relates past values of a time series to present and future values—a necessity for even the simplest type of forecast.

In this section, we review three additional classes of smoothers—all provided by the `tssmooth` family of commands—that make incrementally stronger assumptions about the process generating the time series. These new smoothers take our analysis to a middle ground between the nonparametric, model-agnostic approach of the `tssmooth nl` command and the fully parametric approach of the time-series models covered in later chapters.

Many academic researchers skip over the methods covered in this and the subsequent chapter and apply only the more formal methods covered later. Other practitioners, particularly in industry, find the tools in these two chapters to be adequate and more readily applied than econometric time-series models, particularly for forecasting. Granger (1980) highlights some of the advantages of the less-formal methods for industry. First, they are low cost; they provide useful results without requiring the services of an expensive statistician. Second, they often are the only feasible tools when many series need to be modeled simultaneously. For instance, an aerospace company that is trying to model and forecast the required inventories of the thousands and thousands of parts used in constructing an airplane may well find it impossible to construct elaborate time-series models for each part, but the techniques presented in this and the next chapter can readily be used, particularly if the company is willing to prespecify some of the tuning parameters we will discuss in a moment. Finally, the methods of these two chapters provide a reasonable summary of the data that is more easily communicated to a nontechnical senior management than the more formal methods. The value of this last point should not be underestimated. In any event, it is worthwhile to be familiar with the full range of tools available for time-series analysis.

In all, `tssmooth` provides four classes of smoothers:

nl: Compound, nonlinear, resistant smoothers plus Hanning These smoothers were discussed above.

ma: Weighted moving averages General weighted moving averages.

exponential and dexponential: EWMAs These smoothers specify an explicit link between the past history of a time series and its future realizations. Recent history is weighted more heavily than the distant past in forming an estimate of the current mean of the series.

hwinters and shwinters: Holt–Winters smoothers These smoothers use multiple EWMAs to model separately the evolution of the trend, cycle, and seasonal.

We discuss the three latter classes in turn.

3.4.1 ma: Weighted moving averages

tssmooth offers a more general moving-average smoother than the Hanning smoother provided by tssmooth nl. Recall that Hanning is the span-3 weighted average

$$y_t^* = (y_{t-1} + 2y_t + y_{t+1})/4$$

In contrast, tssmooth ma allows arbitrary-span moving averages and user-specified weights. Moreover, the smoothers need not be centered over the currently smoothed point.

For uniformly weighted moving averages, the syntax is

tssmooth ma $\left[\, type\,\right]$ *newvar* = *exp* $\left[\, if\,\right]$ $\left[\, in\,\right]$, <u>window</u>($\#_l\left[\,\#_c\left[\,\#_f\,\right]\right]$) $\left[\, \texttt{replace}\,\right]$

The window() option, which is required, specifies the span and centering of the moving average. $\#_l$ specifies the number of lagged observations to include in the moving average, a number between 0 and half the observations in the sample (after applying the *if* and *in* clauses). $\#_c$ is optional. If included, it is either 0, which indicates that the current observation should be excluded from the moving average, or 1, which indicates that it should be included. By default, the current observation is excluded. $\#_f$ also is optional and specifies the number of forward observations, again a number between 0 and half the observations in the sample. A few examples should make this clearer.

window(5): A span-5 lagged moving average. The current observation is not included. In other words,

$$y_t^* = (y_{t-5} + y_{t-4} + y_{t-3} + y_{t-2} + y_{t-1})/5$$

window(0 0 5): A span-5 forward moving average. The current observation is not included.

window(2 1 2): A span-5 moving average centered on, and including, the current observation.

For moving averages with user-specified weights, the syntax is

tssmooth ma $\left[\, type\,\right]$ *newvar* = *exp* $\left[\, if\,\right]$ $\left[\, in\,\right]$, <u>we</u>ights($\left[\, numlist_l\,\right]$ <$\#_c$> $\left[\, numlist_f\,\right]$) $\left[\, \texttt{replace}\,\right]$

The `weights()` option, which is required, specifies the weights and thus implicitly specifies the span of the moving average. *numlist$_l$* and *numlist$_f$* are optional Stata numeric lists that specify the weights to apply to lagged and forward observations, respectively. (See [U] **11.1.8 numlist** for the details on all the ways you can specify a numeric list in Stata.) Observations that are not assigned weights are excluded from the moving average. <#$_c$> is required and specifies the weight to be applied to the current observation. The angle brackets must be typed literally. Stata normalizes the weights to produce an appropriate weighted mean, that is, only the ratio of the weights matters. In other words, the option `weights(1 < 2 > 3)` will generate the same results as the option `weights(100 < 200 > 300)`.

Because the Hanning smoother is a weighted moving average, we can use `tssmooth ma` to (almost) replicate it.

```
. use ${ITSUS_DATA}/trend, clear
. tssmooth nl NLh = rawtrend, smoother(h)
. tssmooth ma MAh = rawtrend, weights(1 < 2 > 1)
The smoother applied was
    (1/4)*[1*x(t-1) + 2 *x(t) + 1*x(t+1)]; x(t)= rawtrend
. describe NLh MAh

              storage  display     value
variable name    type   format     label       variable label

NLh             float   %9.0g                   nl: x(t)= rawtrend: smoother(h)
MAh             float   %9.0g                   ma: x(t)= rawtrend: weights(1 < 2
                                                  > 1)

. compare NLh MAh

                                    ————————— difference —————————
                            count      minimum      average     maximum

NLh<TSh                         1    -1.745006    -1.745006   -1.745006
NLh=MAh                        58
NLh>MAh                         1      3.61306      3.61306     3.61306

jointly defined                60    -1.745006     .0311342     3.61306

total                          60
```

All but two points match because `tsmooth nl` and `tssmooth ma` treat endpoints differently. The Hanning smoother in `tssmooth nl` uses the first and last observations of the raw series as the endpoints of the smoothed series. `tssmooth ma` applies the span-3 weighted average to the first and last points of the raw series but treats y_0 and y_{N+1} as missing values and renormalizes the weights.

```
. format NLh MAh %6.2f

. list rawtrend NLh MAh in f/3
```

	rawtrend	NLh	MAh
1.	3.11	3.11	4.85
2.	8.34	5.97	5.97
3.	4.08	3.45	3.45

```
. /* tssmooth endpoint method */
. display %6.2f = (2*rawtrend[1] + 1*rawtrend[2])/3
  4.85

. list rawtrend NLh MAh in -3/l
```

	rawtrend	NLh	MAh
58.	61.47	59.17	59.17
59.	59.40	62.62	62.62
60.	70.24	70.24	66.62

```
. /* tssmooth endpoint method */
. display %6.2f = (2*rawtrend[60] + 1*rawtrend[59])/3
  66.62
```

Note that the method used by `tssmooth ma` is different from the method used by the `smooth` command's E operator.

```
. tssmooth nl NLhe = rawtrend, smoother(he)

. format She %6.2f

. list rawtrend NLh NLhe MAh in f/3
```

	rawtrend	NLh	NLhe	MAh
1.	3.11	3.11	5.97	4.85
2.	8.34	5.97	5.97	5.97
3.	4.08	3.45	3.45	3.45

```
. /* smooth E endpoint method */
. display %6.2f = 3*Sh[2] - 2*Sh[3]
  11.01
```

The E calculates the first observation of the smooth as

$$y_1^{**} = \text{median}(3y_2^* - 2y_3^*, y_1^*, y_2^*)$$

In this example, $y_1^{**} = \text{median}(11.01, 3.11, 5.97) = 5.97$.

3.4.2   EWMAs

Up to now, we have used smoothers only to filter out the random noise in a time series. However, the `tssmooth ma` smoother introduced the possibility of using a smoother to forecast a time series, a topic we cover in detail in the next chapter.

Prior to the `tssmooth ma` command, all our smoothers were centered over the current point; that is, they included observations both prior to and subsequent to the current point. These smoothers cannot be used for forecasting because they require information about the future. However, the `tssmooth ma` allows the use of lagging moving averages, smoothers that incorporate only past values of the time series. Lagging moving averages can produce forecasts because they do not incorporate any information not yet available. If our only interest is smoothing an already-observed time series, then lagging moving averages are probably not the best choice, because they ignore the information in future observations. If, on the other hand, our goal is to update in real time our estimate of the current mean of an evolving series, a lagging moving average does the trick.

EWMAs are a special type of lagging moving average designed specifically to update our estimate of a time-varying mean. All the time series we have examined so far have time-varying means. Series with trends have means that increase or decrease steadily over time. Series with cycles may have constant unconditional means—for instance, our sine wave example has a global mean of 0—but the local mean at any particular moment in time, t, varies according to whether we happen to be nearer the peak or the trough of the cycle.

EWMAs also accommodate other types of time-varying means. Some series are subject to so-called regime changes, that is, abrupt changes in their behavior. Changes in political administration or significant changes in law often are cited as regime changes. For instance, the passage of an investment tax credit (especially a temporary tax credit) may induce a sudden increase in investment spending by businesses. The TED spread we examined a few pages back provides another example of a series that exhibits regime changes. A EWMA will take some time to recognize a regime change, but the accumulation of observations from the new regime will eventually force the EWMA to catch up.

exponential: EWMAs

EWMAs infer the current mean of the series from the entire available history, but recent observations are weighted more heavily than older observations. The weights decline exponentially the further in the past we go, hence, the name. EWMAs update the local mean of the series according to

$$y_t^* = \alpha y_t + (1 - \alpha)y_{t-1}^*$$

where y_t is the raw series and y_t^* is the mean in period t. α, called the smoothing parameter, lies between 0 and 1 and determines the amount of inertia in the local mean. Values of α near 0 generate very smooth series with slowly changing means. Values near 1 produce more volatile series with rapidly changing means.

The updating formula is recursive; to apply it, we need an estimate of y_{t-1}^*, last period's mean. Substituting repeatedly for this unobserved mean, we have

$$
\begin{aligned}
y_t^* &= \alpha y_t + (1-\alpha)[\alpha y_{t-1} + (1-\alpha)y_{t-2}^*] \\
&= \alpha y_t + (1-\alpha)\alpha y_{t-1} + (1-\alpha)^2 y_{t-2}^* \\
&= \alpha y_t + \alpha(1-\alpha)y_{t-1} + \alpha(1-\alpha)^2 y_{t-2} + \cdots + (1-\alpha)^t y_0^*
\end{aligned}
$$

In all the smoothers we examined previously, the formulas and weights were completely specified either by the definition of the smoother or in options typed by the user. Here, however, we are left with two unknown quantities: α and y_0^*, the mean of the series at time 0. These unknowns can either be estimated from the data or be specified a priori. Note that the influence of y_0^* declines exponentially at the rate $(1-\alpha)$, so for any reasonably long time series, the results will be more affected by the setting of α than of y_0^*.

Stata's `tssmooth` command provides EWMAs. The syntax is

`tssmooth exponential [type] newvar = exp [if] [in] [, replace parms(#`$_\alpha$`)`
  `samp0(#) s0(#) forecast(#)]`

The `forecast()` option adds forecasted values to the end of the *newvar*, expanding the dataset if necessary. We will discuss the properties of these forecasts in the next chapter.

By default, Stata sets y_0^* to the mean of the first half of the sample and estimates α by minimizing the sum-of-squared forecast errors. The `samp0()` and `s0()` options are mutually exclusive ways of overriding Stata's default estimate of y_0^*. `samp0(#)` specifies that y_0^* should be set to the mean of the first # observations in the sample. Alternatively, `s0(#)` specifies y_0^* directly. `parms(#`$_\alpha$`)` overrides Stata's default and sets α to #$_\alpha$ instead.

There are some commonsense guidelines for setting y_0^* and α:

- First, if interest centers only on the mean of the most recent observations and the observed time series is reasonably long, almost any choice of y_0^* will do because the weight, $(1-\alpha)^t$, declines rapidly. The table below lists the observation number t at which this weight first drops below 0.005—half of one percent—for selected values of α.

α	t
0.1	51
0.3	15
0.5	8
0.7	5
0.9	3

- Second, Stata's default calculation of y_0^* as the mean of the first half of the sample is a poor choice for series with pronounced trends. A better choice is the first observation of the series.

- Finally, Stata's default setting of α does not always produce a useful estimate. EWMAs are simple structures that capture some important intuitions about time series, but a EWMA maps directly to a more formal time-series model only in special cases. Granger (1980) quotes a rule of thumb that α should be set to 0.7, 0.5, or 0.3 based on the apparent volatility or inertia in the level of the series.

We can readily see some of the characteristics of EWMAs by applying `tssmooth` to our artificial cycle and comparing the results with a nonparametric smoother.

```
. use ${ITSUS_DATA}/cycle, clear
. tssmooth nl S3rssh = rawcycle, smoother(3rssh)
. tsset month, monthly
        time variable:  month, 1960m2 to 1965m1
                delta:  1 month
. count
  60
. summarize rawcycle in f/30
```

Variable	Obs	Mean	Std. Dev.	Min	Max
rawcycle	30	-.0628047	1.64148	-2.618357	2.941716

```
. tssmooth exponential default = rawcycle
computing optimal exponential  coefficient (0,1)
optimal exponential coefficient =        0.6256
sum-of-squared residuals         =      123.04288
root mean squared error          =        1.4320317
. label variable default "alpha = 0.6"
. list rawcycle cycle in f
```

	rawcycle	cycle
1.	1.51	0.81

```
. tssmooth exponential ewma = rawcycle, parms(0.3) s0(0.26)
exponential coefficient  =        0.3000
sum-of-squared residuals =      140.77
root mean squared error  =        1.5317
. label variable ewma "alpha = 0.3"
```

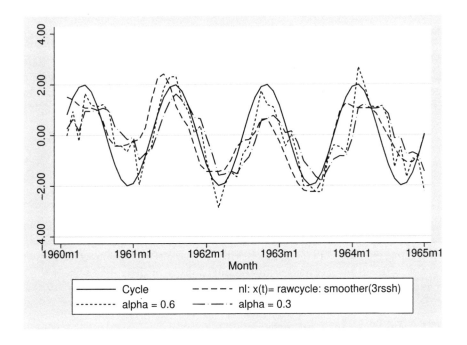

Figure 3.19. Nonparametric and EWMA smoothers applied to cyclical data

tssmooth exponential sets the starting value, y_0^*, to -0.063, the mean of **rawcycle** in the first half of the sample, and estimates α as 0.6256. We stored the resulting smoothed series in a variable named **default**. For comparison, we created another EWMA, named **ewma**, setting y_0^* to 0.26—the initial value of the raw series, which by chance is far from the 0.81 value of the underlying cycle—and setting α to 0.3, a value that produces a much smoother series. We also used **smooth** to generate the 3RSSH smooth of **rawcycle**.

As the graph shows, the nonparametric smoother does a reasonably good job of tracking the underlying cycle. **tssmooth**'s default setting of α weights the current observation heavily and produces a choppier series. Setting α to 0.3 reduces the volatility of the EWMA, and the resulting smoothed series is qualitatively similar to the 3RSSH smooth. Note, however, that this EWMA is "phase shifted"—its peaks and troughs are shifted to the right compared both with the underlying cycle and with the 3RSSH smooth. EWMAs are backward looking; they incorporate information only about the present and past. When the mean of the series changes, as in a series with a pronounced trend or cycle, a EWMA will never quite catch up to the current mean. In contrast, **smooth** calculates centered smoothers that incorporate equal amounts of past and future observations in forming their estimates of the local mean and median. This lagging

characteristic of EWMAs makes them less than optimal for ex post analysis of a time
series, where future values are readily available to improve the estimate of the signal.

dexponential: Double-exponential moving averages

Double-exponential smoothing, provided by the `tssmooth dexponential` command,
is designed to handle series with local trends. Double-exponential smoothing applies
EWMA smoothing twice, using the same smoothing parameter α in both stages. In other
words, the single EWMA smooth is

$$y_t^* = \alpha y_t + (1 - \alpha)y_{t-1}^*$$

and the double-exponential smooth is

$$y_t^{**} = \alpha y_t^* + (1 - \alpha)y_{t-1}^{**}$$

The double-exponential smoother requires two initial values: y_0^* and y_0^{**}. By default,
Stata calculates these from a regression of the first half of the sample on $t - t_0$. The
length of the regression can be overridden with the `samp0(#)` option. Alternatively, the
two initial values can be specified with the `s0(`y_0^* y_0^{**}`)` option. The rest of the syntax
of the `tssmooth exponential` command is identical to that of the single-exponential
version.

The TED spread we analyzed earlier exhibits pronounced local trends during periods
of financial market stress. Let's apply the double-exponential smoother to this series.
The `tssmooth` command treats missing observations, including gaps in the series, as
missing values. Because the TED spread is recorded only on business days, we can
either repeat the last-observed value on weekends and holidays or create a fake daily
date variable that does not have gaps (as we did above in our earlier example of the
TED spread). For simplicity, we use the latter method.

```
. use ${ITSUS_DATA}/ted, clear
. generate int newdate = _n
. tsset newdate, daily
        time variable:  newdate, 02jan1960 to 02apr1976
                delta:  1 day
. tssmooth dexponential dewma = ted
computing optimal double-exponential coefficient (0,1)

optimal double-exponential coefficient =        0.4479
sum-of-squared residuals                =     329349.59
root mean squared error                 =     7.4487194
```

```
. tsline ted smooth dewma if date>mdy(8,31,2008) & date<mdy(1,1,2009),
> lpattern(solid dash shortdash)
```

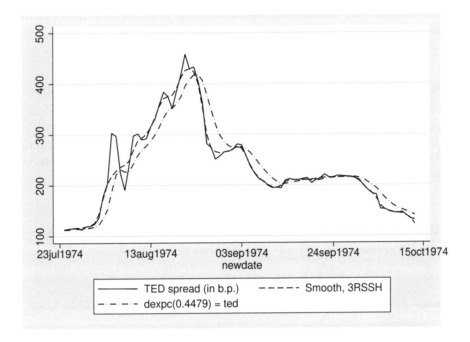

Figure 3.20.   Nonparametric and double-exponential smoothers applied to the TED spread

Note that like the single-exponential smoother, the double-exponential smoother lags changes in the direction of the series. The double-exponential smoother provides a smoother representation of the TED spread than the nonparametric 3RSSH smoother, which tracks the raw series more closely.

3.4.3   Holt–Winters smoothers

hwinters: Holt–Winters smoothers without a seasonal component

Holt (1957) and Winters (1960) proposed adaptive smoothers that model the trend, cyclical, and seasonal components of a time series with separate EWMAs. To simplify the explanation, we will begin with a version that omits the seasonal component. This version is used frequently because most economic data already are seasonally adjusted.

Consider a time series, y_t, that follows a linear trend with random shocks to both the trend and the mean. Let a_{t-1} be the estimated mean of the series in period $t-1$ and b_{t-1} be the estimated trend, that is, the expected increment in the series between periods $t-1$ and t. We calculate the smoothed value, y_t^* in period t, as our estimate of y_t given the information available through period $t-1$,

$$y_t^* = a_{t-1} + b_{t-1}$$

In other words, we just add our best guess of the current trend to our best guess of the last period's mean.

To calculate smoothed values of the trending series y_t, we need to update both a_t and b_t every period to incorporate the information provided by new observations of y_t. We use a EWMA of the currently observed value of y_t and the prior estimate $(a_{t-1}+b_{t-1})$ to update the estimated mean:

$$a_t = \alpha y_t + (1 - \alpha)(a_{t-1} + b_{t-1})$$

To update the estimated trend, we use a separate EWMA that combines this period's increment in the mean of y and the prior estimate of the trend:

$$b_t = \beta(a_t - a_{t-1}) + (1 - \beta)b_{t-1}$$

The updated values, a_t and b_t, allow us to calculate y_{t+1}^*, the next term in the Holt–Winters smooth.

Note that we have performed a neat trick here: we have updated two quantities, a_t and b_t, using only a single new observation, y_t. This two-for-one bargain on information derives from the structure we have assumed for the two EWMAs. However, we will see that trying to extract this much information from a single time series sometimes can make it difficult to bootstrap the smoother by also estimating a_0, b_0, α, and β from the same time series.

The `tssmooth hwinters` command provides Holt–Winters nonseasonal smoothing. The syntax of this command is similar to the syntax of the EWMA smoothers:

`tssmooth` `hwinters` $\big[$*type*$\big]$ *newvar* = *exp* $\big[$*if*$\big]$ $\big[$*in*$\big]$ $\big[$, `replace` `parms`$(\#_\alpha\ \#_\beta)$
   `samp0`($\#$) `s0`($\#_{\text{cons}}\ \#_{\text{lt}}$) `forecast`($\#$) `diff`$\big]$

We will discuss the `forecast()` option in the next chapter.

As before, the `parms()` option allows you to specify values for α and β, the smoothing parameters in the two EWMAs. And the `samp0()` and `s0()` provide alternative ways to set the initial values, a_0 and b_0. If neither option is specified, the initial values are obtained from a linear regression of *exp* on a constant and a time trend by using the observations in the first half of the sample. `samp0()` specifies that the regression include the first $\#$ observations. Alternatively, you can use the `s0()` option to set a_0 and b_0 directly.

The Holt–Winters smoother is designed for time series with linear trends, so it is not a good choice for the TED spread data, which are relatively flat over time with occasional, short-lived blowouts. Instead we will use the U.S. GDP data to demonstrate this smoother.

```
. use ${ITSUS_DATA}/quarterly, clear
(Quarterly U.S. GDP, annualized, BEA, as of 4/27/2012)
. tssmooth hwinters hw = gdp
computing optimal weights

Iteration 0:   penalized RSS = -1317826.7
Iteration 1:   penalized RSS = -659185.86  (not concave)
Iteration 2:   penalized RSS = -647840.59  (not concave)
Iteration 3:   penalized RSS = -633540.88
Iteration 4:   penalized RSS = -623527.41
Iteration 5:   penalized RSS = -622727.19
Iteration 6:   penalized RSS = -618929.41
Iteration 7:   penalized RSS = -618894.33

   (output omitted )

Iteration 186: penalized RSS = -618884.25  (backed up)
Iteration 187: penalized RSS = -618884.25  (backed up)
—Break—
r(1);
```

Not a very auspicious beginning. As we warned, multiple combinations of α and β can sometimes produce much the same fit, so we might as well specify reasonable values.

```
. tssmooth hwinters hw = gdp, parms(0.7 0.5) samp0(10)
Specified weights:
                    alpha = 0.7000
                     beta = 0.5000
sum-of-squared residuals = 719751.8
 root mean squared error = 52.51352
. label variable hw "HW(0.7 0.5) = gdp"
```

We modified the variable label that `tssmooth` applied to `hw` to make it fit more readily in the graph legend below.

For comparison, let's calculate a EWMA. The simple EWMA adapts to changes in the mean of the time series without decomposing the changes into changes in the trend and the overall mean.

```
. tssmooth exponential ewma = gdp
computing optimal exponential  coefficient (0,1)

optimal exponential coefficient =      0.9998
sum-of-squared residuals         =   2385188.3
root mean squared error          =     95.5963
. label variable ewma "EWMA(0.9998) = gdp"
```

The strong trend in GDP forces the estimate of α toward the limiting value of 1.

```
. tsline gdp hw ewma, lpattern(solid dash dash_dot)
```

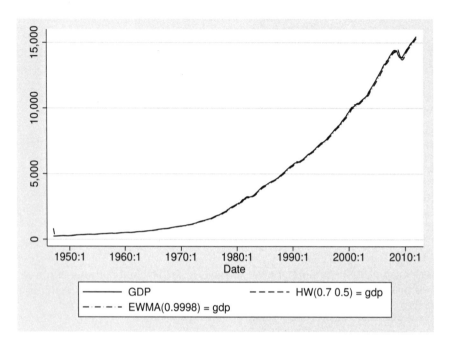

Figure 3.21. U.S. GDP with Holt–Winters and EWMA smoothers

It's hard to see any difference between the three lines. Let's zoom in on the post-2000 performance.

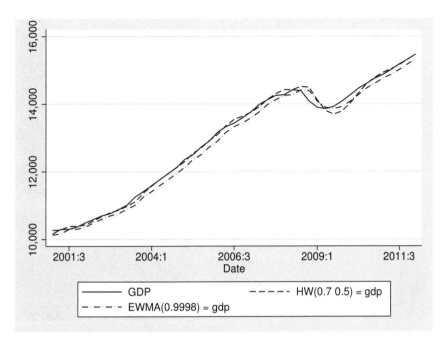

Figure 3.22. U.S. GDP with Holt–Winters and EWMA smoothers

As before, the backward-looking nature of the EWMA forces it to lag any series with a persistent trend. The Holt–Winters smooth is also backward looking, but it relies on constantly updated estimates of the trend as well as on the local mean in forming the smooth. As a result, the Holt–Winters smooth lags the underlying series only when the trend changes, as it does at the onset of the recession that is visible at the end of the GDP series.

Comparing the residuals of the two smoothers highlights this difference.

```
. generate Rhw = hw - gdp
. generate Rewma = ewma - gdp
. tsline Rhw Rewma if year>2000, yline(0) lpattern(solid dash)
```

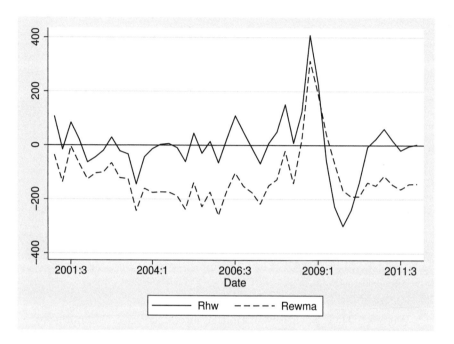

Figure 3.23. U.S. GDP with Holt–Winters and EWMA smoothers

The Holt–Winters residuals appear well behaved until the onset of the recession, while the EWMA residuals are consistently negative until the recession reverses the trend, and the EWMA lag begins to produce positive residuals.

The Holt–Winters smoother is designed for a series with a linear trend. As we noted in section 3.2, GDP appears to follow an exponential trend, and thus the log of GDP follows a linear trend. When smoothing the log of GDP instead of GDP in its original units, `tssmooth hwinters` has better luck finding optimal values of α and β.

```
. generate lgdp = log(gdp)

. tssmooth hwinters lhw = lgdp, samp0(10)
computing optimal weights
Iteration 0:   penalized RSS = -.06552307  (not concave)
Iteration 1:   penalized RSS = -.02894495
Iteration 2:   penalized RSS =  -.0289069
Iteration 3:   penalized RSS = -.02889882
Iteration 4:   penalized RSS = -.02889738
Iteration 5:   penalized RSS = -.02889703
Iteration 6:   penalized RSS = -.02889701

Optimal weights:
                            alpha = 1.0000
                             beta = 0.5441
        penalized sum-of-squared residuals = .028897
                 sum-of-squared residuals = .028897
                 root mean squared error = .0105222
```

The smoother has chosen to use 100% of the current information to set the local mean, a_t ($\alpha = 1.0$), but the estimated trend, b_t, appears to have substantial inertia ($\beta = 0.5441$).

shwinters: Holt–Winters smoothers including a seasonal component

The Holt–Winters smoother can also accommodate time series with seasonal components, and the tssmooth shwinters command provides two different versions. In the default version, the seasonal component enters multiplicatively, which is appropriate for cases where the amplitude of the seasonal component grows (or, less commonly, shrinks) over time. This version makes sense where seasonal variation is a constant percentage deviation in a series that is trending. For example, retail sales peak every year during the Christmas shopping season. But, recessions aside, overall retail sales grow steadily every year. A Christmas increase that was modeled additively as a constant dollar increment would eventually represent a negligible percentage increase in retail sales.

Let S_t be the seasonal component, and let there be s periods in a complete seasonal cycle. For example, in monthly data, s typically equals 12. So S_t is a sequence of values that repeats every s periods. In the case of multiplicative seasonality, the smoothed value of y_t is

$$y_t^* = (a_{t-1} + b_{t-1})S_t$$

The EWMA for updating the estimated mean is

$$a_t = \alpha \frac{y_t}{S_{t-s}} + (1 - \alpha)(a_{t-1} + b_{t-1})$$

In the first term, y_t is deseasonalized by dividing it by the most recent estimate of the seasonal component for period t. The EWMA for updating the trend is unchanged; that is,

$$b_t = \beta(a_t - a_{t-1}) + (1 - \beta)b_{t-1}$$

The EWMA for updating the seasonal component is

$$S_t = \gamma \frac{y_t}{a_t} + (1 - \gamma)S_{t-s}$$

The syntax of the `tssmooth shwinters` command is

tssmooth <u>sh</u>winters [*type*] *newvar* = *exp* [*if*] [*in*] [, replace <u>add</u>itive
  <u>pa</u>rms(#$_\alpha$ #$_\beta$ #$_\gamma$) <u>samp</u>0(#) s0(#$_{cons}$ #$_{lt}$) <u>f</u>orecast(#) <u>pe</u>riod(#)]

The `additive` option specifies that seasonality enters additively rather than multi-plicatively.

3.5   Points to remember

- Prior to analyzing fresh data, make sure you fully understand the definitions of the variables, the relationship between the data at hand and the phenomenon you aim to study, and the processes that generated the observed data.

- Time-series data present additional questions you need to answer prior to analysis. Of particular importance: are the data seasonally adjusted, and are the data revised?

- Exploratory analysis of cross-section data typically begins with an examination of the univariate distributions of the variables in the data. However, there is no meaningful univariate time-series analysis. So-called univariate time-series analysis actually is the analysis of the bivariate relationship between the variable of interest and some measure of time.

- In describing this bivariate relationship, we traditionally split a time series into as many as four components: trend, cycle, seasonal, and random noise. The sum of the first three of these components comprises the signal, that is, the systematic portion of the time series. Trend denotes a persistent tendency for a series to increase or decrease. In common parlance, trends occasionally can shift in magnitude and even direction. For instance, we speak of recent trends in stock prices. Cycle denotes aperiodic oscillations around a trend or, if there is no trend, a mean. The distinction between local trends and phases of a cycle is fuzzy and more a matter of convention than of statistics. Seasonals comprise a special type of cycle; however, they need not be smoothly rising and falling. As a result, they can pose a challenge for exploratory time-series analysis.

- Smoothers are used to filter raw data, that is, to reduce the contribution of the random noise and thus to make it easier to recognize the signal. The simplest smoothers assume only that the signal evolves smoothly over time (which may not be true if the series includes a volatile seasonal) and that the noise component is uncorrelated over time. Many formal time-series models contain correlated error terms, but in this stage of the analysis, that predictable portion of the error is attributed to the signal rather than to the noise.

- Simple smoothers replace the raw data with means or medians of adjacent observations. Elementary smoothers are defined by their span (the number of adjacent observations included) and the type of location estimate used (mean, weighted mean, or median). Filter performance can be improved by constructing compound smoothers, that is, sequences of elementary smoothers. The influence of outliers can be minimized by applying one or more median smoothers prior to any mean-based smoothers. Other refinements include special treatment of the endpoints of the series, splitting of repeated values, repeated application of a smoother until the results converge, and application of the smoother to the estimated residual to recapture any signal that was misattributed in the first pass.

- Smoothers designed specifically for time series can be used to decompose the signal into its trend, cycle, and seasonal components and to generate forecasts of future values. Smoothers that include future values of the time series can produce superior filters but are not feasible for forecasting.

Stata commands and features discussed

foreach ([P] **foreach**): Loop over items; section 3.3

rnormal() ([FN] **Random-number functions**): Standard normal random-number generator; section 3.3

set seed ([R] **set seed**): Specify starting point of random number sequence; section 3.3

tssmooth ([TS] **tssmooth**): Smooth and forecast univariate time-series data; section 3.4

4   A first pass at forecasting

Chapter map

4.1 Forecast fundamentals. The types of forecasts covered in this book. Other types of forecasts. Measuring the quality of a forecast. The elements of a forecast.

4.2 Filters that forecast. And ones that do not. Trailing moving averages. Forecasts based on exponentially weighted moving averages (EWMAs). Forecasting a trending series with a seasonal component. Matching the filter to the type of forecast needed. Pros and cons of each approach.

4.3 Points to remember. Three models. Different filters for different situations.

4.4 Looking ahead. Moving from an intuitive approach to forecasting to a formal statistical approach. Short-horizon versus long-horizon forecasts.

Forecasting is perhaps the most familiar and certainly one of the most important uses of time-series models. Accordingly, much of this book is concerned with aspects of forecasting. We have not introduced modern time-series models yet; nonetheless, some useful forecasting can be done with tools we have already discussed. Indeed a great deal of forecasting in industry uses these methods.

This chapter provides a gentle introduction to forecasting. First, we introduce some of the fundamentals of forecasting—the different types of forecasts, the elements of a forecast, and the criteria for choosing one forecasting method over another. Next we explain how some of the filters discussed in the previous chapter can be used for forecasting and the situations to which they are suited. Most of the information we need was discussed in detail in the last chapter, so this first pass at practical forecasting will be brief. More advanced methods for forecasting will be introduced—and compared with these simpler methods—in later chapters.

4.1   Forecast fundamentals

A forecast is a statement about the outcome, timing, or probability distribution of some event whose realization is not known currently. That said, there are many different kinds of forecasts, and the time-series techniques discussed in this book apply to only a subset of them.

4.1.1   Types of forecasts

For the most part, this book explores stochastic, linear, dynamic time-series models for forecasting continuous variables. Let's consider the import of these adjectives one at a time.

Stochastic: We cannot forecast perfectly because the variables we are interested in are influenced by random disturbances in addition to any systematic determinants.

Linear: Linear models provide adequate, tractable time-series representations of many variables. As a practical matter, nonlinear models are much more difficult to analyze and understand, and there are fewer established techniques for nonlinear models.

Dynamic: This feature is the defining characteristic of time-series models—the influence of the past on the present and future values of a variable. Analyzing the dynamic relationships among variables comprises the bulk of this book.

Continuous: Modeling discrete variables such as the outcomes of elections or sporting events requires specialized statistical techniques, even in a cross-section setting. Dynamic relationships—the essence of time-series models—already add significant complexity to statistical analysis. Indeed this complexity justifies the writing of a book such as this one. Adding discrete variables to the mix increases the complexity beyond the scope of this book.

The time-series models we consider can do a lot. The most common application is the generation of point forecasts, that is, the likeliest values at each future date. However, these models also can produce confidence intervals and complete probability distributions of future events. As with static (non-time-series) regression models, they also can be used for policy experiments: to calculate the likely impact of changes in an explanatory variable on a dependent variable. Moreover, these time-series models are well suited to analyzing the relationships among jointly dependent time series. These models also can produce forecasts of the timing of events, although the forecasts typically are expressed in terms of rates of decay of random disturbances to the time path of a variable or the expected time until a variable crosses a specified threshold.

Before we start generating forecasts, let's review some of the types of forecasts that arise in practice and point out whether the time-series models we discuss are well adapted to producing these types of forecasts.

Although we are restricting our attention to models of continuous variables, predictions of events with discrete, qualitative outcomes may be the most familiar.[1] Which candidate will win the presidential election? Which football team will win this Sunday's contest? If I ask Cindy to the prom, will she say yes? Some of these forecasts concern unique, one-time events. In these cases, time-series models, which emphasize the influence of the past on the present and future, generally are not applicable. In these situations, handicapping can be a useful approach. Instead of attempting to construct a model for the probability distribution of an event—say, the winner of this year's Kentucky Derby—we may base our forecast on the opinion of an expert (the racing form), the empirical distribution of the opinions of a panel of experts (a group of professional handicappers), or the opinions of the public at large (the betting odds that reflect the volume of betting on each horse).

Some forecasts concern events that have occurred already but whose outcome is unknown at present. The most familiar examples arise in gambling. Does my opponent's hole card give him a higher-valued poker hand than mine? Many government and industry statistics are published only with a significant lag, and participants in financial markets forecast these not-yet-disclosed measurements of past events to anticipate their likely impacts on the prices of stocks, bonds, and other financial assets. Time-series models can be used for these types of forecasts as long as dynamic relationships play a role in the determination of the variable of interest.

Some forecasts are focused primarily on the timing of an event whose eventual occurrence is assumed to be nearly certain. For example, market participants generally agreed that the rapid increases in house prices in the years prior to 2007 were unsustainable and that an eventual collapse of this housing bubble was inevitable. However, there was no consensus on the timing (or, as it turns out, the magnitude) of the housing collapse. Time-series models can be used for this type of forecast, but frequently some other type of analysis is better adapted to this task. Survival analysis and the failure analysis techniques of statistical quality control often provide convenient methods.

A final class of forecasts—clearly beyond the scope of this book—includes speculations about alternative histories (how would the United States be different today if the South had won the Civil War?) and the forecasts of futurists (what will computers be like in 30 years, or how will human society adapt to climate change in the 21st century?) These types of forecasts are fascinating. However, time-series models are of little use in producing and evaluating forecasts with such global and open-ended scope.

---

1. Stata provides a rich set of tools for models with discrete outcomes (see the Stata Press book *Regression Models for Categorical Dependent Variables Using Stata* [3rd ed., 2014] or type `help logistic` from within Stata if you are in a hurry). Survival analysis—also well covered by Stata— provides models for the timing of discrete events. See the Stata 16 manual *Survival Analysis Reference Manual* or the Stata Press book *An Introduction to Survival Analysis Using Stata* (rev. 3rd ed., 2016) for more information.

4.1.2   Measuring the quality of a forecast

As we develop different approaches to forecasting, we need some way to measure the quality of each approach to choose among them. The primary measure of the quality of a forecast is a loss function, that is, a function that assigns a value—the loss—to the difference between the forecast of a variable, $\widehat{y}_t$, and its realization, y_t:

$$\text{Loss} = L\left(y_t - \widehat{y}_t\right)$$

A model that produces forecasts with smaller expected loss is preferred, other things being equal, to a model that produces forecasts with larger expected loss. Except in special cases, we measure loss by the square of the forecast error,

$$L(y_t - \widehat{y}_t) = \left(y_t - \widehat{y}_t\right)^2$$

Expected loss is not the only statistical criterion for choosing a forecasting method. Unbiasedness generally is a desirable characteristic, and the forecast method with minimum expected loss may not be unbiased, forcing us to choose between objectives.

Pragmatic criteria can be as important as statistical ones. Some forecast methods are more difficult to apply and understand, more computationally intensive, or more expensive to use than others. These costs may outweigh any statistical advantages of these methods. Fitting and interpreting modern time-series models requires both a significant degree of statistical sophistication and a considerable amount of practical experience—in other words, a firm may have to employ an experienced statistician to use these techniques. In industry, the additional precision of advanced forecast techniques may not justify the cost of employing a statistician—a simpler method that can be applied by an analyst may be adequate. Moreover, it often is challenging for statisticians to communicate effectively to business leaders the advantages and limitations of various statistical approaches.

The forecast methods covered in this chapter are relatively simple to implement and understand and do not require the assistance of a statistician. As a consequence, these techniques are the workhorses of a great deal of forecasting in industry. Moreover, their short-term to medium-term accuracy often is quite reasonable. In settings where large numbers of forecasts are needed routinely—for example, in a company that must forecast the inventory of hundreds or thousands of specialized parts—these methods may be the only feasible ones. For now, we will evaluate the performance of forecast methods on pragmatic and intuitive grounds. We will introduce more sophisticated techniques in later chapters. At that point, we will focus on expected loss and similar measures of forecast quality.

4.1.3   Elements of a forecast

The elements of a forecast consist of an information set; a projection date or time (usually the latest date or time for which the information set is available); a forecast horizon (the amount of time between the projection date and the event being predicted);

and a model of some sort that relates the information set available at the projection date to the event at the forecast horizon. These elements are not fixed by the phenomenon under analysis. You can choose to invest in an extensive information set in an effort to maximize the accuracy of your forecast. Alternatively, you may choose to save money and effort by limiting the information set at the cost of a somewhat greater error variance in your forecast. The forecast horizon is determined solely by your requirements. You may need to forecast the value of a variable next period but may not care about longer-horizon forecasts. Alternatively, you may need to predict the value at a later period—for example, you may need to forecast your available cash at a distant date when you are required to make a large payment—but you may not care about the value of the variable in the intervening periods. To a certain extent, even the model is your choice. All models are approximations to reality, and your choice of model may reflect your trade-off between the closeness of the approximation and other objectives. Of course, there are limits: not all possible model choices will be supported by the data.

The minimum-loss information set includes all the information that can improve the forecast. A reduced information set will produce forecasts with a higher-expected loss. However, it may be costly to construct a model that includes every potentially useful predictor. It also may be costly to collect all the variables in the ideal information set, especially if the forecast is updated frequently. As a result, forecasts frequently are based on reduced information sets.

Univariate time-series models are based on the most commonly used reduced information set. Univariate time-series models predict a variable based solely on its past values (and, potentially, past values of random disturbances). In fact, the defining characteristic of a time-series model is the inclusion of past values of the dependent variable.

The univariate time-series model is a benchmark. More complex models are gauged by their improvement (loss reduction) relative to a univariate time-series model (in much the same way that R^2 measures regression performance by the reduction in error variance relative to a model based solely on the sample mean). It can be surprisingly difficult to improve on the forecasts of a well-specified univariate model. As a result, univariate time-series models are used frequently in practical forecasting. All the filter-based forecasts illustrated in this chapter are types of univariate time-series models.

The most common forecast horizon is one-step-ahead, that is, the first unknown value, and models often are compared based on their one-step-ahead performance alone. The longer the forecast horizon, the greater the number of random disturbances that affect the outcome. As a result, the error variance of a forecast generally increases with the forecast horizon.

> **Note:** A K-step-ahead forecast is a prediction of events in period $t + K$ conditional on information in period t. When $K > 1$, we have a multistep-ahead forecast.

It is important to distinguish between in-sample and out-of-sample forecasts and between static forecasts and dynamic forecasts. Consider the forecast equation

$$y_{t+1} = \mu + \rho y_t + \epsilon_{t+1}$$

An in-sample forecast of y_{t+1} is given by

$$\widehat{y}_{t+1} = \widehat{\mu} + \widehat{\rho} y_t$$

where the "hats" ($\widehat{}$) indicate estimates, and both y_{t+1} and y_t are part of the sample used to calculate $\widehat{\mu}$ and $\widehat{\rho}$. Out-of-sample forecasts apply the same formula to y's that were not included in the estimation sample. In-sample forecasts do not provide the most rigorous test of a model, because the model is fit to optimize in-sample performance.

Static forecasts are based on the actual values of period t information. Sticking to the same forecast equation, we write a static, one-period-ahead forecast of y_{t+1} as

$$\widehat{y}_{t+1} = \mu + \rho y_t$$

that is, we use the observed value of y_t in forecasting y_{t+1}. In contrast, a dynamic,[2] one-period-ahead forecast of y_{t+1} is

$$\widehat{y}_{t+1} = \mu + \rho \widehat{y}_t$$

In this case, we do not observe the actual value of y_t, so we must base our forecast of y_{t+1} on a prior forecast of y_t. Dynamic forecasts generally are less accurate than static forecasts because they include the error in predicting y_t in the forecast.

4.2 Filters that forecast

In the previous chapter, we introduced a wide variety of filters that help in highlighting patterns in time series. Some of these filters also can be used for forecasting. These methods have the advantage of low cost, both in the time and effort required for estimation and in the amount of computation required for forecasting. This low cost comes at the expense of some of the sophistication, flexibility, and (sometimes) accuracy provided by formal time-series models.

To generate a forecast, a filter must estimate future values as a function of current and past information (or, equivalently, current values as a function of past information).[3]

---

2. We use the word "dynamic" in two senses in this book. Sometimes "dynamic" indicates the calculation of distant predictions based on earlier predictions, as described here. Earlier in the chapter, we used "dynamic" to denote a formula or model that describes the evolution of a variable over time. Related, but not identical, ideas. Unfortunately, accepted terminology in time-series analysis can be confusing, as you will see in later chapters. We will do our best to make our meaning clear from context.

3. When trying to forecast the value of a variable that has already been realized but is not yet known, we may include any information available when the forecast is calculated. Some of this information may postdate the variable we are forecasting.

This condition eliminates all the smoothers provided by the `tssmooth nl` command because they combine information from the past, present, and future.

Filters that can produce forecasts are provided by the `tssmooth` command. The four flavors of EWMAs—`tssmooth exponential`, `tssmooth dexponential`, `tssmooth hwinters`, and `tssmooth shwinters`—include a `forecast(#)` option that appends predicted values to the end of the smoothed series.[4] Each of these methods operates differently and is suitable for a specific type of forecasting task. The table below lists the types of forecasts produced by each of these `tssmooth` subcommands.

Table 4.1. Forecasting with `tssmooth`

Subcommand	Able to generate forecasts?	Nature of forecast	Uses
`nl`	No	—	—
`ma`	No	—	—
`exponential`	Yes	All forecasts are equal to the last in-sample EWMA	Noisy but nontrending series
`dexponential`	Yes	Forecasts follow the last estimated trend	Series with a linear secular trend
`hwinters`	Yes	Forecasts follow the last estimated trend	Series with a linear secular trend
`shwinters`	Yes	Forecasts follow the last estimated trend and seasonal pattern	Series with a regular seasonal pattern on top of a linear secular trend

To compare the forecasts produced by these methods, we will construct several forecasts of the U.S. civilian unemployment rate, a series we introduced in the prior chapter on filters.

---

4. Technically, the `tssmooth ma` command can be used to forecast as long as the `window` option includes only past values, that is, as long as we specify a trailing moving average. However, this smoother offers no advantage over `tssmooth exponential`, so Stata does not provide a `forecast` option for `tssmooth ma`.

4.2.1   Forecasts based on EWMAs

Let's start by reexamining the time series of unemployment rates.

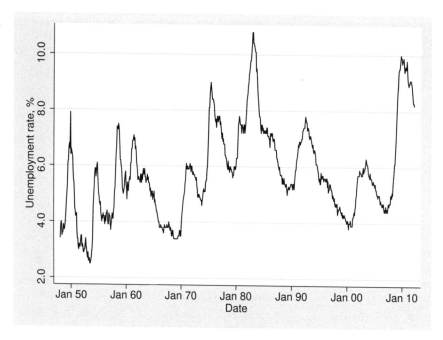

Figure 4.1. U.S. civilian unemployment rate, January 1948 through March 2012

In chapter 3, we used the unemployment rate as an example of a series with a prominent cyclical component. Unlike the artificial cyclical series we constructed to demonstrate the performance of alternative smoothers, the cycle in the unemployment rate is irregular in frequency and shape. It exhibits sharp, inverted, V-shaped peaks associated with recessions and slightly skewed, U-shaped troughs associated with economic expansions.

From this inspection, it is clear that none of our smoothers can produce useful long-run forecasts. Simple EWMAs set all future values to a constant equal to the final estimate of the local mean. The other smoothers project constant trends. None of these smoothers project future oscillations, a key feature of the unemployment rate series. Nonetheless, these forecasts may be useful for short periods, especially when the unemployment rate is hovering near a peak or trough. And, in practice, we usually are most interested in fairly short-term forecasts of the unemployment rate—often just the forecast of next month's rate.

The U.S. unemployment rate began increasing sharply in May 2008 and peaked in October 2009 at 10%. To highlight the strengths and weaknesses of these forecasting approaches, we will generate forecasts from two different starting dates—April 2009, when unemployment was 8.9% and still trending up, and December 2009, just after the peak. We will begin with the `tssmooth exponential` command. We have set α, the weight on the current observation, to 0.9, higher than Granger's recommended maximum of 0.7 but lower than Stata's estimate of 0.9998.[5] We could generate these two forecasts by running the `tssmooth exponential` command twice, once through April 2009 and once through December 2009, and use the `forecast()` option to append as many projections as we need.[6] For this example, though, there is an easier way. The forecasts of `tssmooth exponential` all are equal to the final smoothed value, so we can smooth the entire series once and generate two forecast series equal to the smoothed values in April and December 2009.

```
. use ${ITSUS_DATA}/monthly, clear
(Monthly data for ITSUS)

. keep date unrate

. keep if !missing(unrate)
(420 observations deleted)

. tsset
        time variable:  date, Jan 48 to Mar 12
                delta:  1 month

. tssmooth exponential ewma = unrate, parms(0.9)

exponential coefficient  =      0.9000
sum-of-squared residuals =      39.882
root mean squared error  =     .22744

. label variable ewma "EWMA(0.9)"

. list date if date==tm(2009m4) | date==tm(2009m12)
```

	date
736.	Apr 09
744.	Dec 09

```
. generate apr = ewma[736] if _n>736
(736 missing values generated)

. label variable apr "April 2009 forecast"

. generate dec = ewma[744] if _n>744
(744 missing values generated)

. label variable dec "December 2009 forecast"
```

---

5. The Stata estimate suggests that $\alpha \approx 1$. In other words, the best estimate of tomorrow's unemployment rate is today's rate. That random walk view of the unemployment rate clearly does not hold up over an extended horizon, but, as we shall see below, it works pretty well for a one-step-ahead forecast.

6. The `forecast()` option will extend the current dataset if necessary.

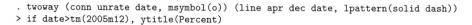

```
. twoway (conn unrate date, msymbol(o)) (line apr dec date, lpattern(solid dash))
> if date>tm(2005m12), ytitle(Percent)
```

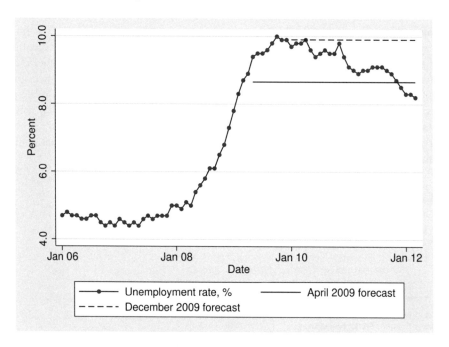

Figure 4.2. EWMA forecasts with different projection dates

The April forecast—when unemployment is still rising—performs very poorly. The December forecast fares better, at least for a few months, because the unemployment rate remained within 0.2% points of 9.9% (the December forecast) until May 2010. Of course, it was impossible to be certain in April 2009 that unemployment was still trending upward or to be certain a few months later in December that the rate was flattening out for a while.

The double exponential smoother projects future values along the last estimated local trend. We will repeat this exercise with the `tssmooth dexponential` command to see how that approach performs.[7]

```
. tssmooth dexponential dewma = unrate if date<tm(2009m5), parms(0.5)
> forecast(35)
double-exponential coefficient  =      0.5000
sum-of-squared residuals        =      32.311
root mean squared error         =      .20952
```

---

7. Now we need the `forecast()` option because the projections of the double exponential smoother are not constant.

```
. generate dapr = dewma if date>tm(2009m4)
(736 missing values generated)

. label variable dapr "April 2009 forecast"

. drop dewma

. tssmooth dexponential dewma = unrate if date<tm(2010m1), parms(0.5)
> forecast(27)
double-exponential coefficient  =        0.5000
sum-of-squared residuals        =        32.703
root mean squared error         =        .20966

. generate ddec = dewma if date>tm(2009m12)
(744 missing values generated)

. label variable ddec "December 2009 forecast"

. twoway (conn unrate date, msymbol(o)) (line dapr ddec date, lpattern(solid dash))
> if date>tm(2005m12), ytitle(Percent)
```

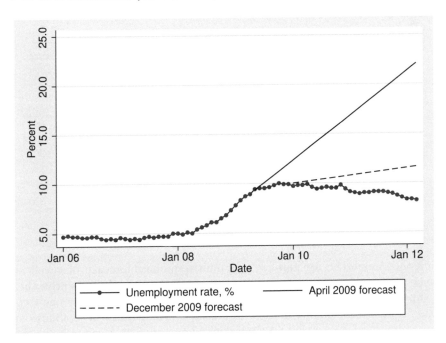

Figure 4.3. DEWMA forecasts with different projection dates

By projecting a fixed trend, both forecasts end up overpredicting by increasing amounts as the forecast horizon increases. Again the December forecast is a little luckier because the forecast begins after the turning point in the unemployment rate.

The scale of this graph makes it difficult to compare the relative performance of the exponential and double exponential smoothers. The graph below compares the forecast errors of all four forecasts. As we saw above, both methods fare better in December than April. Somewhat surprisingly, the EWMA forecasts perform better in both cases, even though they simply project a constant value.

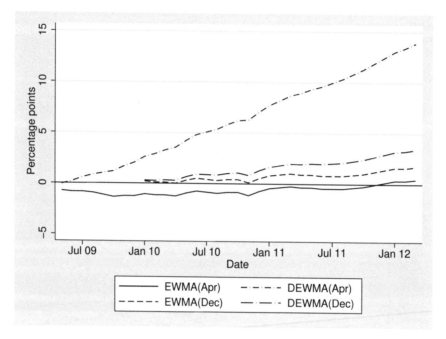

Figure 4.4. EWMA and DEWMA forecast errors

While these smoothers are unreliable for multistep-ahead forecasts of a cyclical series, they may perform acceptably for one-step-ahead forecasts. Conveniently, the values of the exponential and double exponential smooths are, in fact, their one-step-ahead forecasts. To assess the quality of these forecasts, we compare them to a naïve forecast, that is, a simplistic forecast that does not depend on a model. A common benchmark is the "no change" forecast, that is, the assumption that all future values will be the same as this period's. The very high EWMA estimate of α suggests that this naïve forecast will be hard to beat at one step ahead.

We start by recalculating the EWMA and double exponentially weighted moving-average (DEWMA) one-step-ahead forecasts over the entire sample (January 1948–March 2012). For both forecasts, we also calculate the one-step-ahead forecast errors—the arithmetic difference between the forecast and the actual.

```
. /*
>           EWMA one-step-ahead forecast
> */
. drop ewma
. tssmooth exponential ewma = unrate, parms(0.9)

exponential coefficient   =        0.9000
sum-of-squared residuals  =        39.882
root mean squared error   =        .22744

. generate oneewma = ewma - unrate
. /*
>           DEWMA one-step-ahead forecast
> */
. drop dewma
. tssmooth dexponential dewma = unrate, parms(0.5)

double-exponential coefficient   =       0.5000
sum-of-squared residuals         =       54.369
root mean squared error          =       .26555

. generate onedewma = dewma - unrate
```

We will calculate the no change forecast error by hand.

```
. generate onenoch = unrate[_n-1] - unrate
(1 missing value generated)
```

Figure 4.5 displays the one-step-ahead forecast errors for the EWMA, DEWMA, and no change forecasts from January 2006 through March 2012. We have defined the errors as the predictions minus the actuals, so the negative values in 2008 and 2009 indicate that all three methods underpredicted the unemployment rate as it was increasing. This pattern of errors is expected because all three methods lag trends (or changes in trends in the case of the double exponential smoother) by construction.

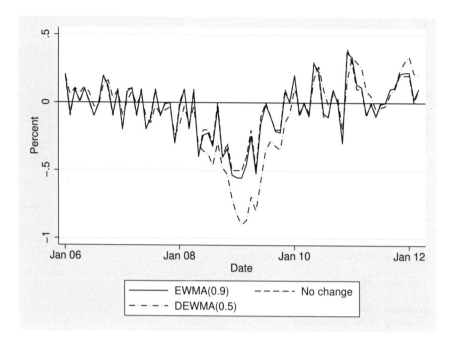

Figure 4.5. One-step-ahead forecast errors for three methods

From figure 4.5, it looks as if all three methods produced roughly equivalent one-step-ahead forecasts when the unemployment rate was in the neighborhood of a peak or trough. However, the double exponential smoother significantly underperformed the other two methods when the rise in unemployment accelerated.

We have zoomed in on the performance of these alternative forecasts over a few years to make it easier to see the differences between them. However, we should assess their relative performance over the full sample before we reach any conclusions. As we mentioned in section 4.1.2, the loss function used to measure forecast quality generally relies on some measure of the squared forecast errors. The most frequently used metric is the root mean squared error (RMSE), that is, the square root of the average squared forecast error. The `tssmooth exponential` and `tssmooth dexponential` commands report RMSE, but we have to calculate it for ourselves for the no change forecast.

```
. generate sqnoch = onenoch^2
(1 missing value generated)

. summarize sqnoch
```

Variable	Obs	Mean	Std. Dev.	Min	Max
sqnoch	770	.046	.1283165	0	2.25

```
. display "RMSE(No change) = " %6.2f = sqrt(`r(mean)')
RMSE(No change) =   0.21
```

The mysterious term '`r(mean)`' in the `display` command represents Stata's method of providing access to the results from the prior `summarize` command. Stata stores selected results from each command in local macros, that is, in character strings assigned to names. In this instance, Stata attached the name `r(mean)` to a string that contains the mean of `sqnoch` (= 0.046).[8]

This name is supposed to look like a "results" function. So, for example, the name `r(sd)` is attached to a string that contains the standard deviation of `sqnoch`. We tell Stata to use the value that is attached to `r(mean)` by surrounding it with a back tick (') and a forward tick ('). In Stataspeak, this is called expanding a local macro.

For easy comparison, the table below displays the RMSEs from all three approaches over the entire sample (January 1948 through March 2012). The naïve, "no change" forecast slightly outperforms the EWMA approach, while the DEWMA method trails the other approaches noticeably.

Table 4.2. Comparison of forecasts

Method	RMSE (%)
No change	0.21
Exponential	0.23
Double exponential	0.27

In practice, what do these differences in RMSEs mean? Well, let us imagine that we are setting up shop as an economic forecaster, perhaps as the chief economist of a major corporation. For the most part, we will keep our forecasts in the middle of the pack of professional forecasts. We want to avoid embarrassing misses that could damage our credibility, although we may occasionally take a flyer on an extreme position because one lucky guess can sustain a career for five or ten years.[9] Let's say that a respectable forecast error is anything within 0.2 percentage points of the actual. We will convert the units of the forecast errors to tenths of a percent for convenience in tracking performance against this threshold.

---

8. Actually, the macro contains the mean of `sqnoch` at the precision used by Stata internally. Thus
   ` . display "'r(mean)'"`
   `.0460000052116811`
9. This method describes a popular and fruitful approach to the ubiquitous "March Madness" office pools, where winners are predicted for all the games in the NCAA invitational basketball tournament. If, like me, you wish to appear savvy despite knowing nothing about any of the teams, select mainly the teams favored by the pretournament seeds. However, pick one or two upsets—I always pick Duke to lose in the first or second round. If you get lucky, act as though you have extraordinary powers of divination. If not, you will still have a respectable number of teams in the later rounds.

```
. generate int etenth = 10*oneewma
. list etenth oneewma in f/5
```

	etenth	oneewma
1.	17	1.755584
2.	-2	-.2244415
3.	-2	-.2224443
4.	0	.0777555
5.	4	.4077756

```
. generate int dtenth = 10*onedewma
. generate int ntenth = 10*onenoch
(1 missing value generated)
```

Now we will tabulate the distributions of these "tenths" variables to see what share are less than two-tenths of a percent. We will use the `tab1` command, which applies the `tabulate` command independently to a list of variables.

```
. tab1 ntenth etenth dtenth
-> tabulation of ntenth
```

ntenth	Freq.	Percent	Cum.
-13	1	0.13	0.13
-10	1	0.13	0.26
(output omitted)			
-2	52	6.75	14.94
-1	71	9.22	24.16
0	397	51.56	75.71
1	87	11.30	87.01
2	54	7.01	94.03
(output omitted)			
7	1	0.13	99.87
15	1	0.13	100.00
Total	770	100.00	

```
-> tabulation of etenth
```

etenth	Freq.	Percent	Cum.
-12	1	0.13	0.13
-10	1	0.13	0.26
(output omitted)			

For the no change method, the share of forecasts with errors less than ± 0.2 is 72% ($9.22 + 51.56 + 11.30$, or equivalently, $87.01 - 14.94$).[10]

---

10. There is more than one way to do it. We could have calculated this share without creating the "tenths" variables by typing
```
count if abs(onenoch)<0.2
local numerator "'r(N)'"
count
display 100 * 'numerator'/'r(N)'
```
We created the "tenths" variables because we want to use them below to look at the complete distribution of forecast errors.

Table 4.3 collects these results in a convenient format. The "no change" and EWMA methods do about equally well in protecting our reputation. Even the DEWMA smoother has a similar success rate.

Table 4.3. Share of "respectable" forecasts

Method	|Errors| < 0.2
No change	72
Exponential	71
Double exponential	68

What about the distribution of "unsuccessful" forecasts? Figure 4.6 displays a histogram of the forecast errors of all three methods.[11] Again not a lot to choose between. The distribution is tighter for the "no change" forecast, but the other two approaches do not do too badly. The case for using the "no change" forecast is a little stronger if our loss function is based on the absolute distance between our forecast and the actual rather than the number of acceptably wrong forecasts, but, again, the difference is not striking.[12]

---

11. At many golf courses, you are asked to pick up your ball and move on after 10 strokes per hole to avoid slowing down play. Similarly, we have capped the absolute forecast errors at 10 tenths (one percent) to keep these histograms readable.

12. Except at turning points.

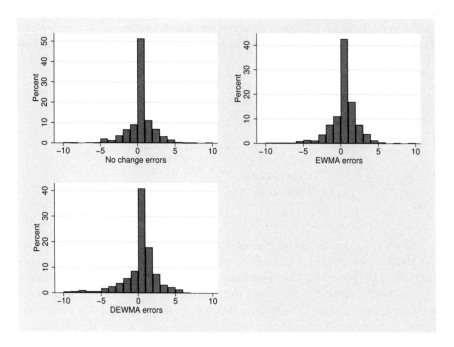

Figure 4.6. Distribution of one-step-ahead forecast errors (in tenths of a percent)

What are we to make of these results? Before you jump to the conclusion that we can dispense with these smoothers and just use the last observed value as our best forecast, remember today's unemployment rate is particularly highly correlated with yesterday's rate (correlation coefficient = 0.9917). Other series are not as highly correlated as the unemployment rate, and for those other series, the EWMA forecast is likely to outperform the naïve forecast. And for a trending series, such as gross domestic product, the DEWMA forecast has a built-in advantage over the EWMA and "no change" forecasts.

Nonetheless, this example serves as a useful reminder of how difficult it can be to beat a naïve forecast by time-series methods alone. In other words, to construct a better forecast, we may need to know more about the factors that influence the predicted variable above and beyond its past values. In subsequent chapters, we will look at some more sophisticated time-series forecasting methods, but we will continue to find that pure time-series methods work best over the short term and that it can be difficult to outperform a naïve benchmark.

Despite these caveats, smoother-based forecasts can be useful tools, especially in an industry setting where many forecasts are required quickly and little is known about the process generating future values.

Finally, we have not forgotten the Holt–Winters smoother. In this example, however, it performs very similarly to the double exponential smoother. Figure 4.7 compares the forecasts of the unemployment rate produced by these two smoothers. They are virtually indistinguishable.

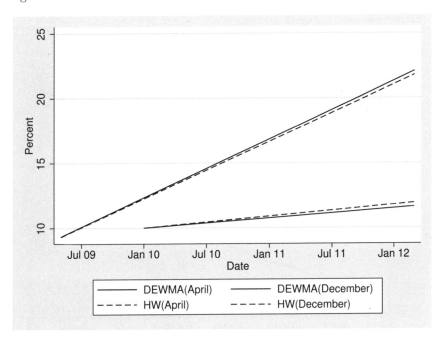

Figure 4.7. Comparison of DEWMA and Holt–Winters forecasts of the unemployment rate

4.2.2 Forecasting a trending series with a seasonal component

Before closing this chapter, let's filter and forecast a trending series that has not been seasonally adjusted. This example will shed additional light on some of the features of forecasting with the simple filters discussed so far and lay the groundwork for the statistical time-series models discussed in subsequent chapters.

For this example, we will analyze weekly measurements of the currency component of M1, a measure of the money supply. There are many different measures of the quantity of money in the economy—M1, M2, M3, etc. While the details of each measure can be complicated (and can differ across countries), the measures with higher numbers generally include more things. M1 is a relatively narrow measure. With some useful imprecision, you can think of M1 as consisting of currency (paper money and coins) and checking accounts. Essentially M1 is "readily spendable" money.[13]

---

13. For precise definitions of U.S. money stock measures, see the H.6 release of the Board of Governors of the Federal Reserve (https://www.federalreserve.gov/releases/h6/current/default.htm).

In the United States, the Federal Reserve publishes a host of statistics on various measures of the money supply, among them a weekly series of the currency component of M1 that is not seasonally adjusted. Let's begin by looking at the data.

```
. use ${ITSUS_DATA}/currcomp, clear
. tsline currcomp, ytitle($ billions)
```

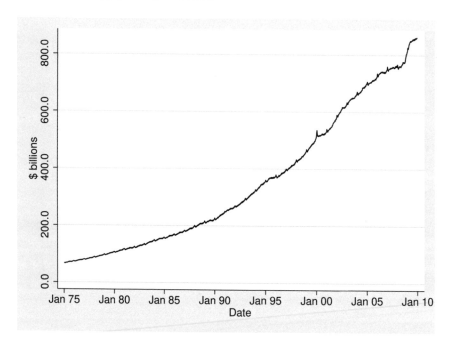

Figure 4.8. Weekly currency component of M1

The amount of currency in circulation appears to grow exponentially. This pattern makes sense because the drivers of currency growth—population and gross domestic product growth—also grow exponentially. There also is a distinctive seasonal pattern to this measure, but it is difficult to see in this chart, so let's zoom in on the post-1997 data.

```
. tsline currcomp if year>1997, ytitle($ billions)
```

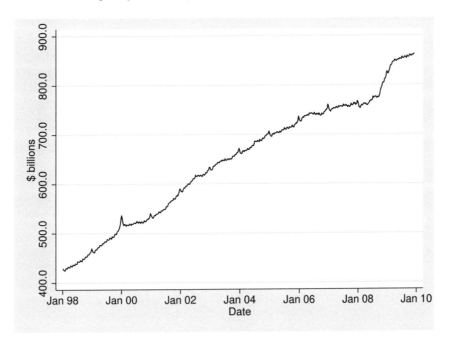

Figure 4.9. Currency component in recent years

Several features stand out in this view. First, there are repetitive "spikes" in the series, corresponding to the end-of-year holiday seasonal uptick in currency holdings. Second, the spike at the end of 1999 is much larger than usual, evidence of the public's concerns about Y2K and the Federal Reserve's actions to accommodate the temporarily higher demand for cash. Third, there is an unprecedented increase in the currency component between September 2008 and January 2009, reflecting the public's response to the financial crises and the uncertainty about the condition of even the largest banks and financial services companies. Unlike the seasonal and Y2K spikes, this increase in currency has not been reversed. Instead something like the prior trend growth rate has resumed, but starting at a higher level.

Because we suspect that trend growth in the currency stock is exponential, we will take the log of this measure before filtering it to convert the growth to a linear trend. Then we will **tsset** the dataset, so we can use the **tssmooth shwinters** command to apply the seasonal Holt–Winters filter.

```
. generate lcc = log(currcomp)
. label variable lcc "Log(currency component of M1)"
. generate wdate = wofd(date)
. label variable wdate "Date"
```

```
. tsset wdate, weekly
repeated time values in sample
r(451);
```

Something has gone wrong. Stata has detected duplicate values of the `wdate`, the weekly date variable, and refused to complete the `tsset` command. Let's see where the problem is occurring. Because the data already are sorted in date order, we can list the observations where the value of `wdate` is the same in successive observations.

```
. list if wdate==wdate[_n-1]
```

	date	currcomp	lcc	wdate
261.	31dec1979	106.3	4.666265	1039
522.	31dec1984	157.7	5.060695	1299
835.	31dec1990	249.0	5.517453	1611
1148.	30dec1996	400.1	5.991714	1923
1409.	31dec2001	589.0	6.378426	2183
1722.	31dec2007	765.4	6.640399	2495

```
. list in 260/261
```

	date	currcomp	lcc	wdate
260.	24dec1979	107.9	4.681205	1039
261.	31dec1979	106.3	4.666265	1039

```
. count if year(date)==1979
    53
```

Fifty-two weeks each with seven days would produce 364-day years. But there are actually 365 (and sometimes 366) days in a year. So occasionally, we end up with 53 "weekly" observations in a year. While it is not an ideal solution, for this example, we will just drop one of each pair of duplicates.[14]

---

14. We checked each of these cases to guarantee we were not dropping an outlier or otherwise significant observation.

```
. drop if wdate==wdate[_n-1]
(6 observations deleted)

. tsset wdate, weekly
        time variable:  wdate, 1975w1 to 2009w49
                delta:  1 week

. tssmooth shwinters shw = lcc, forecast(104)
computing optimal weights

Iteration 0:    penalized RSS = -.28659299   (not concave)
Iteration 1:    penalized RSS = -.06303158
Iteration 2:    penalized RSS = -.05250373   (not concave)
Iteration 3:    penalized RSS = -.04020031
Iteration 4:    penalized RSS = -.03588873   (not concave)
Iteration 5:    penalized RSS = -.03519133   (not concave)
Iteration 6:    penalized RSS = -.03358745
Iteration 7:    penalized RSS = -.03352175
Iteration 8:    penalized RSS = -.03352127
Iteration 9:    penalized RSS = -.03352127

Optimal weights:
                            alpha = 0.4584
                             beta = 0.0030
                            gamma = 0.5158
penalized sum-of-squared residuals = .0335213
        sum-of-squared residuals = .0335213
        root mean squared error  = .0042952

. label variable shw "Seas. HW forecast"

. twoway line lcc shw wdate if year(date)>2006, lpattern(solid dash)
```

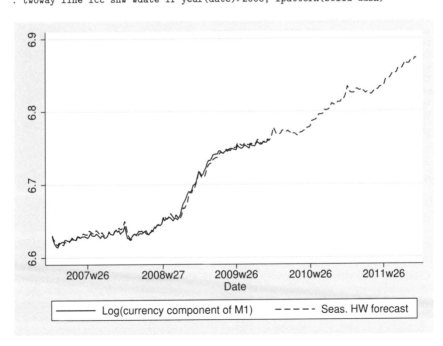

Figure 4.10. Seasonal Holt–Winters forecast

Stata's estimate of α, the weight on the current observation in updating the local mean, is close to 0.5, one of Granger's rule-of-thumb values. Because the stock of currency is a steadily growing quantity, we have applied the multiplicative version of seasonal adjustment.

For this example, we have projected two years (104 weeks) of values, comparable to the forecasts of the unemployment rate in section 4.2.1. Just as with the double-exponential and nonseasonal Holt–Winters filters, the seasonal Holt–Winters filter extends the most recent trend estimate indefinitely into the future. The seasonal version also adds the most recent estimate of seasonal variation (notice the repetition of the holiday spike). Overall, the shape of this forecast looks reasonable.

How confident should we be in this forecast method? A standard way of gauging the reliability of a forecast model is backtesting. We apply the model to an earlier period and compare the projections of the model with the actual outcome. To illustrate this technique, we will estimate the seasonal Holt–Winters filter on data through April 2000, then use that estimate to project the next 500 weeks (the maximum allowed by the `tssmooth shwinters` command).

```
. tssmooth shwinters backtest = lcc if wdate<=tw(2000w17), forecast(500)
computing optimal weights

Iteration 0:    penalized RSS = -.05969274   (not concave)
Iteration 1:    penalized RSS = -.03788054
Iteration 2:    penalized RSS = -.03231528   (not concave)
Iteration 3:    penalized RSS = -.02773728
Iteration 4:    penalized RSS = -.02746347
Iteration 5:    penalized RSS = -.02631589
Iteration 6:    penalized RSS = -.02616972
Iteration 7:    penalized RSS = -.02616718
numerical derivatives are approximate
nearby values are missing
Iteration 8:    penalized RSS = -.02616695
Iteration 9:    penalized RSS = -.02616686

Optimal weights:
                              alpha = 0.4217
                               beta = 0.0000
                              gamma = 0.3851
penalized sum-of-squared residuals = .0261669
        sum-of-squared residuals = .0261669
        root mean squared error  = .0044574

. label variable backtest "Back test"

. local x1 = tw(2000w17)
```

```
. tsline lcc shw backtest if wdate>=tw(1998w1) & wdate<=tw(2009w52),
> lpattern(solid dash dash_dot) xline(`xl´)
```

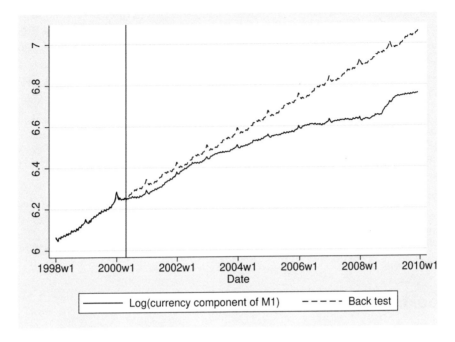

Figure 4.11. Backtesting the seasonal Holt–Winters forecast

For this sample, the estimates of all three parameters are lower than in the full sample. The chart compares the actual values of the currency stock with this backtest. The vertical line indicates the beginning of the backtest forecast. The backtest forecast consistently overshoots actuals. As we noted before, the forecast holds trend growth constant and repeats the last estimate of seasonal variation. But in this case, the trend rate of growth in currency declined after April 2000, particularly after 2003. Not surprisingly, the longer the forecast horizon, the likelier it is that something fundamental has changed in a way that alters the dynamic behavior of our time series.

4.3 Points to remember

- This book explores stochastic, linear, dynamic time-series models for forecasting continuous variables. These models cover many forecasting situations, but they generally are not applicable to forecasting discrete events (the winner of an election); to forecasting the outcome of a unique, unprecedented situation (the collapse of the recent U.S. housing bubble); or to analyzing counterfactual or futuristic scenarios.

- In general, the quality of forecast methods, like other statistical techniques, are measured primarily on their expected loss, a quantitative measure of the importance of different types of forecast errors. Different forecasters can have different loss functions—and hence, rationally choose different methods of forecasting— for the same phenomenon. For instance, one forecaster may care only about the accuracy of the one-step-ahead forecast, while another may care only about the forecast at longer horizon.

- Filter-based forecasts are chosen primarily for their low cost of production and ease of use and interpretation rather than their expected loss.

- The different filters are applicable in different situations. There is no one filter that is "best" in some sense for all purposes. Trailing moving averages and EWMAs generate constant forecasts; double-exponential and nonseasonal Holt–Winters filters generate forecasts with constant trends; and the seasonal Holt–Winters filter generates forecasts with constant trend and seasonals.

- Filters are examples of univariate time-series models—the influence of other more fundamental determinants of the time series are not explicitly modeled. If those fundamental relationships change, the univariate model also may shift. Backtesting is one way of highlighting the sensitivity of a time-series model to this type of disturbance.

4.4 Looking ahead

This chapter provided an informal, intuitive introduction to time-series forecasting and illustrated forecasting using some of the filters introduced in the previous chapter. This informal treatment allowed us to introduce many of the most important topics in forecasting—types of forecasts, elements of forecasts, measures of forecast quality, etc.—without much in the way of formal statistics. The examples were chosen to highlight the essential practical differences between techniques.

We shift gears from this point on and introduce increasingly sophisticated parametric models of time series. The next chapter introduces regression models with autocorrelated disturbances. Subsequent chapters introduce the modern univariate time-series model—a compact and elegant way of capturing a wide variety of time-series behavior— followed by a host of extensions and variations of this model—models with time-varying error variance, with jointly dependent variables, and with other characteristics that require special treatment. These chapters necessarily will include a greater amount of statistical material, but the principles discussed in this chapter and the previous one still apply to these more elaborate models.

Stata commands and features discussed

tsset ([TS] **tsset**): Declare data to be time-series data; section 4.2

tssmooth ([TS] **tssmooth**): Smooth and forecast univariate time-series data; section 4.2

5 Autocorrelated disturbances

Chapter map

5.1 Autocorrelation. What is autocorrelation? Reasons for autocorrelation. Example of autocorrelation in data on mortgage rates.

5.2 Regression models with autocorrelated disturbances. How autocorrelated disturbances violate classical regression assumptions. First-order autocorrelation. A regression model for mortgage rates.

5.3 Testing for autocorrelation. Durbin's alternative test applied to the mortgage rate data. Other tests.

5.4 Estimation with first-order autocorrelated data. Four estimation strategies for regressions with autocorrelated disturbances—ordinary least squares (OLS), transformation, feasible generalized least squares (FGLS), and instrumental variables (IV). Three different regression models and the strategies that work for each model. Examples with simulated data.

5.5 Estimating the mortgage rate equation. Choosing a strategy for the mortgage rate equation. Comparing the results of alternative strategies.

5.6 Points to remember. Why autocorrelation? Serial correlation. Stata's `estat durbinalt` command. Using strategies that match the model.

The prior two chapters on filtering and forecasting covered an eclectic assortment of methods for detecting patterns in time-series data and for extrapolating those patterns into the future. Some of the methods can be grounded in theory, but the approaches are recommended more for their pragmatic virtues than for their formal justifications.

This chapter begins our study of regression-based approaches to time-series analysis. We begin by adding one time-series wrinkle to the classical regression model—autocorrelated disturbances.[1] While more sophisticated approaches to time-series regression have been introduced in recent years, the issues and techniques discussed here anticipate many of the more advanced topics we cover in later chapters.

We begin by discussing some of the intuitive reasons for the prevalence of autocorrelation, that is, correlation among regression disturbances in different periods. We

---

1. Chapter 2 includes a brief review of the classical regression model and introduces the notation used here.

illustrate autocorrelation using data on mortgage interest rates. We introduce first-order autocorrelation—the most common specification—and develop formal tests for autocorrelation in the next two sections. Finally, we discuss some of the methods used to fit regression models with autocorrelated errors.

5.1 Autocorrelation

Not that long ago, serial correlation (also called *autocorrelation*) would comprise the principal—if not the only—time-series topic in an econometrics textbook. Autocorrelation denotes the correlation of a random variable with its past and future values. In particular, time-series econometrics focused on the possibility that the error term in a regression might be autocorrelated. This correlation violates the classical assumption that errors are independent. As a consequence, classical least-squares estimates of the regression coefficients are inefficient, and hypothesis tests are affected.

Much of early time-series econometrics was cast as dealing with the "problem" of autocorrelation. Serial correlation introduced nuisance parameters, and "correcting" for serial correlation was a necessary first step on the way to the real job of testing specific hypotheses about regression parameters. Modern econometricians have rejected this older view in favor of an approach that makes these time-series parameters a central feature of study. Despite this shift, the older techniques remain in use, and it is important to be familiar with them.

In practice, many time series are autocorrelated. Some of the more common reasons for autocorrelation include

- seasonal-adjustment methods applied to many time series, especially economic statistics published by the government, which sometimes induce serial correlation;

- "natural" correlations in seasonally unadjusted data (for example, the summer months are all hotter than average);

- differences in the frequencies of observed data and of the factors that determine the random disturbances. For example, a recession typically endures for several months, while many economic quantities (unemployment, industrial production) are recorded monthly. If the impact of the recession is not explicitly modeled, its influence will enter through the disturbance term, and regressions of monthly data will exhibit serial correlation.

5.1.1   Example: Mortgage rates

To illustrate autocorrelation, we will analyze the relationship between two measures of mortgage interest rates. The Freddie Mac Primary Mortgage Market Survey (PMMS) provides a consistent measure of the mortgage rates offered by lenders.[2] The Fannie Mae current coupon is a market-based measure of the interest rate demanded by investors in the secondary market, that is, the market for mortgage-backed securities. As a rough analogy, the PMMS rate represents the "retail" mortgage rate available to homeowners, while the current coupon represents the "wholesale" rate available to institutional investors. Like retail and wholesale prices on ordinary commodities, there should be a close relationship between the PMMS rate and the current coupon.

Let's look at the relationship between the primary and the secondary mortgage rates for 30-year fixed rate mortgages. Our file of monthly time series includes pmms30, the Freddie Mac PMMS rate (primary rate), and fncc30, the Fannie Mae current coupon (secondary rate), from January 1985 through April 2009. We start by dropping the variables we are not using for this example, and we drop any observations for which we do not have values of either the PMMS rate or the current coupon.

```
. use ${ITSUS_DATA}/monthly, clear
(Monthly data for ITSUS)

. keep date fncc30 pmms30

. describe

Contains data from /Users/sbecketti/Data/itsus1r/data/monthly.dta
  obs:         1,191                          Monthly data for ITSUS
  vars:            3                          11 Jul 2012 07:43
-------------------------------------------------------------------------------
              storage   display    value
variable name   type    format     label      variable label
-------------------------------------------------------------------------------
date           float    %tmm_Y                Date
fncc30         double   %6.2f                 FNMA 30-year current coupon
pmms30         float    %9.0g                 PMMS30 mortgage rate
-------------------------------------------------------------------------------
Sorted by: date
     Note: Dataset has changed since last saved.

. keep if !missing(fncc30,pmms30)
(899 observations deleted)

. tsset
        time variable:  date, Jan 85 to Apr 09
                delta:  1 month
```

---

2. The survey is restricted to rates on so-called conforming mortgages, that is, mortgages within the conforming balance limits of the housing agencies. Larger mortgages are called jumbo mortgages. The conforming limit is updated once a year to keep pace with changes in house prices.

As we did in chapter 4, we begin by looking at a graph of the two rates over time
and some summary statistics of the rates and month-over-month changes.

```
. tsline pmms30 fncc30, legend(cols(1))
. summarize pmms30 fncc30 D.pmms30 D.fncc30
```

Variable	Obs	Mean	Std. Dev.	Min	Max
pmms30	292	7.980132	1.831624	4.81	13.195
fncc30	292	7.539473	1.867227	4.151175	13.10971
pmms30					
D1.	291	−.0284021	.2304231	−.8014998	.7975006
fncc30					
D1.	291	−.0292689	.2550928	−.7326483	1.114377

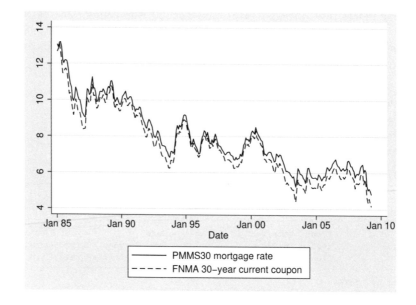

Figure 5.1. Primary and secondary mortgage rates

Figure 5.1 shows that mortgage rates trended significantly lower over this 24-year
period. The primary mortgage rate topped 13% in January 1985 and averaged a sample-
low value of only 4.81% in April 2009. However, this decline in rates was anything
but smooth. Changes in interest rates are commonly stated in basis points. A basis
point (abbreviated b.p.) is 0.01 of a percent. In other words, 100 b.p. = 1%. The
standard deviation of the one-month change in the monthly average primary mortgage
rate (D.pmms30) was 23 b.p. In at least one month, the mortgage rate dropped 80 b.p.,
and in another, it jumped by 80 b.p.

Of particular interest is the primary–secondary spread, the difference between the PMMS rate and the current coupon. Like changes in interest rates, the spread typically is measured in basis points. Let's construct that spread and take a look at it.

```
. generate pssp30 = 100 * (pmms30 - fncc30)
. label variable pssp30 "30-year primary-secondary spread"
. summarize pssp30
```

Variable	Obs	Mean	Std. Dev.	Min	Max
pssp30	292	44.0659	15.7645	6.028579	93.94764

```
. tsline pssp30, yline(44)
```

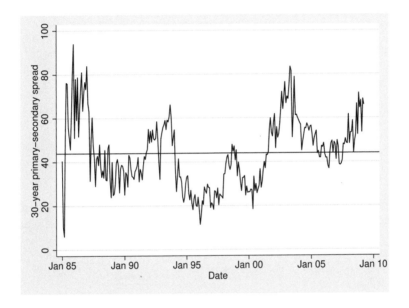

Figure 5.2. The spread between the primary and secondary mortgage rates

The primary–secondary spread exhibits clear evidence of autocorrelation. The spread tends to remain above (below) its mean for many months. Observing an unusually wide (narrow) spread this month makes it highly likely that next month's spread will also be wide (narrow).

The wide and persistent swings in the primary–secondary spread reflect the speed of shifts in the demand for mortgage relative to the ability of the industry to adjust processing capacity. When the general level of interest rates drops sharply, many borrowers take advantage of the low rates to refinance their mortgages. Lenders add staff to help process the surge in loan applications, but it takes time to train the new employees, and at least temporarily, industry capacity is fully utilized, a phenomenon known as pipeline congestion. During these periods, lenders have no incentive to lower primary mortgage rates further, even if secondary market rates continue to fall, because the demand for new loans temporarily exceeds the industry's ability to supply them. As a consequence, the primary–secondary spread widens and remains wide until either loan demand falls or processing capacity catches up.

In periods of high and rising interest rates, the situation is reversed. Lenders find themselves with idle processing capacity. Lenders are slow to reduce capacity too quickly in case interest rate trends reverse unexpectedly. Instead, they compete more aggressively for borrowers by narrowing their margins, that is, by narrowing the primary–secondary spread.

5.2   Regression models with autocorrelated disturbances

Intuitively, autocorrelation should be a helpful property to forecasters. Today's realization of an autocorrelated random variable provides useful information about tomorrow's likely value. In the next chapter, we will introduce some methods for exploiting the predictive information provided by autocorrelation.

On the other hand, autocorrelated regression disturbances pose some estimation challenges to econometricians.

A simple time-series regression can be written as

$$y_t = \beta_0 + \beta_1 x_{1t} + \cdots + \beta_K x_{Kt} + \epsilon_t$$

where y_t represents the value of the dependent variable in period t, the x_t's are variables that influence y_t, the β's are the regression coefficients that quantify the influence of each x_t—β_0 is the constant term in the regression—and ϵ_t is the random error or disturbance. In classical regression, we assume that the disturbances have an expected value of 0 given the values of the regressors

$$E\left(\epsilon_t | x_{1t}, \ldots, x_{Kt}\right) = 0, \forall t$$

and a constant variance (the assumption of homoskedasticity),

$$E\left(\epsilon_t^2 | x_{1t}, \ldots, x_{Kt}\right) = \sigma^2, \forall t$$

and that disturbances in different periods are uncorrelated:[3]

$$E\left(\epsilon_s \epsilon_t\right) = 0, s \neq t$$

This last assumption does not hold when the regression disturbances are autocorrelated.

---

3. These conditions are frequently strengthened to the assumption that the errors are identically and independently distributed (i.i.d.). Even stronger, we sometimes assume that the errors are i.i.d. normal.

5.2.1   First-order autocorrelation

First-order autocorrelation is the most-studied case of serial correlation. In this specification, the current period's disturbance is a multiple ρ of the previous period's disturbance plus a serially uncorrelated error

$$\epsilon_t = \rho\epsilon_{t-1} + \eta_t$$

where

$$E\left(\eta_t^2\right) = \sigma_\eta^2$$

and

$$E\left(\eta_s\eta_t\right) = 0, \forall s \neq t$$

This specification is called first-order autocorrelation because only a one-period lag of ϵ_t appears in the equation above. This specification also is called an AR(1) model, where "AR" stands for autoregression. Higher-order autocorrelation can also occur, where the order of the autocorrelation denotes the longest lag in the equation. For example, a pth-order autocorrelation is written as

$$\epsilon_t = \rho_1\epsilon_{t-1} + \rho_2\epsilon_{t-2} + \cdots + \rho_p\epsilon_{t-p} + \eta_t$$

In later chapters, we will analyze models with higher-order autocorrelation, AR(p) models. For now, however, we will stick to AR(1) models.

To determine the mean and variance of the regression disturbance ϵ_t in the AR(1) model, we substitute to eliminate ϵ_{t-1} (and earlier terms):

$$\epsilon_t = \rho(\rho\epsilon_{t-2} + \eta_{t-1}) + \eta_t = \eta_t + \rho\eta_{t-1} + \rho^2\eta_{t-2} + \cdots$$

So the expected value of the error is

$$E\left(\epsilon_t\right) = \sum_{i=0}^{\infty} \rho^i E\left(\eta_{t-i}\right) = 0$$

The expected value of ϵ_t is 0, just as in the classical regression model. Note, however, that ϵ_t depends on an infinite series of current and past η disturbances. The impact of each η_s on ϵ_t depends on the value of ρ. If $|\rho| > 1$, the impact of η is larger for distant past values of η than for recent values. In fact, the impact of past η's grows without bound as the distance in time grows. This possibility seems counterintuitive. On the other hand, if $|\rho| < 1$, the influence of past values of η diminishes geometrically as they recede in time—the most important influences are the most recent ones. This possibility corresponds most closely to our intuition about real-world processes, and we will assume $|\rho| < 1$ in the remainder of this chapter.

The possibility that $|\rho| = 1$ is an interesting one, although we will have to wait until later chapters to explore it. In this case, each η represents a permanent shock to current and future values of ϵ. The influence of these shocks neither diminishes over time nor grows explosively. Instead today's ϵ represents the simple accumulation of the current and all past η's. Efficient market theories of asset prices incorporate this type of process, as do some dynamic models of the macroeconomy.

The variance of ϵ_t is the expected value of ϵ_t^2. Because the η's are serially uncorrelated, the expectation of all cross terms is 0, and

$$E\left(\epsilon_t^2\right) = \sigma_\eta^2 + \rho^2 \sigma_\eta^2 + \rho^4 \sigma_\eta^2 + \cdots$$

Because we are assuming that $|\rho| < 1$, this infinite series converges, and the variance is

$$E\left(\epsilon_t^2\right) = \frac{\sigma_\eta^2}{1 - \rho^2}$$

A time series with a constant (not a function of time) mean and variance is called a stationary process.[4] In this simple case of first-order autocorrelation, the series is stationary if and only if ρ is less than 1 in absolute value.

Note that first-order autocorrelation does not imply that only adjacent observations of ϵ are correlated. In fact, all the ϵ's, no matter how far apart in time, are correlated. For instance, the covariance between ϵ_t and ϵ_{t-1} is

$$\begin{aligned}
E\left(\epsilon_t \epsilon_{t-1}\right) &= E\left\{\left(\eta_t + \rho\eta_{t-1} + \rho^2\eta_{t-2} + \cdots\right)\left(\eta_{t-1} + \rho\eta_{t-2} + \rho^2\eta_{t-3} + \cdots\right)\right\} \\
&= E\left(\eta_t \epsilon_{t-1}\right) + E\left(\rho\epsilon_{t-1}^2\right) = \rho\sigma_\epsilon^2 = \frac{\rho\sigma_\eta^2}{1 - \rho^2}
\end{aligned}$$

By similar reasoning, the covariance between ϵ_s and ϵ_t, where $s = t - k, k > 0$, is

$$E\left(\epsilon_s \epsilon_t\right) = \rho^k \sigma_\epsilon^2 = \frac{\rho^k \sigma_\eta^2}{1 - \rho^2}$$

In other words, the covariance of ϵ_s and ϵ_t decreases geometrically with the distance (in time) between s and t, a direct implication of our assumption that $|\rho| < 1$. This pattern makes sense intuitively: events (disturbances) that occur at close to the same time are affected by some of the same underlying factors, while distant events bear only a weak relationship to each other. The implication that covariances decline geometrically may be too rigid a specification in some situations, but it seems like a reasonable and tractable approximation.

When $\rho > 0$, the covariance between ϵ_t and ϵ_{t-1} is positive, and the errors are said to be positively autocorrelated. Conversely, when $\rho < 0$, the covariance between ϵ_t and ϵ_{t-1} is negative, and the errors are said to be negatively autocorrelated. As a practical matter, we rarely run across negatively autocorrelated errors, and we will not consider that alternative in the remainder of this chapter.

---

4. More precisely, these conditions define a weakly or covariance stationary process. Strong stationarity requires a time-invariant distribution, not just a time-invariant mean and variance.

5.2.2   Example: Mortgage rates (cont.)

In the mortgage rate example in the previous section, we saw that the spread between the primary mortgage rate, the Freddie Mac PMMS30 survey of rates offered to retail borrowers, and the secondary mortgage rate, the Fannie Mae 30-year current coupon, shows signs of autocorrelation.

Most lenders sell their mortgages to be packaged as mortgage-backed securities. As a consequence, the secondary mortgage rate, the rate demanded by institutional investors, plays a large part in determining the primary rate that lenders offer to borrowers. Let's try to quantify this influence by estimating a regression of the primary rate on the secondary rate.

```
. regress pmms30 fncc30

      Source |       SS           df       MS              Number of obs   =       292
-------------+----------------------------------           F(1, 290)       =  41836.65
       Model |  969.539395          1  969.539395          Prob > F        =    0.0000
    Residual |  6.72057681        290  .023174403          R-squared       =    0.9931
-------------+----------------------------------           Adj R-squared   =    0.9931
       Total |  976.259972        291  3.35484526          Root MSE        =    .15223

------------------------------------------------------------------------------
      pmms30 |      Coef.   Std. Err.      t    P>|t|     [95% Conf. Interval]
-------------+----------------------------------------------------------------
      fncc30 |   .9775502   .0047793   204.54   0.000     .9681438    .9869567
       _cons |   .6099185    .037118    16.43   0.000     .5368636    .6829734
------------------------------------------------------------------------------

. predict resid, resid
```

`. scatter resid L.resid`

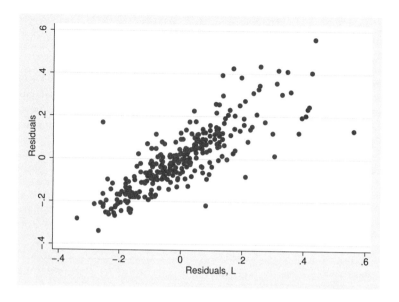

Figure 5.3. Current residuals versus lagged residuals

The regression makes a rough sort of sense—on average, the primary mortgage rate moves almost one for one with the secondary mortgage rate, and judging by the F statistic and R^2, the fit seems very good. The regression estimates raise some interesting questions, but we will hold these for later.

As we expected, the regression residuals are strongly autocorrelated. The pattern in the scatterplot indicates that a formal significance test is unnecessary in this instance.

5.3   Testing for autocorrelation

In our mortgage rate example, statistical tests for autocorrelation seem beside the point. In many instances, though, we cannot tell so easily whether apparent autocorrelation is statistically significant. Durbin's alternative test provides a formal test of the null hypothesis of serially uncorrelated disturbances against the alternative of autocorrelation of order p.[5] We can use Stata's `estat` command to apply Durbin's alternative test to our mortgage rate data.

---

5. This test is called Durbin's *alternative* because it supplants an earlier test proposed by Durbin and Watson. The original Durbin–Watson test was a workhorse of early time-series analysis and appears in almost all commercial statistical packages—including Stata—even though preferable tests are available.

The `estat` command is Stata's unified way of accessing a wide variety of postestimation statistics. The `regress` command, like Stata's other estimation commands, always displays the most commonly used statistics along with the parameter estimates. The regression table displayed by `regress` includes—in addition to the estimated coefficients—the standard errors, t statistics and their p-values, and the 95% confidence intervals (CIs) of the coefficients. Above this regression table, `regress` displays an analysis of variance table (on the left) and a compact table of other statistics (on the right). At last count, there were approximately a zillion additional regression-based statistics that are useful in specific cases. Rather than displaying all of them (and making the regression output pretty much unreadable), Stata provides the `estat` command as a convenient method to display just the information that is needed.

`estat durbinalt` displays the test statistic for Durbin's alternative test.

```
. estat durbinalt

Durbin's alternative test for autocorrelation
```

lags(p)	chi2	df	Prob > chi2
1	632.077	1	0.0000

```
                 H0: no serial correlation
```

As expected, the null hypothesis of no serial correlation is strongly rejected.

By default, the alternative hypothesis in this test is first-order autocorrelation of the regression disturbance. However, Durbin's alternative test also can be used to test for higher-order autocorrelation. For example, a pth-order serial correlation process, in which ϵ_t depends on lagged values of ϵ up to p periods ago, can be written as

$$\epsilon_t = \rho_1\epsilon_{t-1} + \cdots + \rho_p\epsilon_{t-p} + \eta_t$$

To test the null hypothesis of no autocorrelation, that is,

$$H_0 : \rho_1 = \cdots = \rho_p = 0$$

against the alternative that at least one of the ρ_i's is not 0, type

```
. estat durbinalt, lags(p)
```

where p is replaced with the order of the process.

5.3.1 Other tests

Durbin and Watson proposed one of the first tests for autocorrelation. The Durbin–Watson test is based on some restrictive assumptions: the regression disturbances follow a first-order autocorrelation process, there are no lagged dependent variables in the regression,[6] and the regression contains a constant term. In addition, the Durbin–

---

6. More precisely, all regressors must be strictly exogenous.

Watson test does not produce a clear yes or no answer. Their test statistic can fall into a "zone of uncertainty" where we neither reject nor accept the null hypothesis.

Durbin's alternative test does not suffer from these limitations and has largely supplanted the older Durbin–Watson test. Durbin's alternative test can accommodate lagged dependent variables. In addition, the test statistic in Durbin's alternative test is distributed asymptotically χ^2, so there is no zone of uncertainty, and p-values are readily calculated.[7]

Stata offers another modern test for serial correlation, the Breusch–Godfrey test. Its motivation is similar to Durbin's alternative, and the test is asymptotically equivalent to Durbin's alternative test.

5.4 Estimation with first-order autocorrelated data

Estimating regressions with first-order autocorrelated disturbances takes careful analysis. There is no "silver bullet" technique that is appropriate in all cases. Instead the estimation strategy must be matched to the type of regression model at hand. To adequately cover the variety of model types and estimation strategies and to review some of the problems that may arise in practice, this section of chapter 5 is longer than the previous sections combined.

Regression models with autocorrelated disturbances present two types of challenges. First, when the regression disturbance is autocorrelated, the variance matrix of the residuals is no longer a scalar matrix. This violation of the classical regression assumptions implies that least-squares estimates of the regression coefficients are inefficient. Second, if there are also lagged dependent variables in a regression model with autocorrelated disturbances, the regressors are correlated with the regression error. This second violation of the classical assumptions is more serious. In this case, least-squares estimates are biased and inconsistent. Thus, when the tests described earlier in this chapter reject the null hypothesis of no autocorrelation, a modification of the simple least-squares estimation procedure is needed.

There are four strategies for dealing with these two issues in regression models with autocorrelated disturbances:

OLS Use OLS to estimate the regression coefficients, but use a corrected estimator for the variance–covariance estimator (VCE).

Transformation Transform the original model into an equivalent model with i.i.d. disturbances, and use OLS to estimate the parameters.

FGLS Use FGLS to estimate the parameters.

IV Use IV to estimate the parameters.

---

7. Because some practitioners still use the Durbin–Watson test, Stata supplies it. Typing `estat dwatson` will produce the Durbin–Watson d test statistic.

No one strategy applies to all possible cases, but we usually use the transformation strategy.

In econometrics, as in medicine, it is important to match the statistical "treatment" to the model "disease". The OLS and FGLS strategies are effective treatments for a nonscalar residual variance matrix but do not treat the case of correlation between regressors and the error term. The IV strategy is applicable when regressors are correlated with the error term, but it is not designed for cases where the nonscalar residual variance matrix is the only issue. The transformation strategy is a broad-spectrum statistical drug: it can be used to treat either or both "diseases".

In the next three subsections, we discuss the applicability of these four strategies to three different time-series regression models. In the first model, all the independent variables are strictly exogenous, and the errors are autocorrelated. The second contains a lag of the dependent variable, but the errors are i.i.d. The third model contains a lag of the dependent variable, and the errors are first-order autocorrelated.

5.4.1 Model 1: Strictly exogenous regressors and autocorrelated disturbances

Let's evaluate which strategies make sense for

Model 1:

1. $y_t = \beta_0 + \beta_1 x_t + \epsilon_t$,
2. $\epsilon_t = \rho \epsilon_{t-1} + \eta_t$,
3. η is i.i.d., and
4. x_t is strictly exogenous.

You will recall that strict exogeneity is defined as a lack of correlation between the error term, ϵ_t, and all past, present, and future values of x_t. Thus the assumption that x_t is strictly exogenous means there is no correlation between the regressors and the regression error

$$E\left(x_t \epsilon_s\right) = 0, \quad \forall s, t$$

Thus the IV strategy is not applicable to model 1. However, the variance matrix of the residual is nonscalar, so the OLS, transformation, and FGLS strategies can be applied to fit model 1.

Let's look more closely at the variance matrix of the residual before we discuss each of the applicable strategies. We begin by rewriting model 1 in matrix form:

$$y = X\beta + \epsilon$$

Because x_t is strictly exogenous, the OLS estimator of the coefficient vector in model 1 is still unbiased.

$$E\left(\widehat{\beta}|X\right) = E\left\{(X'X)^{-1}X'y|X\right\} = \beta + E\left\{(X'X)^{-1}X'\epsilon|X\right\} = \beta$$

The difficulties arise when we estimate the variance of $\widehat{\beta}$, which depends crucially on the variance of ϵ. In the classical regression model without autocorrelation, the variance of ϵ is

$$E\left(\epsilon\epsilon'|X\right) = \sigma^2 I$$

This scalar covariance matrix is a compact way of specifying both homoskedasticity (the diagonal elements—the variances of the ϵ_t—are all equal to the scalar σ^2) and lack of autocorrelation (the off-diagonal elements—the covariances of ϵ's in different periods—are all equal to 0).

In model 1, the variance of ϵ is no longer a scalar matrix. Arranging the results we derived in section 5.2 in matrix form, we have

$$E\left(\epsilon\epsilon'|X\right) = \sigma_\eta^2 \Omega = \frac{\sigma_\eta^2}{1-\rho^2}\begin{pmatrix} 1 & \rho & \rho^2 & \cdots & \rho^{T-1} \\ \rho & 1 & \rho & \cdots & \rho^{T-2} \\ \rho^2 & \rho & 1 & \cdots & \rho^{T-3} \\ \vdots & \vdots & \vdots & \ddots & \rho \\ \rho^{T-1} & \rho^{T-2} & \rho^{T-3} & \cdots & 1 \end{pmatrix}$$

The diagonal elements of Ω are

$$E\left(\epsilon^2|X\right) = \frac{\sigma_\eta^2}{1-\rho^2}$$

and the off-diagonal elements are the covariances between ϵ's at different points in time:

$$E\left(\epsilon_s\epsilon_t|X\right) = \omega_{st} = \rho^{|s-t|}\sigma_\epsilon^2$$

The variance of $\widehat{\beta}$ in this case is

$$\sigma_\eta^2(X'X)^{-1}(X'\Omega X)(X'X)^{-1}$$

The OLS estimator of $\widehat{\beta}$ assumes that $\Omega = I$ and uses the incorrect (in this case) formula

$$\widehat{\sigma}^2(X'X)^{-1}$$

To illustrate the applicability of the OLS, transformation, and FGLS strategies to model 1, we will generate some simulated data based on this model. We will create 120 observations—the equivalent of 10 "years" of "monthly" data. To generate data, we set $\beta_0 = 5$, $\beta_1 = 3$, $\rho = 0.9$, and we simulate η as a standard normal random variable.[8]

```
. clear
. set obs 120
number of observations (_N) was 0, now 120
. set seed 762
. generate eta = rnormal()
. generate epsilon = rnormal()*sqrt(1/0.19) in f
(119 missing values generated)
. replace epsilon = eta + 0.9*epsilon[_n-1] in 2/1
(119 real changes made)
. generate x = rnormal()
. generate y = 5 + 3*x + epsilon
. generate int t = _n
. tsset t
        time variable:  t, 1 to 120
                delta:  1 unit
```

The Stata `rnormal()` function generates standard normal (mean 0, standard deviation 1) random numbers. The `set seed` command anchors the starting point of the sequence of random numbers, so you can reproduce these results exactly if you replicate these commands.[9,10] We create a time variable, `t`, and `tsset` the data so that we can use Stata's time-series commands.

Now let's consider the strategies in turn.

---

8. You are probably wondering why the first observation of ϵ is created differently from the rest of the observations. We know, by construction, that each observation of ϵ depends on the previous observation of ϵ plus a random disturbance. However, for the first observation of ϵ in our sample, we have no previous observation to condition on. The solution is to rewrite the equation as

$$\epsilon_t = \rho\epsilon_{t-1} + \eta_t$$

Collecting epsilons on the left-hand side of the equation, we get

$$\epsilon_t - \rho\epsilon_{t-1} = \eta_t$$

where η_t follows a standard normal distribution. In the next chapter of this book, we show how this equation implies that the stationary distribution of ϵ_t is $N\{0, 1/(1 - \rho^2)\}$ or, in this example, $N(0, 1/0.19)$. Thus, we can simulate the first observation of `epsilon` as

```
generate epsilon = rnormal()*sqrt(1/0.19) in f
```

9. Computers do not generate truly random numbers. Instead they generate what are called "pseudorandom" numbers. Close enough for our purposes.
10. Stata has revised its random-number generators occasionally. These results were generated using Stata 16. If you are using an earlier version, you may get different results.

The OLS strategy

In model 1, OLS produces unbiased estimates of the β coefficients, but the OLS estimator of the covariance matrix of β's is no longer appropriate; therefore, significance levels, CIs, and hypothesis tests based on this estimator are incorrect.

For non-time-series data, the `vce(robust)` option to the `regress` command provides an alternative estimate of the variance of $\widehat{\beta}$. This alternative, associated with Huber and White, is appropriate for many types of nonscalar error variance matrices; however, it requires independent regression disturbances. For autocorrelated disturbances such as in model 1, Stata provides the `newey` command, which uses OLS to calculate a point estimate of β but uses the Newey–West estimator for the variance of $\widehat{\beta}$. The Newey–West estimator is an extension of the Huber–White approach for models with autocorrelation.

We will demonstrate the Newey–West approach using our simulated data. We begin by estimating the equation using OLS and testing for first-order autocorrelation.

```
. regress y x
```

Source	SS	df	MS
Model	793.177879	1	793.177879
Residual	400.453725	118	3.39367564
Total	1193.6316	119	10.0305177

Number of obs	= 120
F(1, 118)	= 233.72
Prob > F	= 0.0000
R-squared	= 0.6645
Adj R-squared	= 0.6617
Root MSE	= 1.8422

y	Coef.	Std. Err.	t	P>|t|	[95% Conf. Interval]
x	2.883601	.1886188	15.29	0.000	2.510084 3.257117
_cons	6.092924	.1686455	36.13	0.000	5.75896 6.426888

```
. estat durbinalt
```

Durbin's alternative test for autocorrelation

lags(p)	chi2	df	Prob > chi2
1	304.638	1	0.0000

HO: no serial correlation

Because we "baked in" serial correlation in the residuals, it's no surprise that Durbin's alternative test detected it. The point estimates of the coefficients are tolerably close to the "true" values used to generate the sample. The 95% CI for β_1, the coefficient on x, includes the true value of 3. The CI for β_0, the constant, is a bit higher than the true value of 5.

Now let's use the **newey** command to correct the estimate of the variance of $\widehat{\beta}$.

```
. newey y x, lag(20)
Regression with Newey-West standard errors    Number of obs   =       120
maximum lag: 20                                F(  1,    118) =    429.50
                                               Prob > F       =    0.0000
```

y	Coef.	Newey-West Std. Err.	t	P>|t|	[95% Conf. Interval]
x	2.883601	.1391406	20.72	0.000	2.608065	3.159137
_cons	6.092924	.3372201	18.07	0.000	5.425137	6.760712

The coefficient estimates are unchanged because **newey** uses the OLS estimates of β. But the variances of the coefficient estimates are different. The Newey–West CI for β_1 is 26% narrower than the OLS CI, and the Newey–West confidence interval for β_0 is 100% wider than the OLS CI.[11]

The transformation strategy

The transformation strategy requires only a little algebra and OLS. This ease of use is one reason for the popularity of the transformation strategy, especially in the years before other methods were readily available in software packages like Stata.

To transform model 1, multiply the equation for y_{t-1} by ρ and subtract it from the equation for y_t.

$$y_t - \rho y_{t-1} = (1 - \rho)\beta_0 + \beta_1 x_t - \rho\beta_1 x_{t-1} + \epsilon_t - \rho\epsilon_{t-1}$$

Rearranging terms and replacing $\epsilon_t - \rho\epsilon_{t-1}$ with η_t produces

$$y_t = \beta_0^* + \alpha^* y_{t-1} + \beta_1^* x_t + \beta_2^* x_{t-1} + \eta_t$$

where

$$\beta_0^* = (1 - \rho)\beta_0, \quad \alpha^* = \rho, \quad \beta_1^* = \beta_1, \text{ and } \beta_2^* = -\rho\beta_1$$

The disturbances in this new regression are serially uncorrelated, so least-squares estimates of the starred statistics are unbiased and efficient. However, OLS does not provide efficient estimates of the fundamental (unstarred) parameters, which are non-linear functions of the starred parameters. Note that the regression coefficient on y_{t-1} provides an estimate of ρ, something the OLS strategy does not provide. In fact, this equation provides two estimates of ρ. The negative of the ratio of the estimated coefficients on x_{t-1} and x_t provides an additional estimate of ρ.

---

11. In the real world, we do not know the true order of the autocorrelation. Accordingly, we allow the **newey** command to test for higher-order autocorrelation.

Let's apply the transformation strategy to our simulated data.

```
. regress y L.y x L.x
```

Source	SS	df	MS
Model	1082.58143	3	360.860475
Residual	110.554113	115	.961340115
Total	1193.13554	118	10.1113181

Number of obs = 119
F(3, 115)   = 375.37
Prob > F   = 0.0000
R-squared   = 0.9073
Adj R-squared = 0.9049
Root MSE   = .98048

y	Coef.	Std. Err.	t	P>|t|	[95% Conf. Interval]
y					
L1.	.8641315	.0498592	17.33	0.000	.7653701 .9628929
x					
--.	2.787235	.1011498	27.56	0.000	2.586877 2.987593
L1.	-2.551587	.1741299	-14.65	0.000	-2.896505 -2.20667
_cons	.796708	.3185881	2.50	0.014	.1656463 1.42777

```
. estat durbinalt
```

Durbin's alternative test for autocorrelation

lags(p)	chi2	df	Prob > chi2
1	1.003	1	0.3167

HO: no serial correlation

As we intended, the transformation eliminates the serial correlation of the regression disturbances.

The regression coefficient on x_t is still an estimate of β_1. In the transformed equation, the point estimate of this parameter is 2.79, close to the true value of 3. Note that the 95% CI for $\widehat{\beta}_1$ in the transformed regression is 27% narrower than the CI in the Newey–West regression. The variance of the error term in the transformed equation is σ_η^2. In the OLS regression underlying the Newey–West estimates, the error variance is $\sigma_\eta^2/(1-\rho^2)$—the division by $(1-\rho^2)$ inflates the error variance.

With a little algebra, estimates of all the original parameters—β_0, β_1, ρ, and σ_η^2—can be recovered from this transformed equation. The constant term in the transformed equation is an estimate of $(1-\rho)\beta_0$. To recover an estimate of β_0, we first have to recover an estimate of ρ. As we noted above, we have two point estimates in this example. The regression on y_{t-1} estimates ρ as 0.86, slightly lower than the value we used to generate the simulated data.[12] The negative of the ratio of the coefficients on x_{t-1} and x_t produces an estimate of 0.92, even closer to the true value. However, this small difference in estimates of ρ implies a larger difference in the point estimates of β_0. The two point estimates are 5.86 and 9.42.

---

12. The estimated coefficient on a lagged dependent variable is biased toward 0 even when the equation errors are not autocorrelated. See Shaman and Stine (1988) for details.

This pencil-and-paper approach does not provide significance levels for parameters like β_0, which are nonlinear functions of the estimated coefficients in the transformed equation. As we discussed in section 1.4.2, we can use Stata's `nlcom` command to estimate the significance of this type of implicit parameter. For example,

```
. nlcom _b[_cons]/(1-_b[L.y])
       _nl_1:  _b[_cons]/(1-_b[L.y])
```

y	Coef.	Std. Err.	z	P>|z|	[95% Conf. Interval]
_nl_1	5.863816	.6715668	8.73	0.000	4.54757 7.180063

provides a point estimate of β_0 along with a standard error, t statistic, p-value, and 95% CI. This example uses the estimated coefficient on y_{t-1} as the estimate of ρ.

In the case of model 1, Stata's `nl` command, also discussed in section 1.4.2, estimates the transformed equation by nonlinear least squares, providing unique estimates of the underlying parameters and their standard errors.

```
. nl (y = ({b0}*(1-{rho})) + {rho}*L.y + {b1}*x - {b1}*{rho}*L.x) in 2/l
(obs = 119)
Iteration 0:  residual SS =   410.5237
Iteration 1:  residual SS =   115.2592
Iteration 2:  residual SS =   111.7827
Iteration 3:  residual SS =   111.7822
Iteration 4:  residual SS =   111.7822
Iteration 5:  residual SS =   111.7822
```

Source	SS	df	MS		
				Number of obs =	119
Model	1081.3533	2	540.676647	R-squared =	0.9063
Residual	111.78224	116	.963640042	Adj R-squared =	0.9047
				Root MSE =	.9816517
Total	1193.1355	118	10.1113181	Res. dev. =	330.2615

y	Coef.	Std. Err.	t	P>|t|	[95% Conf. Interval]
/b0	5.937434	.6561381	9.05	0.000	4.637869 7.236998
/rho	.8621282	.0498872	17.28	0.000	.7633203 .9609362
/b1	2.856325	.0806868	35.40	0.000	2.696515 3.016135

```
Parameter b0 taken as constant term in model & ANOVA table
```

Note that the parameter names are enclosed in curly braces (for example, `rho`) so that Stata can distinguish them from variables.

The transformation strategy works the same way regardless of the number of x-variables on the right-hand side of the regression equation. If we extend model 1 to include K regressors, the transformed equation becomes

$$y_t = (1 - \rho)\beta_0 + \rho y_{t-1} + \sum_{i=1}^{K}(\beta_i x_t - \rho\beta_i x_{t-1}) + \eta_t$$

Every added regressor generates an additional estimate of ρ and thus an additional estimate of β_0. Again you can use the `nl` command to obtain unique estimates of the parameters.

The FGLS strategy

In section 2.3.3, we introduced FGLS estimators to handle models, like model 1, with nonscalar error covariance matrices:

$$E\left(\epsilon\epsilon'\right) = \sigma_\eta^2 \Omega$$

Both the transformation strategy and the FGLS strategy transform the original regression equation to an equivalent equation that can be estimated by OLS. However, where the transformation strategy adds lagged variables to the equation and "scrambles" the parameters to guarantee a scalar error covariance matrix, the FGLS strategy uses a consistent estimate of Ω to transform the regression variables, leaving the original structure and parameters intact.

Specifically, the FGLS strategy

1. calculates $\widehat{\Omega}$, a consistent estimate of Ω;

2. uses $\widehat{\Omega}$ to transform model 1 to a model with i.i.d. errors; and

3. estimates the β coefficients by OLS on data from the transformed model.

The FGLS-transformed equation is

$$\widehat{\Omega}^{-1/2}y = \widehat{\Omega}^{-1/2}X\beta + \widehat{\Omega}^{-1/2}\epsilon$$

As we discussed in section 2.3, the probability limit of the error covariance matrix in this transformed equation is scalar

$$\mathrm{plim}\,\widehat{\Omega}^{-1/2}\epsilon(\widehat{\Omega}^{-1/2}\epsilon)' = \sigma_\eta^2 I$$

and the transformed equation can be estimated efficiently by OLS.

In the case of first-order autocorrelation of the regression disturbances, Ω depends on a single parameter, ρ, and

$$\Omega^{-1/2} = \begin{pmatrix} \sqrt{1-\rho^2} & 0 & 0 & \cdots & 0 & 0 \\ -\rho & 1 & 0 & \cdots & 0 & 0 \\ 0 & -\rho & 1 & \cdots & 0 & 0 \\ \vdots & \vdots & \vdots & \ddots & \vdots & \vdots \\ 0 & 0 & 0 & \cdots & -\rho & 1 \end{pmatrix}$$

so all that is needed is an estimate of ρ.

There are many slightly different versions of the FGLS estimator for model 1. They differ in the method they use for estimating ρ. They also differ in their "stopping rules". Some of the estimators calculate an initial estimate of ρ, transform the variables, estimate the parameters by OLS, and stop. Other methods iterate until the estimates of ρ or the β vector converge. Still, others search for a value of ρ that minimizes the sum of squared errors of the transformed equation. All these methods are asymptotically equivalent; however, they may perform very differently in small samples. There is no firm guidance in econometric theory for selecting among these asymptotically equivalent estimators.

The one piece of guidance in the literature concerns the treatment of the first observation. Notice that the transformation strategy added lagged variables to the equation. As a consequence, the equation includes one less observation than previously (119 observations rather than 120; check the regression output above). Some of the FGLS methods also drop the first observation and rely only on adjacent pairs of observations to obtain an estimate of ρ. Small sample studies suggest that these methods tend to perform more poorly than methods that incorporate the first observation in the estimation of ρ and β.

All the calculations in the FGLS strategy—the initial estimate of the regression, the estimate of ρ, the multiplication of the data matrices by the estimate of $\Omega^{-1/2}$, and the estimation of the transformed regression—can be done with Stata commands, but Stata has saved us the trouble by providing the **prais** command, which implements some of the more popular FGLS methods.[13] The Prais–Winsten method multiplies the data matrices by $\widehat{\Omega}^{-1/2}$. The Cochrane–Orcutt method drops the first observation to avoid the special treatment required, because the 0th observation is not available for calculating a first difference. As an option, the **prais** command will iterate over the FGLS process, trying different values of ρ to find the one that minimizes the sum of squared errors.[14]

The Prais–Winsten and Cochrane–Orcutt FGLS estimators are derived specifically for model 1, that is, a model with strictly exogenous regressors and first-order autocorrelated regression disturbances. In particular, the model cannot contain lags of the dependent variable.

---

13. As a footnote to Stata history, during the early days of Stata, two of the original developers of Stata had a running argument about the ability to implement the Cochrane–Orcutt FGLS estimator as a Stata script or program. Whenever insuperable obstacles were discovered, new capabilities were added to Stata to overcome them; thus out of that argument came many of the programming features available in Stata today. Nonetheless, with the introduction of the time-series features of Stata, the developers threw in the towel and implemented **prais** as part of the core Stata package.

14. This approach was first suggested by Hildreth and Lu (1960).

Let's apply the FGLS strategy to our simulated data.

```
. prais y x

Iteration 0:  rho = 0.0000
Iteration 1:  rho = 0.8616
Iteration 2:  rho = 0.8624
Iteration 3:  rho = 0.8625
Iteration 4:  rho = 0.8625

Prais-Winsten AR(1) regression -- iterated estimates

      Source |       SS           df       MS        Number of obs   =       120
-------------+------------------------------        F(1, 118)       =   1281.52
       Model | 1214.29537          1  1214.29537    Prob > F        =    0.0000
    Residual | 111.809881        118  .947541368    R-squared       =    0.9157
-------------+------------------------------        Adj R-squared   =    0.9150
       Total | 1326.10525        119  11.1437416    Root MSE        =    .97342

------------------------------------------------------------------------------
           y |      Coef.   Std. Err.      t    P>|t|     [95% Conf. Interval]
-------------+----------------------------------------------------------------
           x |   2.856335   .0799005    35.75   0.000     2.69811    3.01456
       _cons |   5.901613   .6147842     9.60   0.000    4.684173   7.119053
-------------+----------------------------------------------------------------
         rho |   .8624549
------------------------------------------------------------------------------

Durbin-Watson statistic (original)     0.298126
Durbin-Watson statistic (transformed)  1.777231
```

By default, the `prais` command uses the Prais–Winsten estimator and iterates until the parameter estimates converge. The Durbin–Watson test is applicable to model 1, so the `prais` command reports it for both the original and the transformed models.[15]

Comparison of estimates of model 1

The table below compares the estimates of model 1 that each strategy produced with the "true" values used to generate the simulated data. The table lists the point estimates, the width of the 95% CIs, and whether the CIs included the true values. The transformation strategy estimates come from the nonlinear least-squares estimates of the parameters.

These results do not necessarily generalize to other models or datasets, but they do serve to highlight the potential differences in estimates produced by alternative acceptable strategies. Note also that the results in the FGLS column represent estimates from just one of the many variants of the FGLS estimator for model 1. Finally, note that none of the strategies did a particularly good job of estimating the constant term in our simulated regression. Our simulated dataset of 120 observations would represent a reasonably large sample for many time-series models. The dispersion of estimates in the table below and the difficulty of pinning down β_0 remind us to temper our confidence in the large-sample properties of estimators when applied to real-world problems.

---

15. These are the statistics from the original Durbin–Watson test, not the newer Durbin's alternative test.

		True	OLS	Transform	FGLS
	point	5.00	6.09	5.94	5.90
β_0	CI width	—	1.34	2.60	2.43
	include	—	No	Yes	Yes
	point	3.00	2.88	2.86	2.86
β_1	CI width	—	0.55	0.32	0.32
	include	—	Yes	Yes	Yes
	point	0.90	—	0.86	0.86
ρ	CI width	—	—	0.20	—
	include	—	—	Yes	—
σ_η	point	1.00	—	0.98	0.97

5.4.2 Model 2: A lagged dependent variable and i.i.d. errors

Let's evaluate which strategies make sense for

Model 2:

1. $y_t = \beta_0 + \alpha y_{t-1} + \beta_1 x_t + \eta_t$,
2. η is i.i.d., and
3. x_t is weakly exogenous.

Model 2 is a trick case—OLS produces consistent and efficient estimates for the parameters of model 2.

Because η is i.i.d., its variance is a scalar matrix. Thus no correction is needed for the OLS estimate of the VCE. Even if the variance matrix was nonscalar, the FGLS strategy would not be applicable because model 2 contains a lag of the dependent variable and independent variable that is not strictly exogenous, violating the assumptions of the FGLS estimators. Because the regressors—both y_{t-1} and x_t—are weakly exogenous, the OLS estimates are consistent, and the IV strategy does not apply. (You will recall that weak exogeneity is defined as a lack of correlation between the error term, η_t, and all past and present—but not future—values of x_t.) Because model 2 possesses both a scalar residual variance matrix and weakly exogenous regressors, there is no need for the transformation strategy.

The important point to note here is that lagged dependent variables by themselves do not invalidate the OLS estimator of the VCE. The critical features are the variance of $\widehat{\epsilon}$ and the correlation between the regressors and the regression disturbance.

With real-world data, we could not be certain a priori that η was i.i.d.; hence, we would not be sure that OLS was the appropriate estimator. Let's see what might happen in practice. As with model 1, we will generate simulated data to illustrate the estimation of model 2.[16]

```
. clear
. set obs 120
number of observations (_N) was 0, now 120
. set seed 63
. generate eta = rnormal()
. generate epsilon = eta
. generate x = rnormal()
. generate y = (5 + 3*x)/(1-0.9) + epsilon in f
(119 missing values generated)
. replace y = 5 + 0.9*y[_n-1] + 3*x + epsilon in 2/1
(119 real changes made)
. generate int t = _n
. tsset t
        time variable:  t, 1 to 120
               delta:  1 unit
```

Now we estimate the equation with OLS.

```
. regress y L.y x
```

Source	SS	df	MS		
Model	6887.15227	2	3443.57613	Number of obs	= 119
Residual	139.652446	116	1.2039004	F(2, 116)	= 2860.35
				Prob > F	= 0.0000
				R-squared	= 0.9801
				Adj R-squared	= 0.9798
Total	7026.80472	118	59.5491925	Root MSE	= 1.0972

y	Coef.	Std. Err.	t	P>|t|	[95% Conf. Interval]
y					
L1.	.8905116	.0130438	68.27	0.000	.8646767 .9163466
x	2.925096	.0979873	29.85	0.000	2.73102 3.119172
_cons	5.312237	.6767448	7.85	0.000	3.971859 6.652616

---

16. We have to use a little ingenuity to calculate y_1. Our first observation should obey

$$y_1 = \beta_0 + \alpha y_0 + \beta_1 x_1 + \epsilon_1$$

but y_0 is not available. To overcome this obstacle, we calculate y_1 as the sum of its unconditional expected value and ϵ_1. Thus we have

$$E(y_1) = E(\beta_0 + \alpha y_0 + \beta_1 x_1 + \epsilon_1) = \beta_0 + \alpha E(y_0) + \beta_1 x_1$$

Because the unconditional expectation of y_1 is identical to the unconditional expectation of y_0, we can collect terms and rewrite this expectation as

$$E(y_1) = \frac{\beta_0 + \beta_1 x_1}{1 - \alpha}$$

If this was real-world rather than simulated data, we would not know that the regression disturbances were i.i.d., and we would test for serial correlation.

```
. estat durbinalt
Durbin's alternative test for autocorrelation
```

lags(p)	chi2	df	Prob > chi2
1	0.020	1	0.8876

HO: no serial correlation

The test correctly accepts the null hypothesis of no serial correlation.

What if the test had rejected the null hypothesis? In real-world applications, we do not know the model structure with certainty. And statistical tests do not prove whether a hypothesis is true or false. Instead statistical tests quantify the probability of type I (incorrect rejection of the null) and type II (incorrect acceptance of the null) errors. An unusual run of regression disturbances could produce a misleading test result. To make reliable inferences, we sometimes need to analyze the results of applying different strategies to guide us to the correct specification.

As an example, imagine Durbin's alternative test had rejected the null hypothesis in our sample. In that situation, we might apply the transformation strategy to fit model 2. As with model 1, we would multiply the equation for y_{t-1} by ρ and subtract it from the equation for y_t. After rearranging terms, we would have

$$y_t = (1 - \rho)\beta_0 + (\alpha + \rho)y_{t-1} - \rho\alpha y_{t-2} + \beta_1 x_t - \rho\beta_1 x_{t-1} + (\eta_t - \rho\eta_{t-1})$$

Notice that the error term in the transformed equation—$(\eta_t - \rho\eta_{t-1})$—is now first-order autocorrelated, while the error in the original equation is i.i.d.

$$E\left\{(\eta_t - \rho\eta_{t-1})(\eta_t - \rho\eta_{t-1})\right\} = \rho\sigma_\eta^2$$

We have inadvertently created the condition that we were trying to cure.

Let's estimate this transformed equation with OLS.

```
. regress y L.y L2.y x L.x
```

Source	SS	df	MS
Model	6783.98936	4	1695.99734
Residual	136.029816	113	1.20380368
Total	6920.01917	117	59.145463

Number of obs	= 118
$F(4, 113)$	= 1408.87
Prob > F	= 0.0000
R-squared	= 0.9803
Adj R-squared	= 0.9796
Root MSE	= 1.0972

y	Coef.	Std. Err.	t	P>|t|	[95% Conf.	Interval]
y						
L1.	.9005817	.0933403	9.65	0.000	.7156576	1.085506
L2.	-.0005753	.084372	-0.01	0.995	-.1677314	.1665808
x						
--.	2.936754	.098492	29.82	0.000	2.741623	3.131884
L1.	-.1289145	.2890571	-0.45	0.656	-.7015887	.4437598
_cons	4.850934	.8407461	5.77	0.000	3.185264	6.516603

```
. estat durbinalt
```

Durbin's alternative test for autocorrelation

lags(p)	chi2	df	Prob > chi2
1	0.001	1	0.9815

H0: no serial correlation

In this example, Durbin's alternative test failed to detect the autocorrelation we have induced. If it had, we might realize that our original transformation was unnecessary. On the other hand, we might conclude that we should have allowed for higher-than-first-order autocorrelation of the regression disturbance.

The regression output does contain some other indications that the transformation strategy is not appropriate for these data. The regression coefficients on x_t and x_{t-1} imply $\rho = 0.04$, and the insignificant estimates of the coefficients on y_{t-2} and x_{t-1} suggest that ρ is close to 0 (and α is close to 1).

The picture is even clearer when we use nonlinear least squares to obtain unique estimates of the parameters.

```
. nl (y = ({b0}*(1-{rho})) + ({rho}+{alpha})*L.y - {rho}*{alpha}*L2.y + {b1}*x -
> {b1}*{rho}*L.x) in 3/l
(obs = 118)

Iteration 0:  residual SS =  3802.605
Iteration 1:  residual SS =   227.542

  (output omitted)

Iteration 12:  residual SS =   137.5841
```

Source	SS	df	MS		
				Number of obs =	118
Model	6782.4351	3	2260.8117	R-squared =	0.9801
Residual	137.58409	114	1.20687794	Adj R-squared =	0.9796
				Root MSE =	1.09858
Total	6920.0192	117	59.145463	Res. dev. =	352.9885

y	Coef.	Std. Err.	t	P>|t|	[95% Conf. Interval]	
/b0	5.237608	.6940238	7.55	0.000	3.862752	6.612464
/rho	.0111779	.0942549	0.12	0.906	-.1755403	.197896
/alpha	.8921824	.0133958	66.60	0.000	.8656454	.9187193
/b1	2.936913	.0986176	29.78	0.000	2.741552	3.132273

```
Parameter b0 taken as constant term in model & ANOVA table
```

The estimate of ρ is not significantly different from 0. The estimates of the other parameters are close to their true values.

With real-world data, the clues are rarely clear cut. Nonetheless, this example highlights the usefulness of comparing estimates from different strategies before settling on a final model specification.

5.4.3 Model 3: A lagged dependent variable with AR(1) errors

Let's evaluate which strategies make sense for

Model 3:

1. $y_t = \beta_0 + \alpha y_{t-1} + \beta_1 x_t + \epsilon_t,$
2. $\epsilon_t = \rho \epsilon_{t-1} + \eta_t,$
3. η is i.i.d., and
4. x_t is strictly exogenous.

Now we have both problems simultaneously. As we showed in our discussion of model 1, ϵ has a nonscalar variance matrix. At the same time, we have correlation between a regressor and the error term:

$$E\left(y_{t-1}\epsilon_t\right) = E\left(\beta_0 + \alpha y_{t-2} + \beta_1 x_t + \epsilon_{t-1}\right)\left(\rho\epsilon_{t-1} + \eta_{t-1}\right) = \frac{\rho\sigma_\eta^2}{1 - \rho^2}$$

This correlation rules out the OLS and FGLS strategies. (Even in the absence of this correlation, the lagged dependent variable is a knockout punch for the FGLS strategy.) However, the transformation and IV strategies can be used.

The transformation strategy

The transformation strategy is probably the most frequently used strategy with model 3. The algebra is essentially the same that we used for model 1. To transform model 3, multiply the equation for y_{t-1} by ρ and subtract it from the equation for y_t.

$$y_t - \rho y_{t-1} = \alpha y_{t-1} - \rho \alpha y_{t-2} + (1-\rho)\beta_0 + \beta_1 x_t - \rho \beta_1 x_{t-1} + \epsilon_t - \rho \epsilon_{t-1}$$

Rearranging terms and replacing $\epsilon_t - \rho \epsilon_{t-1}$ with η_t produces

$$y_t = \alpha_1^* y_{t-1} + \alpha_2^* y_{t-2} + \beta_0^* + \beta_1^* x_t + \beta_2^* x_{t-1} + \eta_t$$

where

$$\alpha_1^* = (\rho + \alpha), \ \ \alpha_2^* = -\rho\alpha, \ \ \beta_0^* = (1-\rho)\beta_0, \ \ \beta_1^* = \beta_1, \ \text{and} \ \beta_2^* = -\rho\beta_1$$

OLS estimates of this equation are consistent and efficient, and we can use the standard estimator of the VCE. As in model 1, we can use Stata's `nl` command to estimate this equation by nonlinear least squares and obtain unique estimates of the underlying parameters.

As usual, we start by simulating data consistent with model 3.

```
. clear
. set obs 120
number of observations (_N) was 0, now 120
. set seed 51
. generate eta = rnormal()
. generate epsilon = rnormal()*sqrt(1/0.19) in f
(119 missing values generated)
. replace epsilon = eta + 0.9*epsilon[_n-1] in 2/l
(119 real changes made)
. generate x = rnormal()
. generate y = (5 + 3*x)/0.1 + epsilon in f
(119 missing values generated)
. replace y = 5 + 0.9*y[_n-1] + 3*x + epsilon in 2/l
(119 real changes made)
. generate int t = _n
. tsset t
        time variable:  t, 1 to 120
                delta:  1 unit
```

If this was real data, we would not know that the errors are autocorrelated, so we will apply our usual test.

```
. regress y L.y x
```

Source	SS	df	MS
Model	57248.4084	2	28624.2042
Residual	578.440982	116	4.98656019
Total	57826.8493	118	490.058045

Number of obs	=	119
F(2, 116)	=	5740.27
Prob > F	=	0.0000
R-squared	=	0.9900
Adj R-squared	=	0.9898
Root MSE	=	2.2331

y	Coef.	Std. Err.	t	P>|t|	[95% Conf. Interval]	
y						
L1.	.9397952	.0088592	106.08	0.000	.9222484	.957342
x	2.834387	.2173545	13.04	0.000	2.403889	3.264886
_cons	3.443071	.6147755	5.60	0.000	2.225431	4.660711

```
. estat durbinalt
```

Durbin's alternative test for autocorrelation

lags(p)	chi2	df	Prob > chi2
1	323.582	1	0.0000

H0: no serial correlation

The null hypothesis is soundly rejected, so our next step is to estimate the transformed equation.

```
. regress L(0/2).y L(0/1).x
```

Source	SS	df	MS
Model	53159.125	4	13289.7813
Residual	130.319631	113	1.15327107
Total	53289.4446	117	455.465339

Number of obs	=	118
F(4, 113)	=	11523.55
Prob > F	=	0.0000
R-squared	=	0.9976
Adj R-squared	=	0.9975
Root MSE	=	1.0739

y	Coef.	Std. Err.	t	P>|t|	[95% Conf. Interval]	
y						
L1.	1.794581	.044929	39.94	0.000	1.705569	1.883594
L2.	-.8064204	.0424817	-18.98	0.000	-.8905843	-.7222566
x						
--.	2.936166	.1078646	27.22	0.000	2.722466	3.149865
L1.	-2.874575	.1645064	-17.47	0.000	-3.200492	-2.548658
_cons	.7015882	.3338048	2.10	0.038	.0402607	1.362916

The estimated coefficients on the y's and x's are in line with the values used to generate the data. The estimate of the constant, however, is not very accurate.

Let's check for first-order autocorrelation in the transformed equation. There should not be any.

```
. estat durbinalt

Durbin's alternative test for autocorrelation
```

lags(p)	chi2	df	Prob > chi2
1	0.095	1	0.7576

```
                    H0: no serial correlation
```

And there is not.

Let's use Stata's `nl` command to produce unique estimates of the underlying parameters.

```
. nl (y = ({rho}+{alpha})*L.y - {rho}*{alpha}*L2.y + (1-{rho})*{b0} + {b1}*x -
> {rho}*{b1}*L.x) in 3/1
(obs = 118)

Iteration 0:  residual SS =  2027.769
Iteration 1:  residual SS =  137.9669
     (output omitted)
Iteration 11:  residual SS =  131.2766
```

Source	SS	df	MS		
				Number of obs =	118
Model	53158.168	3	17719.3893	R-squared =	0.9975
Residual	131.27661	114	1.15154923	Adj R-squared =	0.9975
				Root MSE =	1.073103
Total	53289.445	117	455.465339	Res. dev. =	347.4509

y	Coef.	Std. Err.	t	P>|t|	[95% Conf. Interval]	
/rho	.9298964	.0395844	23.49	0.000	.85148	1.008313
/alpha	.8429796	.0458644	18.38	0.000	.7521226	.9338365
/b0	9.15354	2.738819	3.34	0.001	3.727961	14.57912
/b1	2.974388	.0992946	29.96	0.000	2.777686	3.17109

```
  Parameter b0 taken as constant term in model & ANOVA table
```

Again all the parameters except the constant are in line with the true values.

The IV strategy

The IV strategy, discussed in section 2.3.2, depends on finding suitable IV. Finding instruments can be difficult and usually requires compelling nonstatistical information—theoretical or practical arguments that make clear why the instruments are 1) correlated with the endogenous regressors (instrument relevance); 2) uncorrelated with the regression disturbance (instrument exogeneity); and 3) not included in the original regression model.

Fortunately, we have constructed our simulated data to provide just such an instrument. x_{t-1} is a relevant instrument; it is an explanatory variable in the regression for our endogenous regressor, y_{t-1}. x_{t-1} is an exogenous instrument because x is strictly exogenous. Finally, x_{t-1} does not belong in the original regression model by construction. This last condition is more difficult to ensure in real-world applications.

As discussed in section 1.4.2, Stata provides the `ivregress` command for single-equation IV estimation. For this example, we will apply the generalized method of moments with a heteroskedasticity- and autocorrelation-consistent weighting matrix.

```
. ivregress gmm y x (L.y = L.x), wmatrix(hac bartlett 1)
```

Instrumental variables (GMM) regression				Number of obs	=	119
				Wald chi2(2)	=	229.24
				Prob > chi2	=	0.0000
				R-squared	=	0.9796
GMM weight matrix: HAC Bartlett 1				Root MSE	=	3.1476

y	Coef.	HAC Std. Err.	z	P>|z|	[95% Conf. Interval]	
y						
L1.	.8425707	.058925	14.30	0.000	.7270798	.9580615
x	2.880139	.271666	10.60	0.000	2.347684	3.412595
_cons	9.803146	3.766628	2.60	0.009	2.420691	17.1856

Instrumented: L.y
Instruments: x L.x
HAC VCE: Bartlett kernel with 1 lag

The coefficients on y_{t-1} and x_t are in line with the true values. The CI for the constant is wide enough to include the true value, although the point estimate is almost double the true value.

5.5 Estimating the mortgage rate equation

In section 5.1.1, we estimated a simple linear regression of the primary mortgage rate (the "retail" rate) on a constant and the current coupon (the "wholesale" rate):

$$\text{Primary rate}_t = \beta_0 + \beta_1 \text{Current coupon}_t + \epsilon_t$$

In section 5.3, we applied Durbin's alternative test for first-order autocorrelated residuals and rejected the null hypothesis of no autocorrelation.

This mortgage rate equation is an instance of model 1, so the OLS, transformation, and FGLS strategies are applicable. The table below compares the estimates obtained by each strategy.

		OLS	Transform	FGLS
β_0	point	0.61	1.46	1.30
	CI width	0.20	0.60	0.54
β_1	point	0.98	0.86	0.89
	CI width	0.03	0.07	0.06
ρ	point	—	0.94	0.93
	CI width	—	0.07	—
σ_η	point	—	0.08	0.08

Once again, these asymptotically equivalent strategies produce somewhat different results. Note, for example, the significantly higher estimate of β_1 produced by the OLS strategy than the estimates produced by the other two methods.

Finally, figure 5.4 displays the in-sample fits of these three approaches. While the coefficient estimates differ noticeably, the fitted values are not that different.

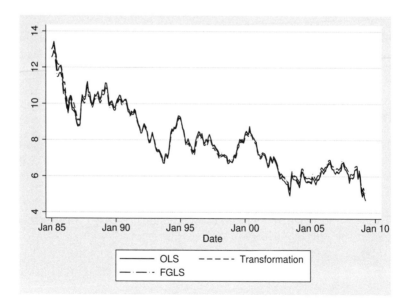

Figure 5.4. In-sample fit of three estimation strategies

5.6 Points to remember

- There are several intuitive reasons why autocorrelation is likely—seasonal adjustment, "natural" correlations in seasonally unadjusted data, and the playing out of influential events over longer periods than our data observation frequency.

- In a regression context, these tendencies toward autocorrelation can show up as serial correlation in the regression disturbance. This type of autocorrelation violates the independence assumption of classical regression and produces nonscalar variance matrices of the residual.

- First-order autocorrelation is very common in practice and is the most analyzed form of serially correlated residuals.

- The preferred test for autocorrelated residuals is Durbin's alternative test because it can be used for autocorrelation of any order and in models with endogenous regressors (such as lagged dependent variables). The Stata command estat durbinalt calculates Durbin's alternative test.

- Depending on the model structure, regressions with autocorrelated disturbances can pose different statistical challenges. It is important to use an estimation strategy that is matched to the type of model. The OLS and FGLS strategies are appropriate for models with nonscalar residual variance matrices. The IV strategy is appropriate for models with correlation between the residual and one or more regressors. The transformation strategy can be used in both situations.

- Acceptable estimation strategies can provide materially different regression estimates, even when these strategies are asymptotically equivalent. The econometrics literature provides limited guidance for choosing among these strategies. It is the responsibility of the researcher to investigate whether the estimates from a given method make theoretical and practical sense.

Stata commands and features discussed

estat ([R] **estat**): Postestimation statistics; section 5.3

estat durbinalt ([R] **regress postestimation time series**): Calculate Durbin's alternative test of the null hypothesis of no residual autocorrelation; section 5.3

ivregress ([R] **ivregress**): Single-equation IV regression; section 5.4

newey ([TS] **newey**): OLS regression with Newey–West standard errors; section 5.4

nl ([R] **nl**): Nonlinear least-squares estimation; section 5.4

nlcom ([R] **nlcom**): Nonlinear combinations of estimators; section 5.4

prais ([TS] **prais**): Prais–Winsten and Cochrane–Orcutt regression; section 5.4

rnormal() ([FN] **Random-number functions**): Standard normal random-number generator; section 5.4

`set seed` ([R] **set seed**): Specify starting point of random-number sequence; section 5.4

`tsline` ([TS] **tsline**): Plot time-series data; section 5.1

6 Univariate time-series models

Chapter map

6.1 The general linear process. The fundamental time-series model. Building more complex models out of linear combinations of white noise.

6.2 Lag polynomials: Notation or prestidigitation? Some notation with surprising properties.

6.3 The ARMA model. Combining autoregressive (AR) and moving-average (MA) components. Tracing the ARMA model back to the general linear process.

6.4 Stationarity and invertibility. The central role of stationarity. Using invertibility to choose among representations of the disturbance.

6.5 What can ARMA models do? The AR component defines a difference equation. The general linear model defines the impulse–response function.

6.6 Points to remember. White-noise disturbances. Representing the infinite-order linear process. The fundamental characteristic of a time series.

6.7 Looking ahead. Now we are ready to model a real-world time series.

The previous three chapters may have seemed a little scattershot. We introduced several important topics in each chapter, but the connections across chapters may not have been obvious. This chapter marks a change in the style of the presentation—from here on, we will be weaving the loose threads into a (relatively) seamless fabric. To achieve this increase in coherence, this brief chapter introduces the theory and terminology that will carry us through the rest of this book. We will not see Stata in action until the next chapter. It will be worth it though. This foundational material will simplify the exposition in the rest of the book.

We begin by defining the general linear process, a deceptively simple-looking model. From the general linear process, it's a short step to ARMA models, a flexible way to capture complex time-series behavior in models with only a few parameters. But first, to make ARMA models tractable, we introduce lag polynomials, a compact notation for describing time-series structures. With this notational tool in hand, we define ARMA models, describe the roles of stationarity and invertibility in these models, and highlight the types of dynamic behavior ARMA models can exhibit.

6.1   The general linear process

Univariate time-series models are easier to understand if we build them from the ground up, that is, if we start with simple building blocks and combine the blocks to create our model. The initial building block in time-series models is the white-noise process, ϵ_t.[1] To qualify as white noise, the random variable ϵ_t must satisfy three conditions:

$$E\epsilon_t = 0$$

$$E\epsilon_t^2 = \sigma_\epsilon^2$$

and

$$E\epsilon_t\epsilon_{t-j} = 0, \forall j \neq 0$$

In plain English, ϵ_t has mean 0, finite (and constant) variance, and is uncorrelated with past or future values of ϵ.[2]

By itself, a white-noise process is not very interesting. However, we can generate a wide variety of interesting time series by constructing linear combinations of current and past values of a white-noise process:

$$y_t = \mu + \epsilon_t + \psi_1\epsilon_{t-1} + \psi_2\epsilon_{t-2} + \cdots$$

The constant, μ, is included to allow for a nonzero mean.[3] The ellipsis ($\cdots$) at the end of the equation indicates there are an infinite number of terms in the equation; this equation defines an infinite MA.[4] This particular MA is called, interchangeably, the general linear process or the linear filter model. The right-hand side of the equation defines a linear filter, an infinite series that converts an infinite sequence of unobservable ϵ's to the observable time series y_t. The linear filter model is the basis of all the time-series models discussed in this book.

Some fundamental properties of the general linear process are easy to derive. Because each ϵ_t has mean 0, we have

$$Ey_t = E(\mu + \epsilon_t + \psi_1\epsilon_{t-1} + \psi_2\epsilon_{t-2} + \cdots) = \mu$$

---

1. To statisticians, the term process means time series. Using the word "process" emphasizes the sequential nature of the series, that is, the dependence of the present value of the series on its values in prior periods.
2. Why is this called noise, and what makes it white? This terminology is borrowed from spectral analysis, an approach to time-series analysis that represents time series as the sum of contributions at different frequencies. Spectral analysis is a convenient framework for time-series analysis in physics and applied fields such as acoustics and electrical engineering. The term noise derives from actual, physical noise and indicates an unsystematic, random component in contrast to the deterministic signal. (We borrowed this terminology in our discussion of filters in chapter 3.) White noise is composed of equal contributions at each frequency. There are, in fact, other colors of noise: pink or flicker noise, red noise (also called Brown noise after Robert Brown, the discoverer of Brownian motion), and grey noise. In this book, we like our noise white.
3. Frequently, we will ignore μ in discussing the properties of the general linear process. Except in rare instances, μ is a nuisance parameter with no impact on the statistical properties we are discussing.
4. Normally, we expect the weights in an MA—the ψ coefficients here—to sum to 1. Strictly speaking, this equation describes a linear combination, but not necessarily a weighted average, of ϵ_t's. However, the use of the term "MA" to denote this type of model is standard in time-series analysis.

Because ϵ_t has constant variance and is serially uncorrelated, the variance of y_t is

$$
\begin{aligned}
E(y_t - \mu)^2 &= E(\epsilon_t + \psi_1 \epsilon_{t-1} + \psi_2 \epsilon_{t-2} + \cdots)^2 \\
&= (1 + \psi_1^2 + \psi_2^2 + \cdots)\sigma^2 \\
&= \sigma_\epsilon^2 \sum_{i=0}^{\infty} \psi_i^2
\end{aligned}
$$

where

$$
\psi_0 \equiv 1
$$

So y_t has finite variance if and only if the series

$$
\sum_{i=0}^{\infty} \psi_i^2
$$

converges to a finite value.

Finally, the covariance between y_t and y_{t-j} is called the jth autocovariance:

$$
\gamma_j \equiv E(y_t - \mu)(y_{t-j} - \mu) = \sigma^2 \sum_{i=0}^{\infty} \psi_i \psi_{i-j}
$$

The general linear process transforms a sequence of white-noise disturbances into an autocorrelated time series. The nature of the autocorrelation—that is, the dynamic structure of the process—is determined completely by the sequence of ψ_i's.

6.2   Lag polynomials: Notation or prestidigitation?

It may seem pointless to specify a model with an infinite number of terms. For one thing, it is not obvious how to fit such a model. Nonetheless, the statistical theory of time series is more tractable under the assumption that time-series processes can reach back infinitely into the past and evolve forward infinitely into the future. And as it turns out, these models can be recast in forms suitable for estimation with a manageable number of unknown parameters. But first, we need a more convenient way of writing the model.

Fortunately, there is an elegant way to write this model compactly, Let's define a lag operator, L, that decrements the time subscript of any variable. In other words,

$$
L\epsilon_t \equiv \epsilon_{t-1}
$$

To obtain earlier values of y, we simply apply the lag operator repeatedly.

$$
L(L\epsilon_t) = L\epsilon_{t-1} = \epsilon_{t-2}
$$

Because we are applying the lag operator twice, it's intuitive to abbreviate that operation by superscripting the lag operator, just as we would if we were raising a variable to a power.

$$
L^2 \epsilon_t = \epsilon_{t-2}
$$

Using the lag operator, we can rewrite the general linear process as

$$y_t = \mu + \epsilon_t + \psi_1 L \epsilon_t + \psi_2 L^2 \epsilon_t + \cdots$$

So far, we have only made things worse—at this point, the lag operator is just extra notation. But what if we pretended L were an ordinary algebraic variable? We have already taken a step in that direction by superscripting the lag operator, treating repeated applications of the operator as if it were the same as multiplying the operator times itself. Extending this intuition, we can take the common term, ϵ_t, out of the expression and rewrite the model as

$$y_t = \mu + (1 + \psi_1 L + \psi_2 L^2 + \cdots) \epsilon_t$$

The model now is composed of a constant term and a "polynomial" in the lag operator "multiplied" by ϵ_t. We can write this more compactly still by defining the lag polynomial $\psi(L)$, where

$$\psi(L) \equiv 1 + \psi_1 L + \psi_2 L^2 + \cdots$$

so finally we have

$$y_t = \mu + \psi(L) \epsilon_t$$

This is elegant and convenient, but is it legal? Fortunately, the answer is "Yes" under fairly general conditions, so from here on, we will avail ourselves of this tool, treating the lag operator as if it were an ordinary algebraic variable.[5]

Because we are on a roll, let's define an inverse lag operator, the lead operator,

$$F \epsilon_t \equiv L^{-1} \epsilon_t \equiv \epsilon_{t+1}$$

a difference operator,

$$D \epsilon_t \equiv (1 - L) \epsilon_t = \epsilon_t - \epsilon_{t-1}$$

and, to round out the set, a seasonal difference operator,

$$S \epsilon_t \equiv (1 - L^s) \epsilon_t = \epsilon_t - \epsilon_{t-s}$$

where s is the seasonal span (4 for quarterly data, 12 for monthly data, and so on). As you can see from the definitions, the lead, difference, and seasonal difference operators are not strictly necessary—they all can be written in terms of the lag operator—but it is convenient sometimes to use these operators instead.

If you are one of the readers who did not skip chapter 1, you already know that Stata allows you to use these operators in time-series varlists if you have tsset your data. For example, Stata lets you fit the regression model

$$y_t = \mu + \phi_1 L y_{t-1} + \phi_2 L^2 y_{t-2} + \epsilon_t$$

---

5. I consulted with my more mathematically knowledgeable colleagues—that would be all of them—and they inform me that we can thank something called the calculus of operators for all of this convenience. Personally, I regard it as a sort of mathematical voodoo, but I'm not going to turn my nose up at anything that makes life easier.

with the command

```
. regress y L.y L2.y
```

or even

```
. regress L(0/2).y
```

6.3 The ARMA model

We have taken our time getting to this point—we introduced white noise and the general linear process, and we showed how to write the general linear process more compactly by using lag polynomials—but now we are finally ready to use these concepts to describe a very flexible and useful time-series model, the ARMA model. The name "ARMA model" is shorthand for autoregressive moving-average model, and the model is written

$$\phi(L)y_t = \theta(L)\epsilon_t$$

where $\phi(L)$ and $\theta(L)$ are finite-order lag polynomials, y_t is an observable time series, and ϵ_t is white noise.[6] $\phi(L)y_t$ is the AR component of the model, and $\theta(L)\epsilon_t$ is the MA component.

Numbers indicating the last nonzero parameter in each linear combination frequently are appended to the ARMA label to indicate the order of the model. For example,

$$(1 - \phi_1 L - \phi_2 L^2)y_t = (1 - \theta_1 L)\epsilon_t$$

is an ARMA(2,1) model.[7] The model name is shortened in an obvious way if either the AR or MA component has no lags. For example,

$$y_t = (1 - \theta_1 L - \theta_2 L^2)\epsilon_t$$

typically is called an MA(2) model rather than an ARMA(0,2) model.

The name "ARMA" is a trifle confusing. Both the AR and MA components of the model have the same mathematical form—they are linear combinations of present and past random variables. The linear combination of observable variables, $\phi(L)y_t$, is the AR component, while the linear combination of unobservable white-noise disturbances,

---

6. For convenience, we have dropped the constant μ for the moment. In this discussion, μ is a nuisance parameter. If we want to be fussy and eliminate μ more carefully, we can always define

$$y_t^* = y_t - \mu^*$$

where μ^* is the unconditional mean of y_t, and end up with an ARMA model for y_t^* that contains no constant term.

7. It turns out to be convenient to write the lag polynomials $\phi(L)$ and $\theta(L)$, which define the AR and MA components, respectively, with minus signs, while we write the lag polynomial $\psi(L)$, which defines the general linear process, with plus signs. Nothing fundamental is changed by rearranging the signs, and other authors adopt different conventions.

$\theta(L)\epsilon_t$, is the MA component.[8] To make matters worse, in chapter 5, we described regression models with MA disturbances as models with autocorrelated disturbances, and this is the common way to denote these models. I'm afraid that if you intend to get involved in time-series analysis, you are going to have to get used to this type of confusing terminology. The situation is similar to some foreign languages where a single word can have many meanings depending on the speaker's pronunciation and intonation. You just have to figure it out from the context.[9]

You will notice in the sections below that we continue to treat the AR and MA components separately and apply different names to what appear to be similar features. This treatment is not arbitrary. The y_t are observable, real-world quantities. As a consequence, some of the properties of the AR component define essential constraints on the dynamic process that describes the behavior of y_t. On the other hand, the ϵ_t are an unobservable, theoretical construct. As a result, we have more freedom in the specification of the MA component. Thus the same properties that define constraints on the behavior of y_t are used in the MA component to choose among statistically equivalent representations of ϵ_t.

We promised at the beginning of this chapter that we were going to cease jumping from topic to topic and start connecting topics more coherently. With that promise in mind, let's show the connection between ARMA models and the general linear process. To rewrite an ARMA model as a general linear process, just multiply both sides of the ARMA equation by the inverse of $\phi(L)$

$$\phi^{-1}(L)\phi(L)y_t = y_t = \phi^{-1}(L)\theta(L)\epsilon_t = \psi(L)\epsilon(t)$$

where

$$\phi^{-1}(L)\theta(L) \equiv \psi(L)$$

Once again, treating the lag operator as if it were an ordinary algebraic variable pays big dividends. In this case, however, the inverse of $\phi(L)$ is not guaranteed to exist, so we have to be a bit more careful before we forge ahead.

The conditions that guarantee the existence of $\phi^{-1}(L)$ are easiest to see in an AR(1) model:

$$(1 - \phi_1 L)y_t = \epsilon_t$$

---

8. Again the MA component is not, strictly speaking, an MA: the θ_j's in the MA lag polynomial do not sum to 1.

9. Yet another structurally similar linear combination of random variables can be added to the model. A distributed lag is a linear combination of exogenous variables, x_t. In the context of the current model, a distributed lag would appear as

$$\phi(L)y_t = \gamma(L)x_t + \theta(L)\epsilon_t$$

To write the AR(1) model as a general linear process, we substitute repeatedly for lagged values of y_t,

$$
\begin{aligned}
y_t &= \epsilon_t + \phi_1 y_{t-1} \\
&= \epsilon_t + \phi_1 \epsilon_{t-1} + \phi_1^2 y_{t-2} \\
&= \epsilon_t + \phi_1 \epsilon_{t-1} + \phi_1^2 \epsilon_{t-2} + \phi_1^3 \epsilon_{t-3} + \cdots
\end{aligned}
$$

so

$$
(1 - \phi_1 L)^{-1} \equiv (1 + \phi_1 L + \phi_1^2 L^2 + \phi_1^3 L^3 + \cdots)
$$

Note that these substitutions have transformed the finite-order AR(1) model into an infinite-order MA model, which is an example of the general linear model, where, in this case, $\psi_i = \phi_i^i, \forall i$. From the last equation, we can see that the inverse $\phi^{-1}(L)$ is well defined if $|\phi_1| < 1$. Furthermore, y_t has finite variance:

$$
E y_t^2 = \sigma_\epsilon^2 \sum_{i=0}^{\infty} \phi_1^i = \frac{\sigma_\epsilon^2}{1 - \phi_1^2}
$$

The condition $|\phi_1| < 1$ guarantees that the influence of ϵ_{t-j} on y_t diminishes as j increases, that is, the further apart in time y_t and ϵ_{t-j} are. If $|\phi_1| > 1$, then the impact of disturbances grows without bound as the disturbances are further in the past. This type of relationship—where events in the distant past have greater impact on the present than recent events—is counterintuitive and will not play a part in our analyses. However, an interesting real-world case is the random walk ($\phi_1 = 1$), where the impact of distant disturbances neither diminishes nor grows,

$$
y_t = y_{t-1} + \epsilon_t
$$

The general linear model representation of the random walk is

$$
y_t = \sum_{j=0}^{\infty} \epsilon_{t-j}
$$

The unconditional mean of y_t is 0, but the variance is

$$
E y_t^2 = E \sum_{j=0}^{\infty} \epsilon_{t-j}^2 = \sum_{j=0}^{\infty} \sigma^2 = \infty
$$

The first difference of a random walk, $D y_t = y_t - y_{t-1} = \epsilon_t$, is white noise, which is much more tractable.

We also can multiply both sides of the ARMA equation by the inverse of the MA lag polynomial to create a pure AR model:

$$
\theta^{-1}(L)\phi(L)y_t \equiv \omega(L)y_t = \theta^{-1}(L)\theta(L)\epsilon_t = \epsilon_t
$$

Again there is no guarantee that θ^{-1} exists. As an example, consider the MA(1) model

$$y_t = \epsilon_t - \theta_1 \epsilon_{t-1}$$

Repeated substitution for lagged values of ϵ_t and a little rearrangement of terms yields

$$(1 + \theta_1 L + \theta_1^2 L^2 + \theta_1^3 L^3 + \cdots)y_t = \epsilon_t$$

so

$$(1 - \theta_1 L)^{-1} y_t \equiv (1 + \theta_1 L + \theta_1^2 L^2 + \theta_1^3 L^3 + \cdots)y_t = \epsilon_t$$

if $|\theta_1| < 1$. Just as an AR(1) model can be rewritten as an infinite-order MA model, an MA(1) model can be rewritten as an infinite-order AR model.

While we can work out the properties of the AR(1) and MA(1) models fairly easily, it makes sense at this point to tally the properties of the general ARMA model. The key properties are stationarity and invertibility, and they derive from properties of the lag polynomials $\phi(L)$ and $\theta(L)$.

6.4   Stationarity and invertibility

Stationarity—or the lack of it (called, sensibly for a change, nonstationarity)—is the most important property of a time series. Stationary time series have the same unconditional distribution (or, at least, the same mean and autocovariances) at any point in time. Moreover, stationary variables are not "explosive"; they have finite variances and autocovariances. More formally, a time series, y_t, is said to be strictly stationary if its probability distribution does not depend on t. So, for example,

$$\begin{aligned} Ey_t &= \mu \\ E(y_t - \mu)^2 &= \sigma^2 \end{aligned}$$

and so on. The mean, variance, and all higher moments are independent of t. In addition to the time independence of the moments of y_t, the joint distribution of a vector of elements of a strictly stationary time series is also independent of t. This property implies that the autocovariances of y_t are independent of t. If y_t is strictly stationary, the relationship between y_t and, say, y_{t-j} depends on the distance between them in time (that is, on j) but not on their absolute location on the time line (not on t).

A familiar example of a time series that is not strictly stationary is a linear trend with random disturbances,

$$y_t = \mu + \alpha t + \epsilon_t$$

because the mean of y_t is a function of t,

$$Ey_t = \mu + \alpha t$$

Another familiar example is the random walk we analyzed in the previous section. A random walk has infinite variance, so it cannot be stationary.

For most of our purposes, we do not require strict stationarity. Instead we can rely on weak stationarity (also called covariance stationarity). A time series is said to be weakly stationary if its mean, variance, and autocovariances are independent of t, that is,

$$Ey_t = \mu$$
$$E(y_t - \mu)^2 = \sigma^2$$

and

$$E(y_t - \mu)(y_{t-j} - \mu) = \gamma_j$$

Higher moments of a covariance-stationary series need not be independent of t. From here on, we will use the term "stationarity" as a shorthand for covariance stationarity unless we explicitly indicate strict stationarity.

The white-noise process, e_t, is covariance stationary. The general linear process also is covariance stationary if the mean and variance are well defined, that is, if

$$\sum_{i=0}^{\infty} \psi_i^2 \quad \text{converges}$$

Because the sum of a finite number of ψ_i's is finite, any finite-order, pure MA model is covariance stationary. Also, because the distribution of a normally distributed random variable depends only on its mean and variance, normally distributed covariance-stationary time series are strictly stationary.

It turns out that stationarity can be determined solely from $\phi(L)$, the lag polynomial of the AR component. Recall that the AR(1) model can be written as

$$y_t = \phi^{-1}(L)\epsilon_t$$

as long as $|\phi_1| < 1$. There is another way to specify this condition. Replace the lag operator L in $\phi(L)$ with an ordinary algebraic variable, z, and write the equation

$$\phi(z) = 1 - \phi_1 z = 0$$

The root of this equation is $z = 1/\phi_1$. So

$$|\phi_1| < 1 \Longrightarrow |z| > 1$$

A fancy way of stating this condition is to say that the root of $\phi(L)$ lies outside the unit circle. This terminology is overkill for the AR(1) model, but it turns out that this condition is sufficient to guarantee stationarity for any ARMA model. That is, it can be shown that the model

$$\phi(L)y_t = \theta(L)\epsilon_t$$

is stationary if all the roots of $\phi(z) = 0$ lie outside the unit circle.[10]

---

10. Why the unit circle? Because $\phi(z)$ is a polynomial of arbitrary degree, the roots can be complex numbers. The extension of absolute value to complex numbers is the modulus. The unit circle in the complex plane is defined by $|z| = 1$. For more precise statements of the conditions guaranteeing stationarity, see Box and Jenkins (1976) or Hamilton (1994).

In keeping with the confusing time-series tradition of calling similar things by different names, an ARMA model is invertible if all the roots of $\theta(z) = 0$ lie outside the unit circle. In fairness, invertibility plays a different role than stationarity in time-series models. Stationarity is fundamental. A nonstationarity time series must be transformed to a stationary representation prior to estimation. For instance, instead of modeling U.S. gross domestic product (GDP), which is clearly nonstationary, we may choose to model the growth rate of GDP.[11] In contrast, multiple parameterizations of an MA component may have equivalent statistical properties.[12] The invertible representation is the most tractable and is called the fundamental representation.[13]

6.5   What can ARMA models do?

We are almost ready to start fitting ARMA models, but before we do, let's spend some time examining the types of time-series behavior that ARMA models can capture. We already mentioned the flexibility of ARMA models several times. This flexibility represents one of the greatest strengths of this class of models. At the same time, it represents one of the greatest challenges because more than one ARMA specification may provide a good fit to an observed time series.

A pure AR model combines a discrete difference equation with random shocks. If we set the disturbance to 0, we have

$$\phi(L)y_t = 0$$

a deterministic difference equation that controls the dynamic behavior of y_t. The random shocks are supplied by the ϵ_t, but the dynamic reaction to those shocks is completely described by the difference equation summarized in $\phi(L)$.

---

11. In a later chapter, we will discuss extensions to the ARMA framework designed to model nonstationary time series directly.

12. For every invertible representation of an MA process, there is a noninvertible representation with the same first and second moments. See Hamilton (1994) for details.

13. You can think of invertibility as an identification condition that chooses the most useful among equivalent representations.

In the simplest case of the AR(1) model

$$(1 - \phi_1 L)y_t = 0$$

future values of y are given by

$$y_{t+j} = \phi_1^j y_t$$

In other words, future values of y decay exponentially toward 0 (see figure 6.1) in the absence of any random disturbances.[14]   The parameter ϕ_1 controls the rate of decay. Notice that a nonstationary model—one where $|\phi_1| > 0$—is explosive. If ϕ_1 is positive, future values grow exponentially to positive or negative infinity depending on the sign of y_0. If ϕ_1 is negative, the magnitude of y_t again grows without bound, but the sign changes every period.

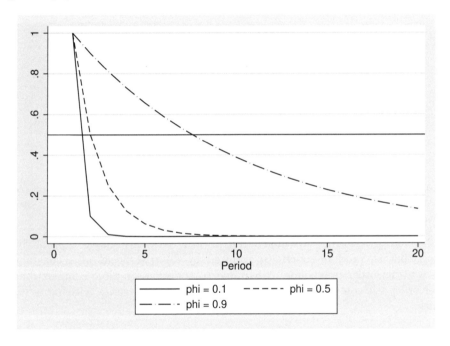

Figure 6.1. Exponential decay in a stationary AR(1) model

_____

14. If the model includes a constant, μ, future values will decay toward the unconditional mean of y_t, $\mu/(1 - \phi_1)$.

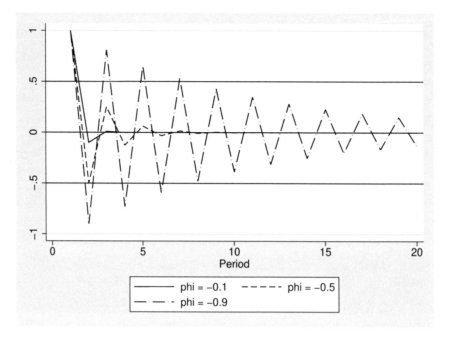

Figure 6.2. Decay when $\phi_1 < 0$

Figures 6.1 and 6.2 display the dynamic responses to a unit disturbance ($\epsilon_0 = 1$, $\epsilon_t = 0, \forall t > 0$) of several stationary AR(1) models. That is, figures 6.1 and 6.2 show how y_t evolves in response to a one-period shock that occurs at time $t = 0$.

A stationary first-order AR model can project only this simple type of dynamic behavior, that is, random shocks (introduced by ϵ_t) that decay at a constant rate. Just as a higher-order polynomial is required to approximate more complex functions, a higher-order AR component is required to model a time series with more complex dynamics.

Consider, for example, the AR(2) model

$$(1 - \phi_1 L - \phi_2 L^2)y_t = 0$$

where, once again, we have set the disturbance to 0 for convenience. In the previous section, we noted that stationarity is determined by the roots of the equation

$$\phi(z) = 0$$

where we have replaced the lag operator L in $\phi(L)$ with an ordinary algebraic variable, z. In addition to stationarity, the roots of this equation also determine the dynamic response of y_t to random disturbances. Quadratic equations have two roots, and those roots are both either real numbers or complex conjugates. If the roots are real, disturbances either decay or grow exponentially, depending on stationarity. If, however, the roots are complex, random disturbances induce oscillations that either decay over time (if the model is stationary) or grow in amplitude without bound (if the model is nonstationary).[15] Figure 6.3 displays an example of an AR(2) model with complex roots.

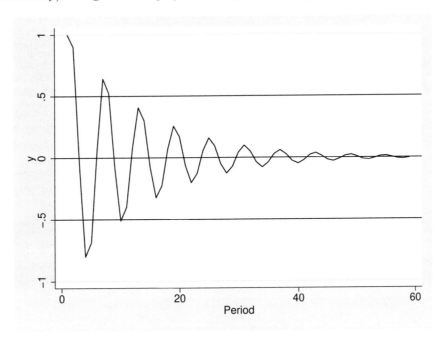

Figure 6.3. Decay in an AR(2) model with complex roots

What about higher-order AR models? Again the roots of $\phi(z) = 0$ determine both stationarity and the shape of the dynamic response to disturbances. Real roots generate either exponential decay or growth, and complex roots generate either damped or explosive oscillations; see table 6.1.

Table 6.1. Dynamic behavior of an AR(2) model

	Stationary	Nonstationary
Real roots	Exponential decay	Exponential explosion
Complex roots	Damped oscillation	Explosive oscillation

15. If the root with the smallest modulus lies exactly on the unit circle, random disturbances accumulate instead of decaying or growing without bound.

So far we have considered only the contribution of the AR component to the dynamic behavior of a time series, but the MA component plays a role as well. Consider the general linear process

$$y_t = \mu + \epsilon_t + \psi_1 \epsilon_{t-1} + \psi_2 \epsilon_{t-2} + \cdots$$

In a pure MA model, only a finite number of the ψ_{t-k}'s are nonzero. If there is an AR component, then $\psi()$ is infinite order and

$$\psi(L) \equiv \phi^{-1}(L)\theta(L)$$

Another name for $\psi(L)$ is the impulse–response function, because the coefficients trace the impact on y_t of a single random impulse, a single, nonzero ϵ_t. Imagine, for example, that $\epsilon_0 = 1$, all other ϵ's are 0, and, for convenience, $\mu = 0$. In this case,

$$y_t = \psi_t, \forall t$$

In other words, the ψ coefficients measure the impact of a single shock working its way through the model over time. In a pure (that is, finite-order) MA model, if the final nonzero coefficient is ψ_{t-k}, then a single shock has no impact after k periods. However, when the model contains an AR component, the impact of a single shock persists forever, although the magnitude of the impact decays over time if the model is stationary.

6.6   Points to remember

- All the time-series models that we consider can be represented as a general linear process, that is, an infinite, weighted sum of white-noise disturbances.

- The infinite-order linear process can be represented as a more tractable combination of finite-order AR and MA components.

- The lag polynomial notation provides a convenient method for representing and manipulating ARMA models and for determining their essential properties.

- Stationarity is the fundamental characteristic of a time series. Random shocks to a stationary time series decay over time, and the time series returns eventually to its unconditional mean. Random shocks to a nonstationary time series generate an explosive response.

- The coefficients of the AR representation of a time series—$\phi(L)$ for a pure AR model and $\theta^{-1}(L)\phi(L)$ for an ARMA model—define a difference equation that determines the dynamic response to random shocks.

6.7   Looking ahead

The ARMA model introduced in this chapter provides a flexible representation for a wide variety of time-series behavior. In the next chapter, we use the example of U.S. GDP to show how to fit an ARMA model to a real-world time series and how to interpret the ARMA estimates. In subsequent chapters, we will extend the ARMA model to multiple time series, nonstationary time series, and series with more flexible error-variance properties.

7 Modeling a real-world time series: The example of U.S. gross domestic product

Chapter map

7.1 Getting ready to model a time series. Summarizing the data with tools we introduced previously. What does it mean to interpret these summaries as "models" of gross domestic product (GDP)? Some daunting advice.

7.2 The Box–Jenkins approach. The Box–Jenkins three-step "recipe": identification, estimation, and forecasting. The role of parsimony.

7.3 Specifying an ARMA model. How an autoregressive moving-average (ARMA) model becomes an autoregressive integrated moving-average (ARIMA) model—or how to handle a nonstationary series. Using autocorrelations and partial autocorrelations to determine the number of autoregressive (AR) and moving-average (MA) parameters.

7.4 Estimation. How to estimate the parameters of an ARMA model. The state-space representation. Diagnosing problems with a fitted model.

7.5 Looking for trouble: Model diagnostic checking. Overfitting, or adding parameters, and tests of the residuals.

7.6 Forecasting with ARIMA models. How we got to this stage. One-step-ahead and dynamic forecasts. Forecasting differences and levels.

7.7 Comparing forecasts. Comparing forecasts from alternative ARIMA specifications. Comparing ARIMA forecasts to simpler methods.

7.8 Points to remember. "Best" specification. The Box–Jenkins approach. Stata's `predict` command.

7.9 What have we learned so far? How the last five chapters are connected after all.

7.10 Looking ahead. The three issues we have not discussed yet.

In the previous chapter, we introduced ARMA models, a flexible representation of a wide variety of time-series behavior, but we did not explain how to fit these models to real-world time series. In this chapter, we correct that oversight using the example of U.S. GDP and the time-series tools in Stata.

This chapter is very long. We will need some additional tools to fit a real-world time series to the statistical models described in the previous chapter. Moreover, we take the time to illustrate the types of puzzles you will need to solve (and how you might solve them) when you fit a time series. So make yourself a cup of coffee, fire up Stata, and follow along as we fit our first ARMA model.

7.1   Getting ready to model a time series

Fitting a univariate time-series model is a little different from other statistical investigations. Most univariate statistical tools are used to characterize the probability distribution of the variable of interest. Means and medians provide estimates of the location of a random variable's distribution. Variances, interquartile ranges, and the like provide measures of the dispersion of the distribution. Letter values, histograms, and kernel densities help to fill in the full picture of the distribution. Diagnostic plots and assorted tests help gauge the conformance between an empirical distribution and an ideal parametric distribution.

Univariate time-series models provide many of these same indicators—means, autocovariances, etc.—for a time-series variable. Intuitively though, the process for a time-series variable seems a little backward. When a researcher analyzes a cross-section variable, univariate statistics guide him or her to a view of the variable's distribution. For a time-series variable, a model first must be fit to the data before some of the key distributional features of the variable can be measured. But it can be difficult to determine the appropriate time-series model to apply, thus making it difficult to assess some of the fundamental characteristics of the random variable. The source of much of the difficulty is the extraordinary flexibility of time-series models—the same data may be fit almost equally well by several different specifications (with different implications). This difficulty led Granger and Newbold (1989), only partly in jest, to recommend that no one should fit a time-series model for the first time. We will illustrate some of these challenges—and ways to overcome them—by fitting a univariate time-series model to U.S. GDP.

So how should we start? Recall that we used GDP in chapter 3 to illustrate a time series dominated by a trend. In that example, we used nominal GDP, a series that combines two dynamic processes, one that determines the growth of the real economy and another that determines the level of prices or inflation. While these processes are related, they are sufficiently different that it makes sense to model them separately. Accordingly, we will analyze a measure of real GDP, that is, GDP in constant dollars.

In addition to statistics on nominal GDP and its constituents, the United States Department of Commerce also publishes GDP expressed in the prices of a specific year.

This series attempts to correct for the changing purchasing power of the dollar and to provide a measure of real output.[1] Currently, this real series is measured in 2005 prices. Recall from chapter 3 that GDP appears to grow exponentially, so the log of GDP follows a linear trend.

```
. use ${ITSUS_DATA}/quarterly, clear
(Quarterly U.S. GDP, annualized, BEA, as of 4/27/2012)

. tsset
        time variable:  date, 1947:1 to 2012:1
                delta:  1 quarter

. generate lrgdp = log(gdp2005)

. label variable lrgdp "Log of real GDP"

. tsline lrgdp
```

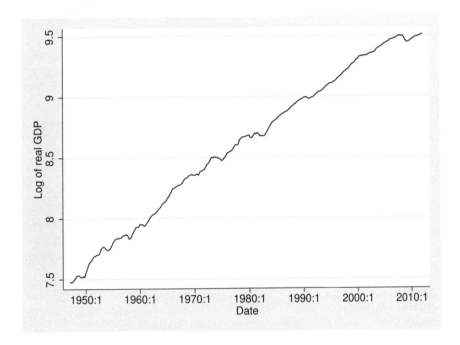

Figure 7.1. Log of United States real GDP

Before we fit an ARMA model to real GDP, let's see what we can glean by applying some of the tools we introduced in earlier chapters.

---

1. The Bureau of Economic Analysis, the agency within the Department of Commerce responsible for the national income and product accounts, provides descriptions of its methodologies online at https://www.bea.gov/resources/methodologies.

Fitting a straight line to this transformation of GDP provides an estimate of trend growth. The slope of the regression line measures the average growth rate of GDP. The regression residuals highlight episodes of above-trend and below-trend growth.

```
. regress lrgdp date

      Source |       SS           df       MS            Number of obs   =       261
-------------+----------------------------------          F(1, 259)       =  34348.33
       Model |  96.9324243         1   96.9324243         Prob > F        =    0.0000
    Residual |  .730908842       259   .002822042         R-squared       =    0.9925
-------------+----------------------------------          Adj R-squared   =    0.9925
       Total |  97.6633331       260   .375628204         Root MSE        =    .05312

-------------------------------------------------------------------------------------
       lrgdp |      Coef.   Std. Err.      t    P>|t|     [95% Conf. Interval]
-------------+-----------------------------------------------------------------------
        date |   .0080885   .0000436   185.33   0.000     .0080026    .0081744
       _cons |   7.977469   .0047329  1685.52   0.000     7.968149    7.986789
-------------------------------------------------------------------------------------
```

The `date` variable is a counter incremented by 1 each quarter and normalized by setting the value to 0 in the first quarter of 1960. These quarterly measures of GDP are annualized, so we can recover the trend rate of real growth by annualizing the regression coefficient on `date`.

```
. display 400 * _b[date]
3.235397
```

From 1947:1 through 2012:1, the U.S. economy as measured by GDP grew at an average annual real rate of 3.2%.

In chapter 3, we applied the Holt–Winters smoother to estimate jointly the trend and cyclical components of a time series. We can use the Holt–Winters smoother here to obtain an alternative characterization of real GDP. In addition, we will use the estimated Holt–Winters smoother to calculate forecasts for the log of real GDP for the three years covering 2012:2 through 2015:1.

```
. tssmooth hwinters hw = lrgdp, forecast(12)
computing optimal weights

Iteration 0:    penalized RSS = -.05181424   (not concave)
Iteration 1:    penalized RSS = -.02592632
Iteration 2:    penalized RSS = -.02592506
Iteration 3:    penalized RSS = -.02592497

Optimal weights:
                              alpha = 1.0000
                               beta = 0.2936
penalized sum-of-squared residuals = .025925
         sum-of-squared residuals = .025925
         root mean squared error = .0099664

. label variable hw "HW smooth"
```

To compare the linear trend and Holt–Winters characterizations, we will add residuals for both models and fitted values from the linear trend model to the dataset.[2] First, the Holt–Winters residuals:

```
. generate Rhw = hw - lrgdp
(12 missing values generated)
. label variable Rhw "HW residual"
```

Now the fitted values and residuals for the linear trend model. Stata remembers the linear trend regression, so we do not have to reestimate it. The `forecast()` option of the Holt–Winters command filled in 12 quarters of future quarterly dates, so the single explanatory variable of the linear trend model has already been filled in for 2010–2012.

```
. predict trend
(option xb assumed; fitted values)
. label variable trend "Trend growth, 1947:1-2012:1"
. predict resid, residual
(12 missing values generated)
. replace resid = - resid
(252 real changes made)
. label variable resid "Trend residual"
```

---

2. While it makes no difference statistically, textbooks typically define residuals as the actual values minus the fitted values. My personal preference is just the reverse. To my simple brain, it seems more intuitive if a positive residual represents a model overprediction and vice versa. Suit yourself.

Figure 7.2 compares the fitted values from both characterizations.

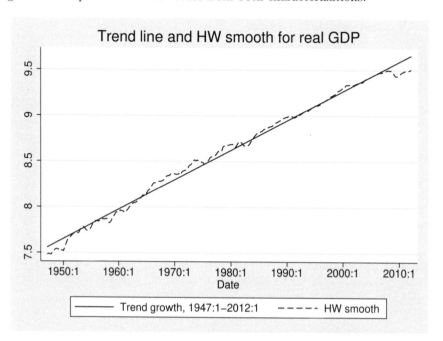

Figure 7.2. Trend line and Holt–Winters smooth for real GDP

We did not include the actual values in this graph because they would be indistinguishable at this scale from the Holt–Winters smooth.

These two smoothers highlight different aspects of the underlying variable. The linear trend characterization highlights the average rate of growth over an extended period, while the Holt–Winters smoother emphasizes the variability of the local trend. Note that the Holt–Winters smoother sets α to its limiting value of 1.0. In essence, for this series, the Holt–Winters smoother does not smooth the data at all. The residuals, displayed in figure 7.3, illustrate these differences. The fixed linear trend produces residuals that are large in magnitude compared with the Holt–Winters residuals.

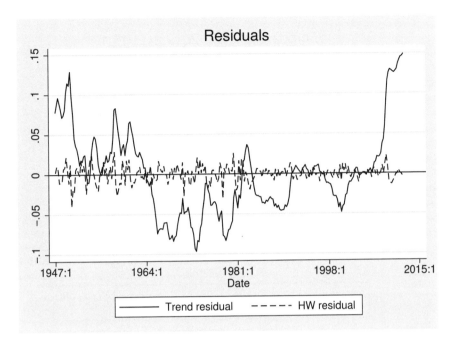

Figure 7.3. Comparison of trend line and Holt–Winters residuals

Note that the residuals from the trend line take extended excursions above and below 0. In other words, they show evidence of autocorrelation.

When we use these approaches to generate forecasts of the data, these techniques become competing models of real GDP—with significantly different implications—rather than just different perspectives on the sample data. Figure 7.4 compares the forecasts from each model for the three years after 2012:1.

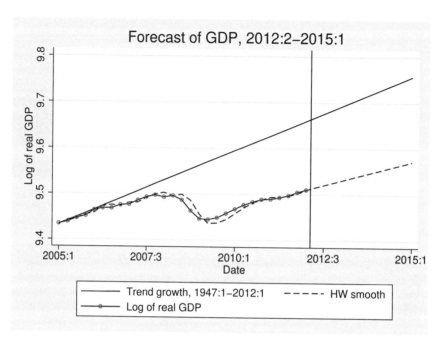

Figure 7.4. Comparison of trend line and Holt–Winters forecasts

We have overlaid these fitted values (for 2005:1–2012:1) and forecasts (for 2012:2–2015:1) with the actual values of the log of real GDP through 2012:1. Both forecasts track the actuals closely through the middle of 2006. At that point, however, they diverge sharply. The growth of actual GDP declines as the recession approaches. The trend forecast ignores this economic weakness, while the Holt–Winters forecast continues to track actual GDP closely. By the end of the three-year forecast period, the Holt–Winters forecast of real GDP is 1% lower than the trend forecast. Moreover, the Holt–Winters forecast projects future real growth at only 2.0%.

How can we make sense of these two very different forecasts? The trend line forecast ignores the information contained in recent GDP growth. If the trend line regression is regarded as a model of real GDP, it incorporates the maintained hypothesis that the growth rate of GDP will average around 3.2%. As a consequence, recent growth below or above that long-run average rate implies faster or slower growth, respectively, in the future to maintain the long-run pace of expansion.

In contrast, the Holt–Winters model is heavily influenced by recent experience. If growth has been slow recently, it is projected to remain slow, and vice versa. If the economy suffers a serious downturn, as it did in 2008 and 2009, the Holt–Winters approach assumes that the economy will not make up the "lost" growth. A recession represents a permanent loss in output. These two views are very different, and they lead to very different macroeconomic policies, so it matters greatly if we believe these tools provide more than convenient characterizations of the data and also can be used as guides to the future.

What if we soften the view of the linear trend model? We can retain the view that real growth will average 3.2% in the future but adopt the Holt–Winters convention that past subpar growth does not imply future above-par growth. This blended approach is similar to a series of one-step-ahead forecasts, where we reset the level of the projection each period as another actual value is realized. Figure 7.5 applies this reasoning.

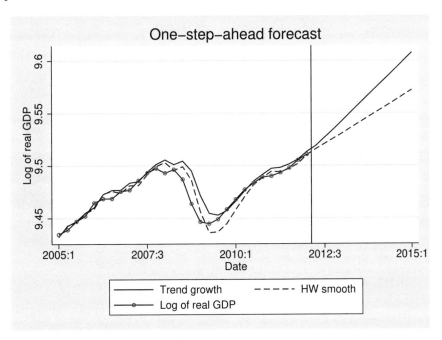

Figure 7.5. Converting the linear trend to a one-step-ahead forecast

Now the two forecasts begin from the same point. The only difference is in the view of the future trend of growth. The linear trend model fixes future growth at 3.2%, while the Holt–Winters model assumes that the recent trend (2.0% in this example) will continue without change.

Note: In case you are curious about the construction of figure 7.5, here are some of the details.

First, how do I know that the Holt–Winters estimate of the recent trend is 2.0%? Remember that the trend is constant in the Holt–Winters forecast period. In this example, the forecast period starts in observation number 262, so I typed

```
. regress hw date in 262/l
. display 400*_b[date]
```

Stata stores the estimated regression coefficient of `date` in `_b[date]`. Because the left-hand-side variable is the log of real GDP, this coefficient is an estimate of the quarterly growth rate (in decimal). The preceding "400" in the `display` command converts this estimate to an annual percentage rate of growth. The `display` command produced the answer 2.0425556.

Adjusting the one-step-ahead forecasts of the linear trend model so that they square with the most recent actual value of log real GDP takes a few extra steps. The original forecasts from the linear trend model are[3]

```
trend = _b[date]*date + _cons
```

In other words, `_b[date]` is the model's forecast of the quarterly increase in `lrgdp`. To apply the in-sample adjustment, I typed

```
. generate ahead1 = _b[date] + L.lrgdp
```

During the out-of-sample forecast, this technique will not work, because `L.lrgdp` is filled with missing values. There are a couple of ways to solve this problem. Here is what I did:

```
summarize resid if date==tq(2012q1)
local adjust = `r(mean)'
replace ahead1 = trend - `adjust' if missing(ahead1)
label variable ahead1 "Trend growth"
```

I calculated the regression error in the final observation and applied that shift throughout the forecast period. There are probably more elegant solutions, but I have a simple mind.

7.2 The Box–Jenkins approach

ARMA modeling of time series is closely associated with the work of George E. P. Box and Gwilym Jenkins, whose 1970 monograph *Time Series Analysis: Forecasting and Control*[4] not only summarized the statistical literature but also outlined a modeling philosophy and provided a practical step-by-step approach—almost a recipe—along

---

3. Of course, I let Stata calculate these forecasts for me. I did not type this formula explicitly.

4. A fourth edition—with Gregory C. Reinsel as an additional coauthor—was published in 2008.

with high-level specifications for computer programs to implement the approach. The appearance of this volume enabled programmers to make up-to-date time-series techniques available, either as a set of stand-alone computer programs or as procedures within larger statistical packages. As a consequence, the Box–Jenkins approach was adopted and applied by many researchers.

Box and Jenkins proposed an iterative approach to time-series modeling with three steps:

1. **Identification:** In Box and Jenkins's terminology, identification refers to the determination of the order of the ARMA model. Is the series in question best modeled as an ARMA(1,1) process, or would an ARMA(2,0) work better?[5]

2. **Estimation:** Once the model has been identified, the parameters—μ, $\phi(L)$, $\theta(L)$, and σ_ϵ^2—must be estimated. Box and Jenkins presented innovative techniques for overcoming some of the estimation challenges involved; however, this aspect of time-series modeling has evolved substantially since the original publication of the Box–Jenkins monograph.

3. **Diagnostic checking:** Statistical tests are applied to determine the adequacy of the model. The detection of model deficiencies raises questions about the initial model identification and restarts the three-step process. This cycle ends when no problems are diagnosed in a candidate model.

The Box–Jenkins approach is best suited to experimental data where fresh samples can be generated for each iteration; in fact, Box and Jenkins are particularly interested in the statistical control of manufacturing and chemical processes, where this type of experimentation is feasible. Nonetheless, the Box–Jenkins iterative approach frequently is applied to nonexperimental data, such as GDP.[6]

Another key concept in the Box–Jenkins philosophy is parsimony. We have already noted the extraordinary flexibility of the difference equations embedded in ARMA models. As a consequence of that flexibility, it often is very difficult to choose the "best" among several reasonable time-series models. Box and Jenkins encourage researchers to select the most parsimonious model, that is, the one with the fewest parameters.

Stata makes it easy to fit ARMA models, so we will not spend much time on that aspect of the Box–Jenkins approach. Instead we will focus on identification and diagnostic checking.[7]

---

5. This use of the term "identification" differs from usual statistical practice, which uses this term to indicate which parameters can, in principle, be pinned down from the sample.

6. Hendry (1995) has criticized many time-series practices, including some aspects of the Box–Jenkins approach, and has proposed an alternative method for developing and fitting time-series models. Part of Hendry's approach emphasizes applying hypothesis tests to a general model that nests the alternatives and applying the tests in an order that preserves their classical interpretation.

7. You may be wondering why we spent so much time in chapter 5 analyzing estimation strategies, yet we virtually spend no time in this chapter doing so, especially because all models with MA components have autocorrelated disturbances. In chapter 5, we were estimating linear relationships by using least squares; hence, autocorrelated disturbances raised potential estimation problems. Stata fits ARMA models by using maximum likelihood, avoiding the issues we confronted in chapter 5.

7.3   Specifying an ARMA model

We want to fit an ARMA(p,q) model to our data on real GDP, but we do not know the appropriate values of p and q. The tools we have used so far—fitting a linear trend and applying a time-series smoother—have not provided any clues so far. In the previous chapter, we highlighted the relationship between the order of the AR component and the dynamic behavior of the observable variable: AR(1) processes generate exponential decay of random shocks, while higher-order processes can produce oscillations if they have complex roots. We could scan a smoothed version of our series to get an impression of the likely order of the AR component, but this approach would be very imprecise. If deviations from the mean decay exponentially without much evidence of oscillations, we might be able to guess that an AR(1) process would suffice or, alternatively, that a higher-order process is needed because the series oscillates noticeably around the trend. However, we could not readily distinguish between an AR(2) process and an even higher-order AR component. In addition, this approach would not help us pin down the order of the MA component. Moreover, evidence of autocorrelation in our series could be generated by either component or by both components.

Box and Jenkins proposed a method for choosing the order of an ARMA model based on sample statistics. The Box–Jenkins method is neither foolproof (hence, the iterative cycle of identification, estimation, and diagnostic checking) nor definitive (you still have to exercise judgment), but it provides structured guidance for a challenging problem. We illustrate the method in the remainder of this section by applying it to our data on GDP.

7.3.1   Step 1: Induce stationarity (ARMA becomes ARIMA)

You may have been wondering how we are going to fit an ARMA model, which requires a stationary observable variable, to the clearly nonstationary real GDP series. Any trending series is nonstationary—the mean of the series is increasing (or decreasing) over time. One way to induce stationarity is to detrend the series, that is, estimate the trend first, then subtract the trend from the observed series. In the Box–Jenkins approach, however, it is more common to difference the observed series one or more times.[8] When differencing is required, we rewrite our ARMA(p, q) model as an ARIMA(p, d, q) model, where d indicates the order of differencing, that is, the number of times y_t is differenced to achieve stationarity.

How does differencing produce stationarity? Let's look first at the linear trend model we fit to the log of real GDP earlier in this chapter. The estimated equation had the form

$$y_t = \alpha + \beta t + \epsilon_t$$

In this specification, the mean of y_t is

$$E(y_t) = \alpha + \beta t$$

---

8. There are important differences between trend-stationary and difference-stationary processes. We will return to this issue in chapter 10.

a function of time, so y_t is nonstationary. In this specification, detrending produces a stationary (in fact, a white noise) series:

$$y_t - \alpha - \beta t = \epsilon_t$$

Differencing also produces stationarity in this model. If we take the first difference of y_t, we obtain

$$y_t - y_{t-1} = (\alpha + \beta t + \epsilon_t) - \{\alpha + \beta(t-1) + \epsilon_{t-1}\} = \beta + \epsilon_t - \epsilon_{t-1}$$

This difference is covariance stationary, meaning that the mean and variance of $y_t - y_{t-1}$ do not depend on time

$$E(y_t - y_{t-1}) = \beta$$

and

$$E(y_t - y_{t-1} - \beta)^2 = E(\epsilon_t - \epsilon_{t-1})^2 = 2\sigma_\epsilon^2$$

It may look as if we have found a specification for our model of real GDP; this example can be written as an ARIMA(0,1,1) model (although the MA component is not invertible). However, as we saw above, the linear trend model is a special case with some strong implications about reversion to the trend path. Let's suspend judgment about the specification for the moment.

While detrending works nicely in the special case above, differencing is a general solution for nonstationarity. Recall that the ARMA model

$$\phi(L)y_t = \theta(L)\epsilon_t$$

is stationary if and only if all the roots of $\phi(z)$ lie outside the unit circle. We can rewrite the AR polynomial as

$$\phi(L) = \prod_{i=1}^{p}(1 - \lambda_i L)$$

where the ith root of $\phi(z)$ is $1/\lambda_i$. So all the roots of $\phi(z)$ lie outside the unit circle if all the ratios, $1/\lambda_i$, lie inside the unit circle.

Suppose that $\lambda_1 = 1$ but $|\lambda_i| < 1$ for $i = 2, 3, \ldots, p$. Now y_t is nonstationary and

$$\phi(L) = (1 - L)\prod_{i=2}^{p}(1 - \lambda_i L) \equiv (1 - L)\phi^*(L)$$

In this case, $(1 - L)y_t$, the first difference of y_t, is stationary because all the roots of $\phi^*(z)$ lie outside the unit circle:

$$\phi^*(L)(1 - L)y_t = \theta(L)\epsilon_t$$

If two λ's equal 1, y_t must be differenced twice to achieve stationarity, and in general, the order of differencing, d, is the number of λ's that are equal to 1.[9],[10]

For many series, there is consensus on the appropriate order of differencing. For example, many economic series exhibit something like exponential growth, especially if, like real GDP, they are closely related to population growth. These types of series generally are modeled as the first difference of their natural logs. For other series, though, you will not know the order of differencing in advance. For these series, you need to look for clues in the sample.

It turns out that the autocorrelations of the series can point to the order of differencing needed for stationarity. As it happens, the autocorrelations of y_t follow exactly the same difference equation—defined by the AR component of the model—as y_t.[11] Because we can calculate estimates of the autocorrelations, we can observe the behavior of the autocorrelations to determine whether differencing is required.

Let's start with the general ARIMA model

$$y_t = \phi_1 y_{t-1} + \cdots + \phi_p y_{t-p} + \epsilon_t - \theta_1 \epsilon_{t-1} - \cdots - \theta_q \epsilon_{t-q}$$

If we multiply both sides of this equation by y_{t-k} and take expectations, we obtain a time-series model in terms of the autocovariances, γ_k. This model is a little messy. The AR part of the model is straightforward, but the MA component comprises a bunch of terms that are expectations of the product of y_{t-k} and each of the ϵ_j. However, y_{t-k} is uncorrelated with any ϵ_j dated later than period $t - k$. Thus if $k > q$—that is, if y_{t-k} is lagged far enough to occur earlier than ϵ_{t-q}, the last MA term—then y_{t-k} is uncorrelated with all the ϵ_j's on the right-hand side of the equation above. So for $k > q$, the model simplifies to just the AR component

$$\gamma_k = \phi_1 \gamma_{k-1} + \cdots + \phi_p \gamma_{k-p}$$

Or, more compactly,

$$\phi(L)\gamma_k = 0$$

And because the autocorrelations are defined as $\rho_k = \gamma_k / \gamma_0$, the autocorrelations also follow the same difference equation

$$\phi(L)\rho_k = 0$$

So how does this help us identify the appropriate order of differencing? It turns out that we can rewrite this difference equation in terms of the roots of $\phi(z)$

$$\rho_k = \alpha_1 \lambda_1^k + \cdots + \alpha_p \lambda_p^k$$

---

9. Box and Jenkins (1976) and Hamilton (1994) provide detailed discussions of the role of differencing in inducing stationarity.

10. You might wonder why we do not consider the possibility of roots that lie within the unit circle ($|\lambda_i| > 0$). These cases are explosive; hence, they will not be observed in practice—at least, not for long.

11. Well, almost exactly. The math below spells out the qualifications.

where the α_i's are fixed coefficients and the λ_i's are the multiplicative inverses of the roots $\phi(z)$. If y_t is stationary, $|\lambda_i| < 1, \forall i$, so the sequence of ρ_k's will "die out" as k increases. However, if y_t is nonstationary, at least one of the λ_i's is equal to 1. Say $\lambda_i = 1$. Then as k grows larger,

$$\rho_k \approx \alpha_i$$

In other words, the sequence of autocorrelations of a nonstationary series will not die out. Thus, to determine the order of differencing, examine the autocorrelation function. If the autocorrelations collapse quickly toward 0, the time series is stationary. If the magnitude of the autocorrelations declines approximately linearly and does not collapse to 0, at least one more difference is needed.

Because we do not know the population autocorrelations, we must rely on the sample autocorrelations to form our conclusion. To illustrate this procedure, we will pretend that we do not know the appropriate order of differencing for real GDP. We will use Stata's `ac` command to calculate and display the sample autocorrelations for $d = 0, 1, 2$. The syntax of the `ac` command is

`ac` *varname* $\big[\,$*if*$\,\big]$ $\big[\,$*in*$\,\big]$ $\big[\,$, `lags(`#`)` `generate(`*newvar*`)` `level(`#`)` `fft` $\big]$

There are additional options, mainly to modify the appearance of the graph, but we do not need them now. By default, `ac` calculates and displays the first 40 autocorrelations of *varname*. The `lags(`#`)` option allows you to specify any number of autocorrelations you prefer. In practice, the first 20 autocorrelations are generally sufficient. If your time series is unusually long or you request a large number of autocorrelations, you can use the fast Fourier transform (`fft`) option to speed the calculations. The `level(`#`)` option lets you specify the significance level of the confidence bands. The default level is 95%. Use the `generate(`*newvar*`)` option to store the autocorrelations. You can add the `nograph` option if you do not wish to see the autocorrelations displayed.

We already know that the log of real GDP is not stationary. Let us see whether the sample autocorrelations can detect the trending behavior as readily as we can.

```
. ac lrgdp
```

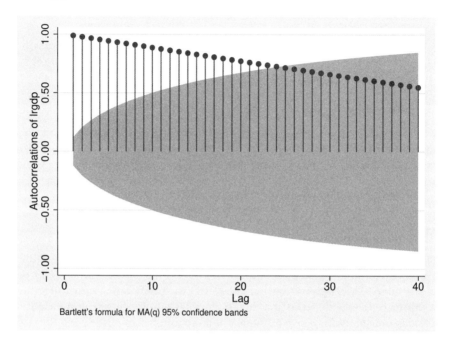

Figure 7.6. Autocorrelations of the log of real GDP

The filled circles indicate the values of the autocorrelations. The dropped lines anchor the autocorrelations to 0. The shaded region indicates the 95% confidence region based on an approximation by Bartlett (1946).

In this example, the autocorrelations clearly decline linearly and do not collapse to 0, indicating that the log of real GDP is not stationary. Now let's recalculate the autocorrelations for the first difference of log real GDP, the growth rate.

```
. generate growth = lrgdp - L.lrgdp
(1 missing value generated)
. label variable growth "Growth rate of real GDP"
. ac growth, lag(20)
```

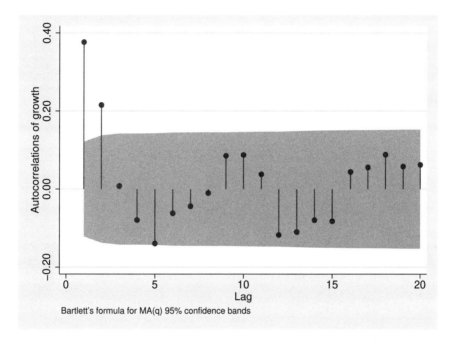

Figure 7.7. Autocorrelations of real GDP growth

Now the autocorrelations quickly collapse to insignificance, indicating the growth rate of real GDP is stationary, at least in this sample.

7.3.2   Step 2: Mind your p's and q's

In the previous subsection, we used the autocorrelations to check for stationarity. In combination with a new statistic, the partial autocorrelation, the autocorrelations can also guide us to appropriate choices of p and q, the order of the AR and MA components, respectively.

The jth partial autocorrelation is defined as the final coefficient in the regression of y_t on j lags of y. In other words, the jth partial autocorrelation measures the correlation between y_t and y_{t-j} after controlling for the correlations between y_t and y_{t-1}, y_{t-2}, ..., and y_{t-j+1}. If we estimate the regression

$$y_t = \beta_0 + \beta_1 y_{t-1} + \cdots + \beta_j y_{t-j}$$

the coefficient β_j is the jth partial autocorrelation. In Stataspeak, if, for example, we run the command

```
. regress L(0/5).y
```

the 5th partial autocorrelation will be stored in _b[L5.y]. Of course, Stata does not make you run a sequence of regressions to calculate the partial autocorrelation function. The pac command does all the work for you.

The syntax of the pac command is

pac *varname* $\big[$ *if* $\big]$ $\big[$ *in* $\big]$ $\big[$, <u>lags</u>(*#*) <u>gene</u>rate(*newvar*) <u>level</u>(*#*) $\big]$

Again there are additional options that we have omitted for now. Most are concerned with the appearance of the graph. All the options above appear as options in the ac command, and their interpretation is the same here.[12]

The patterns of the autocorrelation and partial autocorrelation functions reflect the order of the ARMA model. Let's illustrate this property with some simple examples. First, consider the ARMA(0,0) model, that is, white noise.

$$y_t = \epsilon_t$$

The autocorrelations, γ_k, are all 0 for $k > 1$. The partial autocorrelations are all 0 as well. We will create a white-noise series, then display the autocorrelation and partial autocorrelation functions.

```
. clear
. set obs 120
number of observations (_N) was 0, now 120
. generate int t = _n
. tsset t
        time variable:  t, 1 to 120
                delta:  1 unit
. set seed 95
. generate epsilon = rnormal()
. ac epsilon, name(aceps) lags(20) note("")
. pac epsilon, name(paceps) lags(20) note("")
```

---

12. We have omitted the yw option, which specifies that Stata should solve the Yule–Walker equations to obtain the partial autocorrelations. This method is rarely used anymore.

```
. graph combine aceps paceps, note("95% confidence bands")
```

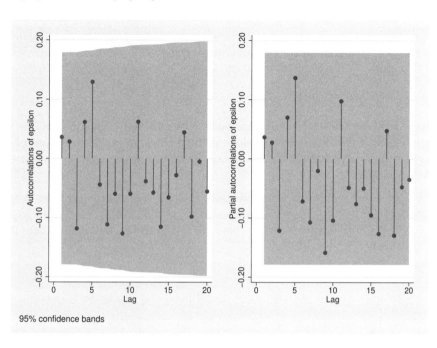

Figure 7.8. Autocorrelation and partial autocorrelation functions of white noise

We used the `name()` option of the `ac` and `pac` commands to attach names to the graphs; then we used the `graph combine` command to display them side by side.[13]

All the autocorrelations and partial autocorrelations lie within the 95% confidence bands, as you would expect with a white-noise series. Most important are the estimates at the left-hand side of each graph, that is, at low lags. To see why this area of the graph is key, we need to think about the autocorrelations and partial autocorrelations of pure AR and MA processes.

We will start with the AR model

$$\phi(L)y_t = \epsilon_t$$

When we discussed differencing to achieve stationarity, we showed that for $k > q - p$, the autocorrelations of y_t follow the same difference equation as y_t, that is,

$$\phi(L)\gamma_k = 0$$

---

13. We specified `note("")` with the `ac` and `pac` commands so that each graph would not have its own note at the bottom. Instead we added a note when we combined them.

Thus if y_t is stationary, the autocorrelation function dies out—exponentially (if all the roots of $\phi(z) = 0$ are real), in damped oscillations (if the roots are complex), or in a combination of exponential decay and damped oscillations (if there are both real and complex roots).

The pattern is different for the partial autocorrelation function of this model. If we rewrite this model as a regression equation, we have

$$y_t = \phi_1 y_t + \cdots + \phi_p y_{t-p} + \epsilon_t$$

The partial autocorrelations should lie outside the confidence bands for the first p lags. For longer lags, the partial autocorrelations should be insignificant. So instead of decaying as lag length increases, the partial autocorrelations should collapse toward 0 sharply after the first p lags.

Let's look at some examples where we know the true order of the process. We reused the white-noise series we generated above and created an AR(1) series with $\phi_1 = 0.9$. Figure 7.9 displays the autocorrelation and partial autocorrelation functions for this series.

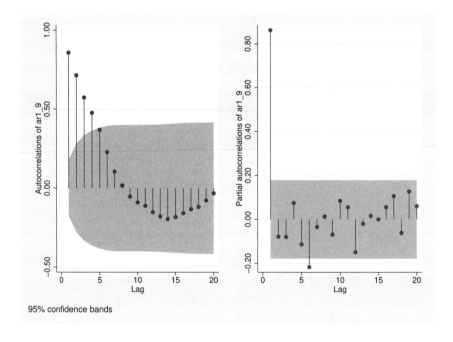

Figure 7.9. Autocorrelation and partial autocorrelation functions of an AR(1) series where $\phi_1 = 0.9$

The influence of the white-noise disturbances adds some noise to the autocorrelation function. Nonetheless, it decays approximately exponentially, as expected. And the first partial autocorrelation stands out, as we expect with an AR(1) process. Also the first partial autocorrelation is 0.84, close to the true value.[14]

The next example highlights one of the challenges in interpreting these statistics. Figure 7.10 displays the autocorrelation and partial autocorrelation functions for an AR(2) process where $\phi_1 = 0.9$ and $\phi_2 = -0.1$. Again we reuse the same white-noise series as before.

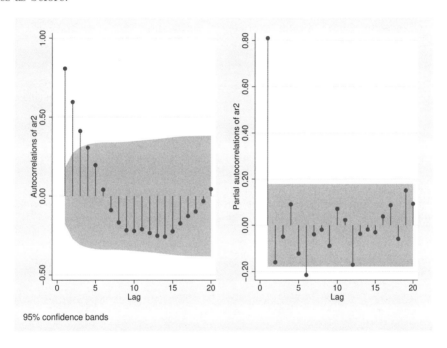

Figure 7.10. Autocorrelation and partial autocorrelation functions of an AR(2) series

The roots of $\phi(z) = 0$ are both real in this example, and the autocorrelation function displays the expected pattern of decay. The estimated values of the first two partial autocorrelations are 0.81 and -0.16, but the second partial autocorrelation lies within the 95% confidence band. As a result, this graph suggests an AR(1) process rather than the AR(2) process we happen to know is the correct specification.

---

14. You can use Stata's `corrgram` command to display a table of the autocorrelations and partial autocorrelations.

The problem lies in the sample size. The conventional approximation to the standard error of the partial autocorrelations is $1/\sqrt{n}$. Our artificial series contains 120 observations. With this sample size, any partial autocorrelations smaller than 0.183 in absolute value will never appear to be significant. We would need a sample size of 400 for an estimated partial autocorrelation of -0.1—the true value of ϕ_2—to be at the edge of significance.

Now let's consider pure MA processes. The autocorrelation and partial autocorrelation functions of pure MA processes are mirror images of those for pure AR processes; the autocorrelations "cut off" after the first q terms, while the partial autocorrelations are dominated by some combination of exponential decay and damped oscillation. Consider the autocovariances of the MA process

$$y_t = \theta(L)\epsilon_t = \epsilon_t - \theta_1\epsilon_{t-1} - \cdots - \theta_q\epsilon_{t-q}$$

The kth autocovariance is

$$\gamma_k = \begin{cases} (-\theta_k + \theta_1\theta_{k+1} + \cdots + \theta_{q-k}\theta_q)\sigma_\epsilon^2 & k = 1, \ldots, q \\ 0 & k > q \end{cases}$$

In simple terms, when $k > q$, the autocovariance is 0 because y_t and y_{t-k} have no ϵ's in common. The derivation of the partial autocorrelation function is complicated, but it can be shown that the partial autocorrelation function of an MA process behaves similarly to the autocorrelation function of an AR process.

Figure 7.11 displays the autocorrelation and partial autocorrelation functions for an MA(1) process with $\theta_1 = 0.9$. (We are still using the same ϵ's.)

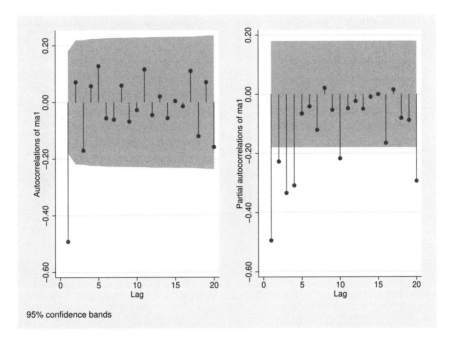

Figure 7.11. Autocorrelation and partial autocorrelation functions of an MA(1) series

The statistics displayed in figure 7.11 suggest an MA(1) process. The autocorrelations "cut off" after the first lag, and the partial autocorrelations die out quickly. In an MA(1) process, the first autocorrelation, ρ_1, is given by

$$\rho_1 = \frac{-\theta_1}{1 + \theta_1^2}$$

For this example, $\theta_1 = 0.9$, which implies $\rho_1 = -0.50$. The sample value in figure 7.11 is -0.49.

Figure 7.12 displays the autocorrelation and partial autocorrelation functions for an MA(2) process with $\theta_1 = 0.9$ and $\theta_2 = -0.1$. As in our AR(2) example, these graphs point to a first-order process rather than the true MA(2) process. Again the problem is the sample size, which is too small for autocorrelations as small as 0.11—our estimate of ρ_2—to be significant.

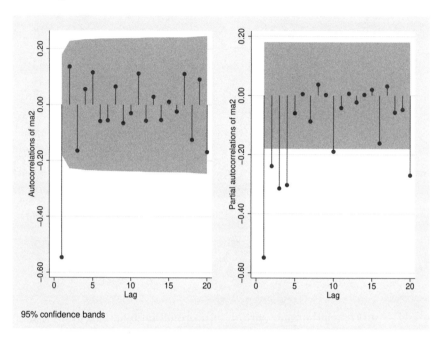

Figure 7.12. Autocorrelation and partial autocorrelation functions of an MA(2) series

To round things out, figure 7.13 displays the autocorrelation and partial autocorrelation functions for an ARMA(1,1) process ($\phi_1 = 0.7, \theta_1 = 0.5$). For an ARMA($p$,$q$) process, the autocorrelation function should "die out" (either in exponential decay or in damped oscillations or in a combination of both) after the first $q - p$ lags, and the partial autocorrelation function should die out after the first $p - q$ lags. In this case, the sample functions send the correct message and suggest an ARMA(1,1) model as a reasonable representation of these data.

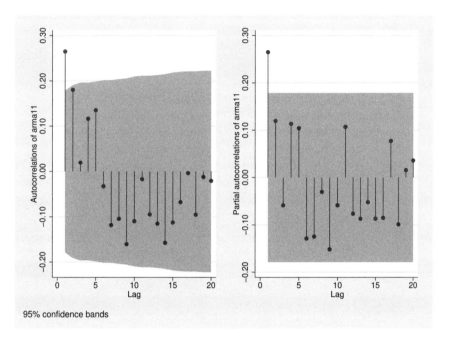

Figure 7.13. Autocorrelation and partial autocorrelation functions of an ARMA(1,1) series

Table 7.1 summarizes the indications the autocorrelation and partial autocorrelation functions provide about the order of the underlying process.

Table 7.1. Indicators of p, d, and q

Process	Autocorrelation function	Partial autocorrelations function
Nonstationary	Autocorrelations do not die out They remain large or diminish approximately linearly	
Stationary	After the first few lags, autocorrelations die out (collapse toward 0 in some combination of exponential decay or damped oscillation)	
AR(p)	Autocorrelations die out	Partial autocorrelations cut off after the first p lags
MA(q)	Autocorrelations cut off after the first q lags	Partial autocorrelations die out
ARMA(p,q)	Autocorrelations die out after first $q - p$ lags	partial autocorrelations die out after first $p - q$ lags

Finally, figure 7.14 displays the autocorrelation and partial autocorrelation functions for real GDP growth. (We already saw the autocorrelation function in figure 7.7 when we determined that a first difference of log real GDP was required for stationarity.) The first two autocorrelations stand out, suggesting a first- or second-order MA process. Only the first partial autocorrelation lies outside the 95% confidence band, suggesting that a first-order AR process may be sufficient. So these sample statistics point us in the direction of either an ARMA(1,1) or an ARMA(1,2) process for real GDP growth.

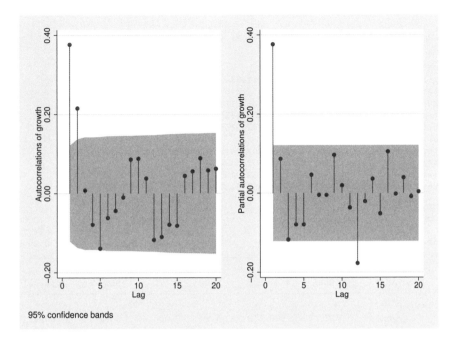

Figure 7.14. Autocorrelation and partial autocorrelation functions of real GDP growth, 1947:2 to 2012:1

7.4   Estimation

We have finally settled on two alternative ARMA models for GDP. To fit them, we will use Stata's `arima` command. The `arima` command is a very powerful and flexible command, but we do not need all of its features just yet. For our present purpose, we will use the simplest version. The syntax of this basic version is

`arima` *depvar*, `arima(#p,#d,#q)`

Let's start by re-creating our data on real GDP growth and fitting an ARMA(1,2) model.

```
. use ${ITSUS_DATA}/quarterly, clear
(Quarterly U.S. GDP, annualized, BEA, as of 4/27/2012)
. tsset
        time variable:  date, 1947:1 to 2012:1
                delta:  1 quarter
. generate lrgdp = log(gdp2005)
. label variable lrgdp "Log of real GDP"
. generate growth = lrgdp - L.lrgdp
(1 missing value generated)
```

```
. label variable growth "Growth rate of real GDP"

. arima growth, arima(1,0,2)

(setting optimization to BHHH)
Iteration 0:    log likelihood =  851.65498
Iteration 1:    log likelihood =  852.84577
Iteration 2:    log likelihood =  852.91291
Iteration 3:    log likelihood =  852.93266
Iteration 4:    log likelihood =  852.93887
(switching optimization to BFGS)
Iteration 5:    log likelihood =  852.94123
Iteration 6:    log likelihood =  852.94295
Iteration 7:    log likelihood =  852.94308
Iteration 8:    log likelihood =  852.94308

ARIMA regression

Sample:  1947:2 - 2012:1                    Number of obs      =        260
                                           Wald chi2(3)       =      59.20
Log likelihood =  852.9431                 Prob > chi2        =     0.0000
```

		OPG				
growth	Coef.	Std. Err.	z	P>|z|	[95% Conf.	Interval]
growth						
_cons	.007777	.0009873	7.88	0.000	.0058419	.0097121
ARMA						
ar						
L1.	.2557184	.2377975	1.08	0.282	-.2103561	.721793
ma						
L1.	.0899396	.2377276	0.38	0.705	-.375998	.5558771
L2.	.1702909	.0818278	2.08	0.037	.0099113	.3306704
/sigma	.0090964	.0002922	31.13	0.000	.0085238	.0096691

```
Note: The test of the variance against zero is one sided, and the two-sided
      confidence interval is truncated at zero.
```

At first glance, this output is a little daunting, so let's work through it methodically. The command itself

```
. arima growth, arima(1,0,2)
```

is straightforward. The option `arima(1,0,2)` tells Stata to fit an ARMA(1,2) model for real growth; the 0 in the middle indicates no differencing.[15] Stata fits the model by maximizing the log of the likelihood function, and the next section of the output traces Stata's choice of optimization method and progress, iteration by iteration. Unless Stata has difficulty maximizing the likelihood function, you do not need to pay attention to this section of the output. In fact, you can suppress it completely (the display, not the calculations) by specifying the `nolog` option.[16]

The results begin after the line that reads `ARIMA regression`. The summary statistics at the top include the sample coverage (the second quarter of 1947 through the first quarter of 2012); the number of observations (260); a Wald test of all the coefficients (except the constant) against the null hypothesis that all of them are 0 (the Wald statistic in this example is 59.20); the p-value of the Wald statistic (0 to 4 decimal places); and the log of the likelihood function (852.9431). Unless I am working on unfamiliar data or I am sorting out some problem, I generally check only the sample period and the number of observations to make sure that I have not inadvertently messed up my data. My focus is on the next segment of output—the table of coefficient estimates and associated statistics.

This table looks more complicated than the usual Stata regression table. The first difference is in the table headings. The column that is usually headed `Std. Err.` is now labeled `OPG Std. Err.` This heading is a reminder that Stata is using the outer product of the gradient to derive the estimates of the standard errors of the estimated coefficients. As with most things in Stata, you can change this method, but in reality, you are unlikely to. Let's move on.

There are three panels in this table of coefficients. This format is used by Stata to display estimates of multiple-equation models. Now we are really confused. We are estimating a single equation,

$$y_t = \mu + \phi_1 y_{t-1} + \epsilon_t - \theta_1 \epsilon_{t-1} - \theta_2 \epsilon_{t-2}$$

but Stata thinks we have two equations, one labeled `growth`, which has a constant term and no regressors, and a second labeled `ARMA`, which has one `ar` term and two `ma` terms. Finally, there is a `sigma` parameter of some sort.

---

15. Instead of fitting an ARMA model for real growth, we could have fit the log of real GDP directly and let Stata do the differencing command by typing `arima lrgdp, arima(1,1,2)`. The results would be identical.
16. But be careful. Convergence problems are common with ARIMA models, especially models with a large number of parameters. If your model runs into these problems, read the Stata manual's description of the many maximization options. For online help, type `help maximize` or see the `maximize` entry in the *Stata Base Reference Manual*.

Stata is not trying to confuse us. It actually rewrote our single-equation model as two equations and for very good reasons. It turns out that a wide range of time-series models—including our ARIMA models—can be recast as state-space models. The state-space representation provides many advantages, especially in estimation, and much modern time-series analysis leverages this representation. Not to worry though. The state-space representation is easily converted back to our intuitive ARIMA form.[17]

A state-space model describes the relationship between one or more observable time series and a vector of unobservable state variables that characterize the "state of the world". The model is composed of two equations—a transition or state equation (stochastic difference equation) that describes the evolution of the state variables and a measurement or observation equation that describes the relationship between the observable variables and the state variables.

In the state-space version of an ARIMA model, the measurement equation is

$$y_t = \mu + \eta_t$$

that is, the observable time series is equal to a constant, μ, plus an unobservable state variable, η_t. The transition equation is

$$\phi(L)\eta_t = \theta(L)\epsilon_t$$

where ϵ_t is the same white-noise error as in the original ARIMA model. It looks as if the difference between the ARIMA model we have been discussing and the state-space model is the location of the dynamics of the model. In an ARIMA model, both the observable y_t and the unobservable ϵ_t follow difference equations defined by the lag polynomials $\phi(L)$ and $\theta(L)$, respectively. In the state-space version, though, all the dynamics are in the transition equation. In fact, in this case, the transition equation is identical to the original ARIMA model except that both η_t and ϵ_t are unobservable.

As it turns out, these differences are apparent, not real. Consider, for example, our ARIMA(1,0,2) model for real GDP. Write the measurement equation as

$$y_t = \mu^* + \eta_t$$

For reasons that will be clear in a moment, we have rewritten the constant as μ^*. The transition equation in this case is

$$\eta_t = \phi_1 \eta_{t-1} + \epsilon_t - \theta_1 \epsilon_{t-1} - \theta_2 \epsilon_{t-2}$$

If we substitute this equation for η_t into the measurement equation, we get

$$
\begin{aligned}
y_t &= \mu^* + \phi_1 \eta_{t-1} + \epsilon_t - \theta_1 \epsilon_{t-1} - \theta_2 \epsilon_{t-2} \\
&= \mu^* + \phi_1(y_{t-1} - \mu^*) + \epsilon_t - \theta_1 \epsilon_{t-1} - \theta_2 \epsilon_{t-2} \\
&= (1 - \phi_1)\mu^* + \phi_1 y_{t-1} + \epsilon_t - \theta_1 \epsilon_{t-1} - \theta_2 \epsilon_{t-2}
\end{aligned}
$$

---

17. Stata provides an **sspace** command that allows you to formulate and estimate a host of state-space models directly. See [TS] **sspace** for details.

But this is just our original ARMA(1,2) model with μ replaced by $(1 - \phi_1)\mu^*$. In other words, we can freely choose to write a univariate time-series model in either ARIMA or state-space form. The choice is a matter of convenience.[18]

One item worth mentioning is the interpretation of the constant term, $\mu*$, in the measurement equation. In an ARIMA model, the constant, μ, is not the mean of y_t. Instead the mean of the observable variable is the product of μ and the inverse of $\phi(L)$.

$$\phi(L)y_t = \mu + \theta(L)\epsilon_t$$
$$\Rightarrow \quad y_t = \phi^{-1}(L)\mu + \phi^{-1}(L)\theta(L)\epsilon_t$$
$$\Rightarrow \quad E(y_t) = \phi^{-1}(L)\mu$$

But in the state-space form, μ^* is the mean of y_t because we can write

$$\phi(L)(y_t - \mu^*) = \theta(L)\epsilon_t$$

This feature makes it easy to compare the mean of real GDP growth implied by the ARIMA model with the estimates we calculated at the beginning of this chapter. Because our model is based on quarterly data, we just annualize the constant term. We prefer to use the stored values of estimated coefficients, but it's not clear how to refer to the constant term. Fortunately, Stata provides a `coeflegend` option to help us out.

```
. arima, coeflegend
ARIMA regression
Sample:  1947:2 - 2012:1                    Number of obs   =       260
                                            Wald chi2(3)    =     59.20
Log likelihood =  852.9431                  Prob > chi2     =    0.0000
```

growth	Coef.	Legend
growth		
_cons	.007777	_b[growth:_cons]
ARMA		
ar		
L1.	.2557184	_b[ARMA:L.ar]
ma		
L1.	.0899396	_b[ARMA:L.ma]
L2.	.1702909	_b[ARMA:L2.ma]
/sigma	.0090964	_b[sigma:_cons]

```
Note: The test of the variance against zero is one sided, and the two-sided
      confidence interval is truncated at zero.
. display 400 * _b[growth:_cons]
3.1107862
```

The average growth rate is 3.1%, similar to our previous estimates.

---

18. So why bother to add an equation and write the model in state-space form? First, as we mentioned above, the state-space model can represent a variety of time-series models, not just the ARIMA model. Second, there are advantages to estimating a time-series model in the state-space form.

One last item to mention is the final panel of the regression table, the part labeled
/sigma. This section reports the estimate of σ_ϵ, the standard error of the white noise
disturbance.

Now that we know how to read the output of the arima command, let's look at the
results.

The only parameter that is highly significant is the constant, μ,[19] which is approximately 0.008. The final coefficient of $\theta(L)$, $\hat{\theta}_2$, is significant at the 5% level. Note that
$\hat{\sigma}_\epsilon$ is roughly 0.009, indicating that the variability of the white-noise disturbances is
large relative to the mean of the process.

For comparison, let's estimate our other candidate specification, the ARMA(1,1)
model.

```
. arima growth, arima(1,0,1) nolog
ARIMA regression
Sample:  1947:2 - 2012:1                      Number of obs   =        260
                                              Wald chi2(2)    =      65.99
Log likelihood =  851.1159                    Prob > chi2     =     0.0000
```

growth	Coef.	OPG Std. Err.	z	P>|z|	[95% Conf. Interval]	
growth						
_cons	.0077788	.0009976	7.80	0.000	.0058236	.0097341
ARMA						
ar L1.	.4904126	.1362777	3.60	0.000	.2233133	.7575119
ma L1.	-.1309908	.1498699	-0.87	0.382	-.4247305	.1627489
/sigma	.0091613	.0002951	31.05	0.000	.0085829	.0097397

---

19. For convenience, we have dropped the asterisk, and we will denote the constant of both the ARIMA
    and the state-space representations as μ. The interpretation should be clear from the context.

The overall fit of the model is about the same as before,[20] but now $\widehat{\mu}$ and $\widehat{\phi}_1$ are significant, while the MA term is insignificant. The AR coefficient, $\widehat{\phi}_1$, is twice as large in this specification as it is in the ARMA(1,2) model, but the coefficients are not significantly different from one another (look at the 95% confidence intervals).

These results are not very encouraging. The data are not pinning down the coefficients very precisely, which implies that the data are not providing a very precise estimate of the dynamic response of GDP growth to economic shocks.

Let's make this last statement more precise. Recall from the previous chapter that the fundamental representation of a time series is a linear stochastic process:

$$y_t = \psi(L)\epsilon_t$$

One way to see how similar or different our two ARIMA models of GDP growth are is to compare the linear stochastic processes implied by the coefficient estimates. In order to make that comparison, we have to calculate the coefficients of $\psi(L)$. Here is one way to carry out that calculation.

Multiply both sides of the linear process above by the AR lag polynomial, $\phi(L)$:

$$\phi(L)y_t = \phi(L)\psi(L)\epsilon_t = \theta(L)\epsilon_t$$

So we can solve for the coefficients of $\psi(L)$ by matching the coefficients of $\phi(L)\psi(L)$ and $\theta(L)$. Let $\psi_1(L)$ be the lag polynomial implied by the estimate of the ARIMA(1,0,2) model and $\psi_2(L)$ be the lag polynomial implied by the estimate of the ARIMA(1,0,1) model. We will use Stata and a little arithmetic to calculate the first four coefficients of these infinite-order lag polynomials.

---

20. You do not have to take my word for it. In addition to the log likelihood and other summary statistics produced by the `arima` command, you can type `estat ic` after the `arima` commands to display the Akaike's information criterion and Bayesian information criterion. This is the same `estat` command we described in chapter 5. In this example, the information criteria confirm the approximate equivalence of the fit.

Model	Degrees of freedom	Akaike's information criterion	Bayesian information criterion
ARIMA(1,0,2)	5	-1695.886	-1678.083
ARIMA(1,0,1)	4	-1694.232	-1679.989

```
. arima growth, arima(1,0,2)
  (output omitted)
. local psi11 = _b[ARMA:L.ar] - _b[ARMA:L.ma]
. local psi12 = (_b[ARMA:L.ar])*(`psi11') - _b[ARMA:L2.ma]
. local psi13 = (_b[ARMA:L.ar])*(`psi12')
. local psi14 = (_b[ARMA:L.ar])*(`psi13')
.arima growth, arima(1,0,1)
  (output omitted)
. local psi21 = _b[ARMA:L.ar] - _b[ARMA:L.ma]
. local psi22 = (_b[ARMA:L.ar])*(`psi21')
. local psi23 = (_b[ARMA:L.ar])*(`psi22')
. local psi24 = (_b[ARMA:L.ar])*(`psi23')
. display as text _n "Psi weights:" _n _n "Model" _skip(7) "1" _skip(9) "2" _n
> _dup(26) _char(208) _n
> as text "psi1" _skip(6) in yellow %6.3f = `psi11' _skip(4) %6.3f `psi21' _n
> as text "psi2" _skip(6) in yellow %6.3f = `psi12' _skip(4) %6.3f `psi22' _n
> as text "psi3" _skip(6) in yellow %6.3f = `psi13' _skip(4) %6.3f `psi23' _n
> as text "psi4" _skip(6) in yellow %6.3f = `psi14' _skip(4) %6.3f `psi24' _n
Psi weights:

Model       1          2
--------------------------
psi1      0.166      0.621
psi2     -0.128      0.305
psi3     -0.033      0.149
psi4     -0.008      0.073
```

To display these coefficients in an easy-to-read table, we stored the calculations in local macros, then used the formatting features of Stata's `display` command to display the table. There are better ways to do this, but for now, let's just focus on the results. The ψ weights implied by the estimates of these two specifications clearly are very different. The ARIMA(1,0,2) estimates imply that a shock to GDP growth persists modestly for one quarter, then is reversed in the succeeding quarters. In contrast, the ψ weights implied by the ARIMA(1,0,1) model paint a very different picture. In this specification, 62% of an economic shock persists into the succeeding quarter, followed by 31% of the original shock, and so on. There is no reversal.

The symptoms we are observing—so-so fit of candidate models, point estimates that embody divergent views of the underlying linear stochastic process—suggest our models may be exhibiting parameter redundancy. To see this problem in its purest form, imagine that a linear stochastic process is consistent with the ARMA(p,q) model

$$\phi(L)y_t = \theta(L)\epsilon_t$$

This model is unchanged if we multiply both sides by the redundant common factor $(1 - \lambda L)$:

$$(1 - \lambda L)\phi(L)y_t = (1 - \lambda L)\theta(L)\epsilon_t$$

This ARMA($p+1$,$q+1$) model will be difficult to estimate, and the estimated parameters will be unstable because an infinite number of values of λ will yield the same results.

In practice, "almost-common" factors—AR and MA parameters that are close but not identical—are almost as problematic, and they can be easy to overlook. The Box–Jenkins principle of parsimony is intended, in part, to guard against parameter redundancy. Let's apply this principle here and prune parameters first from the MA component of the model, then from the AR component.

If we eliminate the MA parameters, we are left with an AR(1) model. In this case, we can use the `ar(#[,#[, ...]])` option to replace the `arima(p,d,q)` option of the `arima` command. With the `ar()` option, we specify the particular parameters we want to include rather than the order of the AR component. For example, `ar(1 4)` specifies the model

$$y_t = \mu + \phi_1 y_{t-1} + \phi_4 y_{t-4} + \epsilon_t$$

In other words, the `ar()` option allows us to impose zero restrictions on some of the AR coefficients. Combining the `ar()` and `ma()` options enables us to specify an ARMA model with zero restrictions.

```
. arima growth, ar(1) nolog
ARIMA regression
Sample:  1947:2 - 2012:1                     Number of obs    =        260
                                             Wald chi2(1)     =      58.51
Log likelihood =  850.5428                   Prob > chi2      =     0.0000
```

growth	Coef.	OPG Std. Err.	z	P>|z|	[95% Conf. Interval]
growth					
_cons	.0077877	.0009309	8.37	0.000	.0059632 .0096121
ARMA					
ar					
L1.	.3762591	.0491894	7.65	0.000	.2798497 .4726685
/sigma	.0091822	.0002939	31.24	0.000	.0086062 .0097583

The model fit is comparable with that of the ARMA(1,0,2) and ARMA(1,0,1) models: all the coefficients are statistically significant, and the estimated value of ϕ_1 lies between the estimates of the more complex ARMA models.

Now let's fit a pure MA model to these data.

```
. arima growth, ma(1 2) nolog
ARIMA regression
Sample:  1947:2 - 2012:1                    Number of obs    =       260
                                            Wald chi2(2)     =     49.63
Log likelihood =  852.2665                  Prob > chi2      =    0.0000
```

growth	Coef.	OPG Std. Err.	z	P>|z|	[95% Conf. Interval]	
growth						
_cons	.007786	.0008921	8.73	0.000	.0060374	.0095345
ARMA						
ma						
L1.	.331545	.0483635	6.86	0.000	.2367543	.4263356
L2.	.2188904	.0496608	4.41	0.000	.1215571	.3162237
/sigma	.0091205	.0002901	31.44	0.000	.0085519	.0096891

Again the fit is reasonable, and the coefficients are statistically significant.

How do we choose between these two specifications? Maybe we do not have to. If we calculate the first four ψ weights of the linear stochastic processes implied by these models, we get

```
Psi weights:
Model       1        2
-------------------------
psi1      0.376    0.332
psi2      0.142    0.219
psi3      0.053    0.000
psi4      0.020    0.000
```

The column labeled 1 reports the ψ weights from the AR(1) model. The column labeled 2 reports the weights from the MA(2) model. The weights are much more similar for these more parsimoniously parameterized models than they were for the ARMA models we estimated originally. The main difference is the exponential decay of the ψ weights in the AR model compared with the discrete drop to 0 in the MA model for all weights after the second lag. Formally, the impact of an economic shock lasts forever in the AR specification, but as a practical matter, the impact can safely be treated as zero after a few quarters. In other words, these two specifications imply roughly the same view of the stochastic process generating GDP growth.

7.5 Looking for trouble: Model diagnostic checking

The next step in the Box–Jenkins approach to time-series analysis is model diagnostic checking. As we saw in the previous section, multiple model specifications may fit a time series about equally well. Model diagnostic checking provides additional information that can help discriminate between candidate models.

Two approaches are commonly used to check the adequacy of ARIMA models: overfitting and tests of the residuals. We discuss each approach in turn.

7.5.1 Overfitting

The Box–Jenkins approach emphasizes choosing the most parsimonious specification, that is, the model with the fewest parameters. As a consequence, there may be a tendency to fit models that are too simple to account for the dynamics of the time series under consideration. One way to guard against this tendency is overfitting, that is, adding parameters to the model. For example, if you have fit an AR(2) model for a time series, you might try fitting a higher-order AR model—AR(3) or AR(4), for instance.

In overfitting, it is important to have a specific notion of how the model may be oversimplified rather than to add parameters blindly. And as we saw in the previous section, you should not add parameters to the AR and MA components of the model simultaneously because of the risk of parameter redundancy.

As an example of overfitting, we will fit AR(2) and MA(3) models to our data.

```
. arima growth, ar(1 2) nolog

ARIMA regression

Sample:  1947:2 - 2012:1                    Number of obs     =        260
                                            Wald chi2(2)      =      58.81
Log likelihood =  851.5176                  Prob > chi2       =     0.0000
```

growth	Coef.	OPG Std. Err.	z	P>|z|	[95% Conf. Interval]	
growth						
_cons	.0077732	.0010251	7.58	0.000	.005764	.0097824
ARMA						
ar						
L1.	.3437587	.0507303	6.78	0.000	.244329	.4431883
L2.	.0863024	.0529982	1.63	0.103	−.0175721	.1901768
/sigma	.0091479	.0002943	31.09	0.000	.0085711	.0097246

```
Note: The test of the variance against zero is one sided, and the two-sided
      confidence interval is truncated at zero.
```

```
. arima growth, ma(1 2 3) nolog
ARIMA regression
Sample:  1947:2 - 2012:1                    Number of obs    =        260
                                            Wald chi2(3)     =      50.28
Log likelihood =  853.1356                  Prob > chi2      =     0.0000
```

| | | OPG | | | |
growth	Coef.	Std. Err.	z	P>|z|	[95% Conf. Interval]
growth						
_cons	.0077765	.000978	7.95	0.000	.0058597	.0096933
ARMA						
ma						
L1.	.3456633	.0518988	6.66	0.000	.2439435	.447383
L2.	.2637154	.0515619	5.11	0.000	.1626559	.364775
L3.	.0862314	.0591503	1.46	0.145	-.0297011	.2021639
/sigma	.0090893	.0002927	31.05	0.000	.0085156	.0096631

Note: The test of the variance against zero is one sided, and the two-sided
      confidence interval is truncated at zero.

In both cases, the additional parameter is insignificant, suggesting the more parsimonious specifications are adequate.

7.5.2 Tests of the residuals

The foundation of time-series models is the linear stochastic process, the weighted sum of present and past values of a white-noise series, ϵ_t. In fact, the process of specifying and fitting a time-series model frequently is described as reducing the observable series to white noise. In other words, the measure of a well-specified and accurately fitted time-series model is evidence that the residuals, the $\widehat{\epsilon}_t$, are white noise.

Earlier in this chapter, we showed how to use the autocorrelation function to help identify the structure of a time-series model, and we displayed an example of the autocorrelation function of a white-noise series in figure 7.8. Thus one obvious check of model adequacy would seem to be a review of the autocorrelation function of the residuals. We could compare the autocorrelations of the residuals with the variance of the autocorrelations of ϵ_t, which is approximately N^{-1}, where N is the sample size.[21] Unfortunately, when the model is misspecified, the true variance of the residuals may be much smaller than N^{-1}, particularly for low lags; hence, this approach is likely to miss many model deficiencies.[22]

It turns out that an effective test can be constructed by using the first k autocorrelations of the residuals taken together rather than individually. The Q statistic

---

21. Box and Jenkins (1976, 290).
22. Examining the autocorrelation function of the residuals may provide evidence about the type of misspecification or parameter misfit; however, this type of evidence can be difficult to disentangle. See Box and Jenkins (1976, 298–299) for an example.

is a weighted sum of the squared autocorrelations that is approximately distributed as $\chi^2(k)$, where k is the number of autocorrelations (equivalently, number of lags) included in the sum. This portmanteau test of the autocorrelations overcomes the weakness of examining the individual autocorrelations for significance, at least if a large enough number of lags are included.

The `wntestq` command in Stata calculates the Q test. The syntax of this command is

wntestq *varname* $\begin{bmatrix} if \end{bmatrix}$ $\begin{bmatrix} in \end{bmatrix}$ $\begin{bmatrix} , \underline{\text{lags}}(\#) \end{bmatrix}$

If you do not specify the number of lags, Stata sets the number of lags to the lesser of 40 or approximately half the sample.

Let's examine the residuals of the MA(2) model we fit in the previous section. We start by rerunning the `arima` command for this model, then use Stata's `predict` command to calculate the residuals.

```
. arima growth, ma(1 2) nolog
(output omitted)
. predict resid, residual
(1 missing value generated)
. wntestq resid
Portmanteau test for white noise
─────────────────────────────────────────
 Portmanteau (Q) statistic =    41.3526
 Prob > chi2(40)           =     0.4114
```

The degrees of freedom in this χ^2 test should be adjusted for the ARMA parameters estimated by the `arima` command. The `wntestq` command does not provide an option for adjusting the degrees of freedom in the test, but the correct p-value can be displayed by typing

```
. display chi2tail(38,41.3526)
.32641631
```

The Q test shows no evidence that the residuals deviate from white noise.

While this book analyzes time series by using time-domain methods, one frequency-domain tool provides a useful diagnostic test. Residuals from poorly specified models can exhibit nonrandom periodicity, another deviation from the assumption of white-noise disturbances.[23] This problem can arise with data that are inadequately seasonally adjusted or in models that do not include appropriate seasonal differencing. Nonrandom periodicity generally will not be detectable by the Q test. An examination of the cumulative periodogram of the residuals can highlight this type of deficiency. For a

---

23. Recall that ideal white noise is composed of equal contributions at each frequency (or period, because the frequency is the inverse of the period). Well-behaved residuals will exhibit random deviations from ideal white noise. Any evidence of a systematic, rather than a random, deviation constitutes evidence of some deficiency in the fitted model.

white-noise series, the cumulative periodogram should be a straight line between the cumulative periodogram value of 0 at a frequency of 0 and 1 at a frequency of 1/2. Marked deviations from this straight line indicate nonrandom periodicity. Note that the period equals the inverse of the frequency. So nonrandom periodicity in seasonal data is likely to produce deviations from the straight line around frequencies corresponding to important periods in the data. For example, in monthly data, one might look for deviations at periods of 12 (frequency = 0.08), 24 (frequency = 0.04), and so on.

The `wntestb` command in Stata calculates and displays the cumulative periodogram along with standard errors under the null hypothesis that the underlying series is white noise. The syntax of this command is

`wntestb` *varname* [*if*] [*in*] [, table level(*#*)]

The `level` option allows you to specify the confidence level of the bands around the cumulative periodogram. The default confidence level is 95%. The `table` option displays the results in a table rather than a graph. Other graph-related options have been omitted.

Again we will use the residuals from the MA(2) specification to illustrate the cumulative periodogram test.

```
. wntestb resid
```

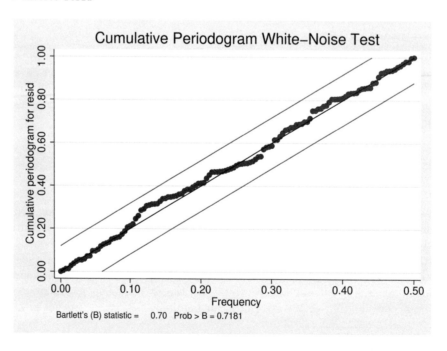

Figure 7.15. Cumulative periodogram of the residuals from the MA(2) specification

In this example, the cumulative periodogram remains close to the 45-degree line and well within the confidence bands; thus the residuals do not exhibit any signs of nonrandom periodicity.

7.6 Forecasting with ARIMA models

We finally are ready to forecast real GDP growth. It took four steps to get to this stage. First, we looked at the data, fitting a trend line and a Holt–Winters smooth to the data to help inform our intuition. Second, we examined the autocorrelation and partial autocorrelation functions of the log of real GDP to determine the degree of differencing and the likely order of the ARMA process. This identification step led us to consider both an ARMA(1,2) and an ARMA(1,1) specification for real growth (that is, the first difference of the log of real GDP). In the third step, we estimated both candidate specifications. The estimates led us to question our original specifications and consider in their place two simpler specifications: an AR(1) and an MA(2) model. These specifications produced reasonable estimates. Finally, in the fourth step, we tested the adequacy of these two models by overfitting them and by calculating two tests of the residuals of the models. These tests failed to detect any model weaknesses.

We use the Stata `predict` command to calculate forecasts from the most recently fitted ARIMA model. This is the same `predict` command used with `regress`, `logistic`, and a host of other Stata estimation commands. Following an `arima` estimation, `predict` allows the following syntax

predict [*type*] *newvar* [*if*] [*in*] [, *statistic options*]

There are several flavors of forecast from ARIMA models, and you have to make sure you ask `predict` for the flavor you really want. To be clear, recall that Stata recasts the ARMA model in its state-space form, with a measurement equation (also called the structural equation)

$$y_t = \mu + \eta_t$$

and a transition (or ARMA) equation

$$\phi(L)\eta_t = \theta(L)\epsilon_t$$

By default, `predict` calculates the one-step-ahead predictions from the fitted model (the `xb` option specifies this statistic explicitly), including the estimated contributions of the ARMA equation. The `structural` option sets all the disturbances, both η's and ϵ's, to 0. In other words, this option tells Stata to calculate forecasts exclusively from the structural equation, ignoring the ARMA equation. At first blush, this option seems pointless because it sets every observation in the forecast equal to μ, the mean of the process, but this option comes in handy in other settings.

The `dynamic(`*time_constant*`)` option specifies a point in the forecast to switch from one-step-ahead to dynamic forecasts. To illustrate, consider the AR(1) model

$$y_t = \phi_1 y_{t-1} + \epsilon_t$$

The one-step-ahead forecast is

$$\widehat{y}_t = \widehat{\phi}_1 y_{t-1} + \widehat{\epsilon}_t$$

that is, the forecast is based on the observed value of y_{t-1}. However, if we are projecting more than one period past the end of the sample, we have not observed y_{t-1} yet. In this situation, we calculate the dynamic forecast

$$\widehat{y}_t = \widehat{\phi}_1 \widehat{y}_{t-1} + \widehat{\epsilon}_t$$

that is, the tth period forecast is conditioned on the forecast in period $t - 1$. The use of actual values of the lagged dependent variable tends to limit the one-step-ahead prediction errors and can lead to a too-optimistic view of the longer-range accuracy of the model predictions. Examining within-sample dynamic forecasts is a good way to assess the horizon over which an ARIMA model is likely to produce useful forecasts.

The final option, t0(*time_constant*), specifies a starting point for the calculation of forecasts. No forecasts are calculated prior to the observation corresponding to the date indicated by the *time_constant*, and all η's and ϵ's prior to that date are set to 0. This assumption changes the initial conditions for the forecasts. As a result, these forecasts will differ from similar forecasts with a different beginning projection date. The t0() option can be combined with either the xb or the dynamic() options.

How different are these forecasts? Let's take a look at four different forecasts from our AR(1) specification. We will annualize the growth rate and the forecasts to make the results easier to read. We begin by reestimating the model to make sure the forecasts are based on the correct model. We will start the dynamic forecasts (options dynamic() and t0()) in the first quarter of 2008.

```
. arima growth, ar(1) nolog

  (output omitted )

. predict arxb
(option xb assumed; predicted values)

. predict arst, structural
(option xb assumed; predicted values)

. predict ardy, dynamic(tq(2008q1))
(option xb assumed; predicted values)

. predict art0, t0(tq(2008q1))
(option xb assumed; predicted values)
(244 missing values generated)

. foreach x in xb st dy t0 {
  2.          replace ar`x´ = 400 * ar`x´
  3. }
(261 real changes made)
(261 real changes made)
(261 real changes made)
(17 real changes made)

. generate agrowth = 400*growth
(1 missing value generated)

. format agrowth ar* %6.1f
```

```
. list date agrowth ar* in f/5
```

	date	agrowth	arxb	arst	ardy	art0
1.	1947:1	.	3.1	3.1	3.1	.
2.	1947:2	-0.6	3.1	3.1	3.1	.
3.	1947:3	-0.3	1.7	3.1	1.7	.
4.	1947:4	6.0	1.8	3.1	1.8	.
5.	1948:1	6.3	4.2	3.1	4.2	.

We selected names for the forecasts that make it easy to refer to them as a group and to use the **foreach** command to apply the same transformation to all of them at once.[24]

Because we calculated the growth rate as the first difference of the log of real GDP, the first observation of the actual growth rate is missing. Even if there had been a nonmissing value of **agrowth** in the first observation, there is no lagged value prior to the beginning of the dataset. In this situation, **predict** fills in the forecast with 3.1%, the mean of the process implied by the model estimates.[25] As it happens, this estimate is very far from the initial actual value of -0.69%. As a consequence, the one-step-ahead forecasts in **arxb** are not very close to the actuals in the first few observations.

The structural forecasts, stored in **arst**, all are equal to the mean of the process, as we noted above. At this point in the data, **ardy** is identical to **arxb**. The dynamic forecasts do not begin until the first quarter of 2008. Similarly, **art0** is filled with missing values until 2008 Q1.

Let's look at the values at the end of the data. Figure 7.16 displays the values from 2007 Q1 through 2012 Q1. The vertical line is drawn at 2008 Q1, the point that the dynamic forecasts (**ardy**) begin and the starting point for **art0**, the version that sets all prior η's and ϵ's to 0. The Stata log below displays the values for each observation.

---

24. The **foreach** command provides a flexible way to execute one or more commands repeatedly, replacing variables or other items in the command each time from a list you specify. In this example, we annualized all four of our predictions with one command repeated (by Stata) four times. Type **help foreach** for details.

25. The "missing" observations prior to the start of the data pose nontrivial challenges to both the estimation of time-series models and the forecasting of time series. See Hamilton (1994) for a detailed, modern treatment of these issues.

```
. list date agrowth ar* if date>tq(2006q4)
```

	date	agrowth	arxb	arst	ardy	art0
241.	2007:1	0.5	3.0	3.1	3.0	.
242.	2007:2	3.6	2.1	3.1	2.1	.
243.	2007:3	2.9	3.3	3.1	3.3	.
244.	2007:4	1.7	3.0	3.1	3.0	.
245.	2008:1	-1.8	2.6	3.1	2.6	3.1
246.	2008:2	1.3	1.3	3.1	2.9	1.3
247.	2008:3	-3.7	2.4	3.1	3.0	2.4
248.	2008:4	-9.3	0.5	3.1	3.1	0.5
249.	2009:1	-6.9	-1.6	3.1	3.1	-1.6
250.	2009:2	-0.7	-0.7	3.1	3.1	-0.7
251.	2009:3	1.7	1.7	3.1	3.1	1.7
252.	2009:4	3.7	2.6	3.1	3.1	2.6
253.	2010:1	3.9	3.3	3.1	3.1	3.3
254.	2010:2	3.7	3.4	3.1	3.1	3.4
255.	2010:3	2.5	3.3	3.1	3.1	3.3
256.	2010:4	2.3	2.9	3.1	3.1	2.9
257.	2011:1	0.4	2.8	3.1	3.1	2.8
258.	2011:2	1.3	2.1	3.1	3.1	2.1
259.	2011:3	1.8	2.4	3.1	3.1	2.4
260.	2011:4	2.9	2.6	3.1	3.1	2.6
261.	2012:1	2.2	3.0	3.1	3.1	3.0

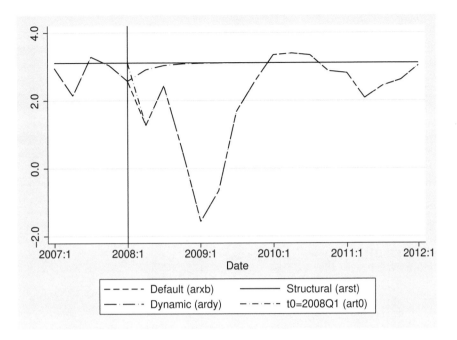

Figure 7.16. Comparing four different uses of the `predict` command

The dynamic forecasts diverge from the default one-step-ahead forecasts starting in the second quarter of 2008, and they converge to the mean of the process by 2008 Q4. Note that the one-step-ahead forecasts that start in the first quarter of 2008 (`art0`) do not match the one-step-ahead forecasts in `arxb` at first, but converge to these values in one quarter. This pattern is determined by the order of the model; the AR(1) model looks back only one quarter, and there are no dynamics in the disturbances to generate persistent deviations.

For comparison, here are the forecasts from the MA(2) specification.

```
. list date agrowth ma* in f/5
```

	date	agrowth	maxb	mast	mady	mat0
1.	1947:1	.	3.1	3.1	3.1	.
2.	1947:2	-0.6	3.1	3.1	3.1	.
3.	1947:3	-0.3	1.8	3.1	1.8	.
4.	1947:4	6.0	1.7	3.1	1.7	.
5.	1948:1	6.3	4.1	3.1	4.1	.

```
. list date agrowth ma* if date>tq(2006q4)
```

	date	agrowth	maxb	mast	mady	mat0
241.	2007:1	0.5	2.8	3.1	2.8	.
242.	2007:2	3.6	2.6	3.1	2.6	.
243.	2007:3	2.9	3.0	3.1	3.0	.
244.	2007:4	1.7	3.3	3.1	3.3	.
245.	2008:1	-1.8	2.6	3.1	2.6	3.1
246.	2008:2	1.3	1.3	3.1	2.8	1.4
247.	2008:3	-3.7	2.2	3.1	3.1	2.2
248.	2008:4	-9.3	1.2	3.1	3.1	1.2
249.	2009:1	-6.9	-1.6	3.1	3.1	-1.6
250.	2009:2	-0.7	-0.9	3.1	3.1	-0.9
251.	2009:3	1.7	2.0	3.1	3.1	2.0
252.	2009:4	3.7	3.0	3.1	3.1	3.0
253.	2010:1	3.9	3.3	3.1	3.1	3.3
254.	2010:2	3.7	3.5	3.1	3.1	3.5
255.	2010:3	2.5	3.3	3.1	3.1	3.3
256.	2010:4	2.3	2.9	3.1	3.1	2.9
257.	2011:1	0.4	2.7	3.1	3.1	2.7
258.	2011:2	1.3	2.2	3.1	3.1	2.2
259.	2011:3	1.8	2.3	3.1	3.1	2.3
260.	2011:4	2.9	2.8	3.1	3.1	2.8
261.	2012:1	2.2	3.1	3.1	3.1	3.1

Much the same as before, with the rate of convergence of `mady` to the mean and of `mat0` to `maxb` determined by the second-order MA specification.

Before we look at these forecasts more closely, let's mention the other statistics `predict` offers. For models estimated, like ours, in terms of first (or higher-order) differences, the `y` statistic will produce forecasts of the level of the time series. For instance, if we estimated our AR(1) with the command `arima lrgdp, arima(1,1,0)` or the command `arima D.lrgdp, ar(1)`, then `predict lrgdphat, y` would generate forecasts of the log of real GDP. This option can be very convenient.

As you would expect, the `residuals` statistic produces residuals in place of forecasts. As an added convenience, the `yresiduals` statistic calculates the residuals associated with the forecasts that would be produced by the `y` statistic. `mse` generates the MSE of the predictions, while `stdp` calculates the standard error of the linear prediction `xb`, ignoring the variation contributed by the ARMA equation.

7.7   Comparing forecasts

We have introduced several different ways to forecast the growth of real GDP. At the beginning of this chapter, we compared forecasts of the log of real GDP based on a trend line regression and on a Holt–Winters smooth of real GDP, and these forecasts can be rewritten as forecasts of real growth. In the previous section of this chapter, we used the `predict` command to calculate forecasts based on our AR(1) and MA(2) models of real growth.

Which of these forecast methods should we use? If we have a correctly specified ARIMA model, the `predict` command will produce minimum mean squared error forecasts and optimal forecasts in the special case when the ϵ's are normally distributed.[26] But we have two ARIMA models. They cannot both be "best". And while forecast accuracy—measured by MSE—is always important, it is not the only feature we might care about.

Let's take a look at some results for each of these approaches to forecasting. For this comparison, we reestimated each of these models—the trend line, the Holt–Winters smoother, and the AR(1) and MA(2) models—by using data through the first quarter of 2002. Then we calculated forecasts of real growth for the remaining 10 years of the sample, that is, from 2002 Q2 through 2012 Q1. We have annualized the quarterly growth rates to make it easier to read the results. Also the residuals are defined as $\widehat{y} - y$ rather than the traditional $y - \widehat{y}$. This convention is a personal idiosyncrasy. I find it more intuitive when a positive residual corresponds to a model overprediction and vice versa.[27]

```
. /*
>         Linear growth trend model
> */
. regress lrgdp date if date<tq(2002q2)
  (output omitted)
. predict trend
(option xb assumed; fitted values)
. predict tres, residual
. label variable trend "Trend"
. label variable tres "Trend"
. generate tg = 400*_b[date]
. generate tgres = tg - growth
. label variable tg "Trend growth"
. label variable tgres "Trend growth"
. /*
>         Holt-Winters model
> */
. tssmooth hwinters hw = lrgdp if date<tq(2002q2), forecast(40)
  (output omitted)
. generate hwg = 400*D.hw
(1 missing value generated)
. generate hwres = hw - lrgdp
. generate hwgres = hwg - growth
(1 missing value generated)
```

---

26. See Box and Jenkins (1976) and Hamilton (1994) for details.

27. For conciseness, I have omitted some "housekeeping" steps from this log—labeling variables and the like.

```
. /*
>          AR(1) model
> */
. arima growth if date<tq(2002q2), ar(1) nolog
  (output omitted)
. predict ardg, dynamic(tq(2002q2))
(option xb assumed; predicted values)
. predict ardgres, residual
(1 missing value generated)
. replace ardgres = -ardgres
(260 real changes made)
. /*
>          MA(2) model
> */
. arima growth if date<tq(2002q2), ma(1 2) nolog
  (output omitted)
. predict madg, dynamic(tq(2002q2))
(option xb assumed; predicted values)
. predict madgres, residual
(1 missing value generated)
. replace madgres = -madgres
(260 real changes made)
```

First, let's look at the actual and predicted growth rates, both within sample (pre-2002:2) and out of sample (2002:2–2012:1).

```
. generate byte period = date>tq(2002q1)
. label define period 0 "In sample" 1 "Out of sample"
. label values period period
. table period, contents(mean growth mean ardg mean madg mean hwg mean tg)
> format(%6.1f) row
```

period	mean(growth)	mean(ardg)	mean(madg)	mean(hwg)	mean(tg)
In sample	3.4	3.4	3.4	3.4	3.4
Out of sample	1.6	3.4	3.4	3.4	3.4
Total	3.1	3.4	3.4	3.4	3.4

The annual rate of real growth was 1.8% points lower on average in the out-of-sample period than between 1947:1 and 2002:1. None of the forecasts predict that shift. The Holt–Winters forecast projects the last estimated trend, which by chance is equal to the 1947:1–2002:1 average. The other three forecasts settle down to the unconditional mean forecast of 3.4% (prior to 2002:2) in at most a few quarters.

The conventional measure of both model fit and forecast accuracy is the dispersion of the residuals, measured here by the standard deviation of the residuals.

```
. table period, contents(sd ardgres sd madgres sd hwgres sd tgres)
> format(%6.1f) row
```

period	sd(ardgres)	sd(madgres)	sd(hwgres)	sd(tgres)
In sample	3.9	3.8	4.8	4.1
Out of sample	2.4	2.4	2.9	2.9
Total	3.7	3.7	4.6	4.0

This table displays several interesting patterns. First, the residual variance is strikingly higher within sample than out of sample, an unusual result but a coincidence. Second, as expected, the ARIMA models have lower residual variances both within sample and out of sample. However, the differences are not that large. For example, the standard deviation of the forecast error from the trend line regression (which translates to a constant real growth rate) is 5% higher than the standard deviation of the forecast error from the AR(1) model within sample and only 21% higher out of sample. Finally, the relative performance of the Holt–Winters forecast improves markedly in the out-of-sample period.

This closeness can be attributed to the character of the forecasts. The trend line forecast is a constant growth rate of 3.368%. The Holt–Winters forecast is a constant 3.440% beginning in the second quarter of 2002. The AR(1) and MA(2) forecasts settle down to a constant 3.388% and 3.387%, respectively, by the fourth quarter of 2002.

The graph below displays the actual growth rates and the forecasts for two years: 2001q2–2002q1 (within sample) and 2002q2–2003q1 (out of sample). The vertical line separates the within-sample and out-of-sample periods. The horizontal line separates positive from negative real growth.

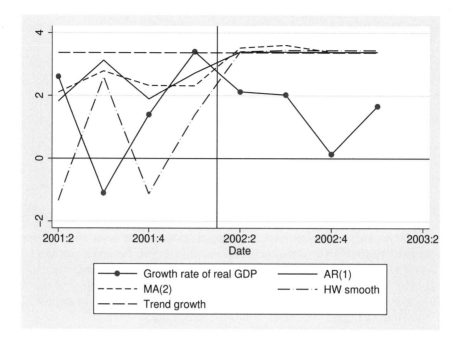

Figure 7.17. Within-sample and out-of-sample forecasts of real GDP, 2001:2–2003:1

Even within sample, none of the models anticipated the drop in real growth in the third quarter of 2001. The decline in the Holt–Winters forecast in the fourth quarter is an "echo effect", that is, a response to the 2001 Q3 drop in growth. The AR(1) and MA(2) models exhibit a more modest response to the Q4 model miss. And as we noted, all the models converge rapidly to their long-run average.

7.8   Points to remember

- The flexibility of time-series models makes it difficult to definitively choose a "best" specification. Be prepared to consider several alternatives and think carefully about the implications of the model specification, not just the apparent goodness of fit.

- The Box–Jenkins approach emphasizes parsimony as a counterbalance to the ability of alternative specifications to provide equivalent fits to a given time series.

- The Box–Jenkins approach is iterative. Models are identified (that is, the order of the model—the values of p, d, and q—is selected), estimated, and tested. If the tests signal deficiencies, the process is repeated until a candidate model is deemed acceptable. This iterative approach is more easily defended in an industrial setting, where independent samples can be drawn for each iteration. Nonetheless, the approach is widely used with nonexperimental data (for example, economic time series).

- Time-series models can produce multiple types of forecasts (one-step-ahead, dynamic, etc.), and Stata's `predict` command can produce each of these types. Be sure to request the type of forecast you need.

- Forecasts from stationary ARMA models settle down to a mean value fairly quickly. As a consequence, long-term forecasts from ARMA models do not differ greatly from the forecasts of simpler trend models.

7.9   What have we learned so far?

We have covered a lot of ground over the last five chapters, and now is a good time to put things in perspective. These chapters may have appeared to be a somewhat scattershot overview of loosely related topics, but, in fact, they comprise a unified summary of most of the essential ideas in time-series analysis, albeit with more of a practical than a theoretical slant. To begin, let's review the issues that we have covered so far.

- Chapter 3 introduced time series in an informal, intuitive way. Instead of plunging directly into a statistical characterization of time series, we framed the questions that a researcher must answer at the beginning of any data analysis. Then we covered some of the terms commonly used to describe the components of a time series and presented some Stata commands that filter out the noise that may obscure the patterns in a time series.

- Chapter 4 highlighted the wide variety of types of forecasts we meet in ordinary life and identified the range of forecasts that we can address with the statistical tools covered in this book. We reused some of the filters introduced in chapter 3 to calculate rough-and-ready forecasts and to identify some of the issues that confront the forecaster.

- Chapter 5 appeared to switch gears completely. Instead of discussing time series, we examined the "problem", encountered frequently in practice, of autocorrelated disturbances in linear regressions when the dependent variable is a time series. While we emphasized the commonsense reasons that residual autocorrelation occurs, this chapter presented formal statistical tests for residual autocorrelation and a collection of estimation strategies for regressions with autocorrelated disturbances.

- Chapter 6 looked like another change of direction. The statistical underpinnings of linear, time-series (ARMA) models were reviewed, and the notation for describing and manipulating these models was covered.

- Chapter 7 completed the topics begun in chapter 6 by identifying and fitting time-series models for real GDP growth and comparing the forecasts of competing models.

So what unifies these disparate topics?

- From the statistical vantage point, the last three chapters formalize the intuitions developed in the first two chapters. Consider the informal concepts of trend, cycle, and seasonal components of a time series. The Holt–Winters filter provides a method for estimating each of these components, is easy to apply, and works well in practice. ARIMA models provide a more formal statistical basis for estimating these components. In the model

$$\phi(L)y_t = \mu + \theta(L)\epsilon_t$$

  the mean (there is no trend) is $\phi^{-1}\mu$, and the cycles, if any, are determined by the roots of $\phi(z)$.[28]

- Implicit in these chapters is a warning. Time-series analysis provides powerful tools for revealing patterns and relationships in data, but the best statistical techniques can only bound, but not eliminate, the irreducible uncertainty we face when analyzing data. Many of our examples relied on simulated data to provide cleaner illustrations of techniques and commands but also to highlight how even well-crafted estimates can sometimes mislead us. There is no substitute for a thoughtful approach to time-series analysis informed by deep subject-matter knowledge and a willingness to apply rigorous tests to every estimate.

  A related implication is the continuing usefulness of older, less formal methods. ARIMA models are more sophisticated than the simpler filters and trend regressions we have used, and under the appropriate circumstances, these more-modern models have superior forecast accuracy. However, as we have seen, the advantages of ARIMA models over smoothers and trend lines are not always as great as we might expect. Moreover, the difficulty of identifying the appropriate ARIMA specification narrows the gap—a misspecified ARIMA model does not necessarily produce more accurate forecasts than a simpler tool.

- It turns out that chapter 5 on autocorrelated disturbances is more closely related to the two subsequent chapters on ARIMA models than it might appear at first. Chapter 5 focuses on (primarily first-order) autocorrelation in the error term. Chapters 6 and 7 concentrate on autocorrelation in the observable dependent variable, although the MA component captures residual autocorrelation of arbitrary degree. But in the state-space representation of the ARIMA model, the time-series properties are represented entirely by an ARIMA structure in the error term of the measurement equation.

---

28. And the seasonals? Did I neglect to mention seasonal components in ARIMA models? We can write

$$\phi(L)\phi^*(L^s)(1-L)^d(1-L^s)^{d_s}y_t = \mu + \theta(L)\theta^*(L^s)\epsilon_t$$

to specify a model with d ordinary differences and d_s seasonal differences plus seasonal AR $\{\phi^*(L^s)\}$ and seasonal MA $\{\theta^*(L^s)\}$ components where s is the seasonal span (4 for quarterly data, 12 for monthly data, and so on). Stata's `arima` command provides the `sarima(p,d,q)`, `mar(numlist,#_s)`, and `mma(numlist,#_s)` options to specify these components. Of course, simply slapping seasonal components on top of an ARIMA specification is not likely to generate meaningful results. A full discussion of the analysis of seasonality is beyond the scope of this book.

7.10   Looking ahead

We have now covered the core of time-series econometrics.[29] Three more topics are required to round out the essential features of modern time-series analysis.

- Real-world time series often exhibit periods of relative calm interrupted by periods of persistent turbulence. The stock market provides a familiar example. Sudden, unexpected world events—natural disasters, political upheavals, wars, disruptions to the supply or price of oil or other essential commodities—generate bursts of uncertainty in financial markets that may take some time to wear off. These patterns suggest a special, time-series type of heteroskedasticity called autoregressive conditional heteroskedasticity.

- So far, we have explored univariate time-series models. But many time series are interdependent, for example, output, inflation, and unemployment. Vector autoregressions models provide a powerful and elegant way of modeling these dependencies.

- Time series may bear a long-run relationship to each other—a long-run "anchor"—despite drifting further apart or closer for extended periods. Such time series are said to be cointegrated, and models of cointegration allow us to estimate short-run dynamics and long-run equilibrium relationships simultaneously.

The next three chapters take up these topics in turn.

Stata commands and features discussed

ac ([TS] **corrgram**): Graph autocorrelations with confidence intervals; section 7.3.1

arima ([TS] **arima**): Fit an ARIMA model; section7.4

corrgram ([TS] **corrgram**): Display autocorrelations, partial autocorrelations, and portmanteau (Q) statistics; section 7.3.2

display ([R] **display**): Display strings and values of scalar expressions; section 7.4

label ([D] **label**): Manipulate labels; section 7.1

local ([P] **macro**): Macro definition and manipulation; section 7.4

pac ([TS] **corrgram**): Graph partial autocorrelations with confidence intervals; section 7.3.2

predict ([R] **predict**): Obtain predictions, residuals, etc., after estimation; section 7.1

regress ([R] **regress**): Estimate linear regression; section 7.1

---

29. At least for linear, time-domain models.

tssmooth ([TS] **tssmooth**): Smooth and forecast univariate time-series data; section 7.1

wntestb ([TS] **wntestb**): Bartlett's periodogram-based test for white noise; section 7.5.2

wntestq ([TS] **wntestq**): Portmanteau (Q) test for white noise; section 7.5.2

8   Time-varying volatility: Autoregressive conditional heteroskedasticity and generalized autoregressive conditional heteroskedasticity models

Chapter map

8.1 Examples of time-varying volatility. An intuitive explanation of time-varying volatility with some examples.

8.2 ARCH: A model of time-varying volatility. Applying Engle's autoregressive conditional heteroskedasticity (ARCH) model to our gross domestic product (GDP) data.

8.3 Extensions to the ARCH model. Using generalized autoregressive conditional heteroskedasticity (GARCH) to limit the order of a model for monthly inflation rates. Asymmetric responses to "news". The impact of changes in the conditional variance on the observed variable. Nonnormal error distributions.

8.4 Points to remember. Conditional variance. The ARCH model and its many extensions.

If you read the previous two chapters carefully, you now know everything there is to know about time-series analysis. Well, almost everything. You learned in chapter 6 how time-series models are built up from a few fundamental building blocks, and you understand how complex dynamic behavior can be modeled by combining low-order autoregressive (AR) and moving-average (MA) components. Chapter 7 taught you to be a time-series "detective", weighing the "clues" provided by the autocorrelations and partial autocorrelations, then grilling your "suspects" by fitting candidate models, examining the fits, residuals, and forecasts to weed out the red herrings until the "culprit" is identified.

So what is left to learn? Well, what we have learned so far is the fundamental approach to time-series models. And econometricians always build upon fundamental models in three ways. First, they generalize a univariate model to handle multiple relationships simultaneously. (We will cover that in chapter 9.) Second, they extend the model to a broader class of phenomena. (That is in chapter 10.) Third, they find ways to handle violations of the classical assumptions. We will cover that topic now.

In the previous two chapters, we assumed the random contribution to our time series, ϵ_t, has constant variance, σ_ϵ^2. In practice, this assumption appears to be violated frequently, especially in time series of financial or economic data. Markets and economies seem subject to periods of "normal" volatility interrupted by periods of unusual turbulence that persist for some time. In other words, the volatility of a time series can itself follow a time-series process.

Nonconstant variance of the error term, or heteroskedasticity, is a familiar problem in cross-section econometrics with familiar solutions (for example, weighted regression when the source of the heteroskedasticity is known and measurable or heteroskedasticity-consistent estimators in more general situations). Time-varying volatility requires a different approach.

We set the stage in the next section by examining some data that exhibit time-varying volatility. The following section covers some of the first models introduced to handle these types of data. The final section reviews generalizations that followed these first models and discusses how to select among the extensive model choices available.

8.1   Examples of time-varying volatility

Financial markets, among other things, produce "price discovery", that is, asset prices that balance the willingness of relatively bullish investors to buy with the willingness of relatively bearish investors to sell. Of course, news arrives every day that shifts the balance of investor opinion about an asset's likely future performance. Some news—the unemployment and inflation rates, the balance of trade, and the like—affects investor views of the market as a whole, while other news—changes in management, updated sales figures, etc.—affects only a specific sector or company. Investors digest this information, update their willingness to buy or sell assets, and their actions change prices. The types of information that affect asset prices are many and varied, and it is impractical to model them all. As a result, their impact on prices is part of the random component of our models, and the typical magnitude of their impact helps determine the magnitude of the residual variance, σ_ϵ^2.

During uneventful periods, little news of import arrives, so investor opinions change relatively little. Moreover, long periods of relatively stable prices tend to narrow the range of investor opinions of asset values. In contrast, sudden and unexpected news can shift investor opinion significantly. More importantly, extraordinary events can sow confusion among investors. Investors may struggle for some time to settle on an updated opinion of value. In these circumstances, investors may exhibit heightened reactions to scraps of information that either increase concern or restore confidence. During these periods, the variance of the random component of prices is increased— at least, that is the way these events are reflected in our models, which, after all, are simplified representations of reality.

As an example of this phenomenon, let's revisit the TED spread that we used in chapter 3 to illustrate the use of some of Stata's smoothers. Recall that the TED spread is the difference in basis points—that is, hundredths of a percent—between the 3-month London interbank offered rate (a rate large banks charge each other) and the yield on the 3-month on-the-run Treasury bill. The TED spread measures the market's opinion of the creditworthiness of the banking sector. United States Treasury securities are regarded as having no risk of default; hence, the yields on Treasury securities provide a benchmark for all other interest rates. When investors become concerned about the strength of the banking sector, interbank interest rates increase relative to Treasury yields. In other words, the TED spread increases.

Figure 8.1 displays the TED spread from 2 January 1987 through 30 December 2011. In chapter 3, we used smoothers to track changes in the level of the TED spread, and we noted the sudden increase in the spread during the financial distress that began in 2007. Here we are more interested in changes in the volatility of the spread than in its level.

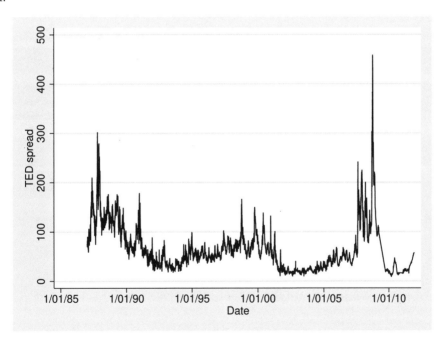

Figure 8.1. The TED spread, 1/2/1987–12/30/2012

You can think of the TED spread as the price the market places on bank risk, and, like many financial prices, the TED spread is reasonably approximated by a random walk.[1]

$$\text{TED spread}_t = \text{TED spread}_{t-1} + \epsilon_t$$

Again, as is common with financial prices, the one-day changes in the TED spread come from a fat-tailed, nonnormal distribution. Figure 8.2 displays a very-squashed box plot of the one-day changes in the TED spread.[2] Half the one-day changes lie within the shaded box in a range of ± 2 basis points. The mass of dots above and below the box and whiskers mark outside values, extreme one-day changes—one as large as 100 basis points—that reflect the fat tails of this distribution.

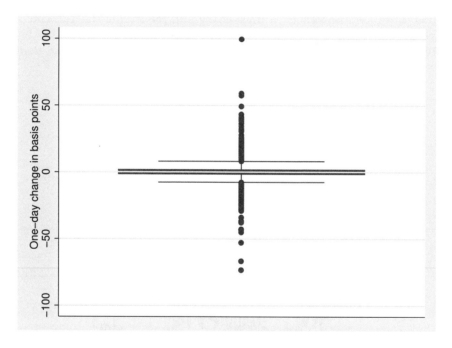

Figure 8.2. The distribution of one-day changes in the TED spread

In this chapter, we are not concerned with the fat tails per se. Rather we are interested in the stability of the dispersion of one-day changes; that is, we are interested

---

1. Actually, the TED spread appears to follow a random walk with drift—the mean of the one-day change in the spread is slightly positive (0.2 basis points) and significant. This drift term is unimportant for our purposes here.

2. Box plots provide convenient summaries of the distributions of random variables. The shaded box depicts the interquartile range; half the observations lie within the box. The horizontal line in the interior of the shaded box marks the median. The "antennae" extending above and below the box indicate the upper and lower adjacent values, which are robust measures of the dispersion in the tails. The dots that appear above and below the adjacent values are outside values, that is, unusually extreme values.

in the behavior of σ_ϵ^2 over time. Figure 8.3 displays the standard deviation of one-day changes in the TED spread by year from 1987 through 2011. Two episodes stand out. The spike in 1987 reflects the uncertainty of investors as they tried to gauge the exposure of financial institutions to the stock market crash in October 1987. Volatility receded in 1988 and continued to decline secularly through 2006. The spike in the volatility of the spread in 2007 and 2008 reflects investor fear and uncertainty as the housing crisis triggered failures and bailouts at large financial firms (Washington Mutual, Countrywide, Bear Stearns, Lehman Brothers, and AIG) and unprecedented government interventions (the TARP program, which injected capital into the banking sector, and the placing of Fannie Mae and Freddie Mac into conservatorship). Even without any formal statistical test, it appears unlikely that a model with constant σ_ϵ^2 will fit these data well.

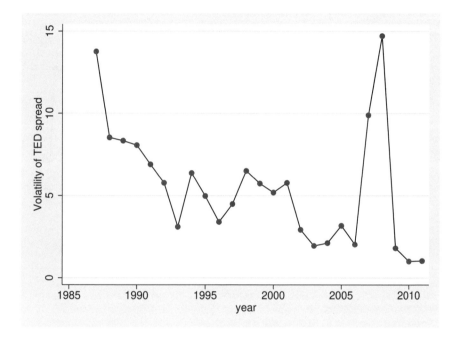

Figure 8.3. Variation in the volatility of the TED spread

While financial time series are particularly prone to time-varying volatility, other time series exhibit similar swings in variance. The real growth rate of GDP, which we analyzed in the previous chapter, provides a good example. Figure 8.4 displays the standard deviation of the quarterly real growth rate of the economy.[3] The graph has a choppy appearance because there are only four quarterly observations in each annual standard deviation. Nonetheless, three eras stand out. The early, post–World War II years (1947–1959) are particularly volatile—the 9.5% standard deviation in 1958 is the highest value in our sample. The variability of real growth drops sharply in 1961, then

---

3. Annualized percentage growth rate.

ratchets up steadily until the turbulent years from the late 1970s through 1982, when the Federal Reserve's so-called monetarist experiment wrung inflation from the economy at the cost of a severe recession. From 1983 through 2007, the volatility of real growth remains muted, an era that has been dubbed the "Great Moderation". This period of calm ends suddenly in 2008 and 2009 as the global financial crisis takes hold.

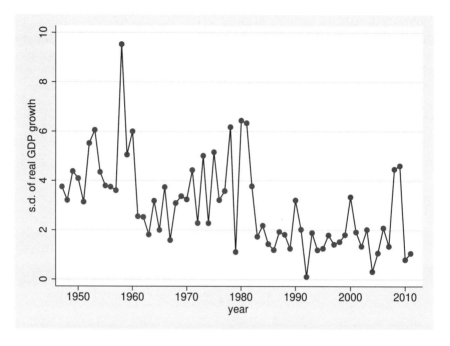

Figure 8.4. Variation in the volatility of real GDP growth

Figure 8.5 provides a clearer picture of those patterns by calculating the standard deviation of real growth over nonoverlapping five-year periods.[4]

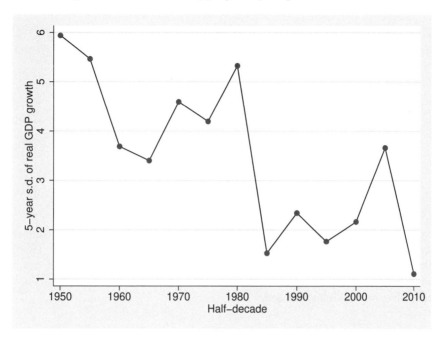

Figure 8.5. Five-year standard deviations of real GDP growth

8.2   ARCH: A model of time-varying volatility

How can we model the empirical observation that many time series exhibit episodes of higher and lower variability? This challenge is a little different, a little less concrete, than what we have been dealing with so far. Up to now, we have built models of observable series, our y_t. To select the appropriate dynamic characterization of y_t, we have relied on statistical summaries—autocorrelations, partial autocorrelations, and the like—but we have been modeling something we can observe, something we can compare directly with model fits and forecasts. Now, however, we are trying to model the evolution over time of an unobservable parameter, σ_y^2. We are at one more degree of separation from observable phenomena.

Engle (1982) introduced the ARCH model to capture time-varying volatility within the structure of standard time-series models. Engle's innovation sparked an explosion of variations of the ARCH approach, and we will look at some of the leading examples a little later. For now, let's walk through Engle's original insight.

---

4. The magnitudes of the estimated five-year and annual standard deviations are not directly comparable, because they cover varying amounts of time.

The autoregressive integrated moving-average (ARIMA) model we have discussed so far is

$$\phi(L)y_t = \mu + \theta(L)\epsilon_t$$

where ϵ_t is white noise with variance equal to σ_ϵ^2. The conditional mean of y_t depends on prior values of y_t (through the coefficients in $\phi[L]$). However, the unconditional mean of y_t is constant:

$$Ey_t = \frac{\mu}{(1 - \phi_1 - \cdots - \phi_p)} \equiv \mu^*$$

Likewise, the unconditional variance of y_t is also constant. In the previous chapter, we showed that this variance—the initial autocovariance, γ_0—is given by

$$\gamma_0 = \phi_1\gamma_1 + \cdots + \phi_p\gamma p - \gamma_{y\epsilon}(0) - \theta_1\gamma\gamma_{y\epsilon}(1) - \cdots - -\theta_q\gamma\gamma_{y\epsilon}(q)$$

where $\gamma_{y\epsilon}(j)$ is the cross-covariance between y_t and ϵ_{t-j}. The expression is messy, but it is composed entirely of constant parameters.

So where does this leave us? How can we incorporate a concept of time-varying volatility? Engle's solution centered on the conditional variance of ϵ_t. Even though the unconditional variance of ϵ_t is the constant σ_ϵ^2, Engle suggested that the conditional variance of ϵ_t could follow an AR process of its own. Specifically, he proposed modeling the square of ϵ_t as an AR(m) process

$$\epsilon_t^2 = \zeta + \alpha_1\epsilon_{t-1}^2 + \cdots + \alpha_m\epsilon_{t-m}^2 + \nu_t$$

or

$$A(L)\epsilon_t^2 = \zeta + \nu_t$$

where ν_t is a new white-noise process. The conditional expectation of ϵ_t^2 is

$$E(\epsilon_t^2|\epsilon_{t-1}^2, \epsilon_{t-2}^2, \ldots) = \zeta + \alpha_1\epsilon_{t-1}^2 + \cdots + \alpha_m\epsilon_{t-m}^2$$

A white-noise series, ϵ_t, satisfying these conditions is described as an ARCH(m) process.

ARCH models require a few more restrictions than ordinary autoregressive moving-average (ARMA) models. ϵ_t^2 can never be negative. This restriction can be guaranteed if $\nu_t \geq -\zeta, \forall t$ and $\alpha_j \geq 0, \forall j$. As with all ARMA models, the roots of

$$A(z) = 0$$

must all lie outside the unit circle. With all $\alpha_j \geq 0$, this condition is equivalent to the condition

$$\alpha_1 + \cdots + \alpha_m < 1$$

When all of these conditions hold, it can be shown that the unconditional variance of ϵ_t is

$$\sigma_\epsilon^2 = E(\epsilon_t^2) = \frac{\zeta}{1 - \alpha_1 - \cdots - \alpha_m}$$

The ARCH model extends the standard ARMA model to incorporate time-varying volatility but at the cost of the additional restrictions required to guarantee that $\epsilon_t^2 \geq 0$. In practice, it can be difficult to impose the additional stationarity and nonnegativity restrictions. Nonetheless, ARCH and its many descendants have proved useful in refining our ability to model many time series.

Let's see how Stata handles ARCH models. We begin with a model for the growth rate of real GDP in the United States. Recall that in chapter 7, we fit an AR(1) model for the growth of real GDP in the United States.[5]

```
. arima growth, ar(1) nolog

ARIMA regression

Sample:  1947:2 - 2012:1                         Number of obs    =       260
                                                 Wald chi2(1)     =     58.51
Log likelihood =  -707.238                       Prob > chi2      =    0.0000
```

growth	Coef.	OPG Std. Err.	z	P>|z|	[95% Conf. Interval]
growth					
_cons	3.115064	.3723441	8.37	0.000	2.385283 3.844845
ARMA					
ar L1.	.3762591	.0491894	7.65	0.000	.2798497 .4726685
/sigma	3.672893	.1175688	31.24	0.000	3.442463 3.903324

As we showed in the previous chapter, this specification provides a reasonable fit to the data. However, our casual inspection above of the variability of real growth suggested the presence of an ARCH component. How can we test this hypothesis?

Stata provides a test for ARCH effects that can be used after a **regress** command. We will demonstrate it by using the AR(1) specification for real growth:

```
. regress L(0/1).growth
```

Source	SS	df	MS			
				Number of obs	=	259
				F(1, 257)	=	42.59
Model	579.230003	1	579.230003	Prob > F	=	0.0000
Residual	3494.96519	257	13.5990863	R-squared	=	0.1422
				Adj R-squared	=	0.1388
Total	4074.19519	258	15.7914542	Root MSE	=	3.6877

growth	Coef.	Std. Err.	t	P>|t|	[95% Conf. Interval]
growth					
L1.	.3764496	.0576814	6.53	0.000	.2628611 .4900381
_cons	1.961878	.2916873	6.73	0.000	1.387477 2.53628

---

5. You may have noticed the difference in the estimate of the constant compared with the estimate reported in chapter 7. We have annualized the quarterly growth rates here to make the units easier to read.

The `regress` estimates are very similar to the `arima` estimates. The root mean squared error in the regression is close to the estimated σ_ϵ calculated by `arima`; the ϕ_1 coefficient estimates are similar; and the estimated constants are similar after the translation to the ARIMA specification is made:

```
. display = _b[_cons]/(1 - _b[L1.growth])
3.1463024
```

The `estat` command, which we introduced in chapter 1, provides a Lagrange multiplier test for ARCH components.

```
. estat archlm, lags(1 2 3 4)
LM test for autoregressive conditional heteroskedasticity (ARCH)
```

lags(p)	chi2	df	Prob > chi2
1	4.789	1	0.0286
2	9.913	2	0.0070
3	9.957	3	0.0189
4	13.541	4	0.0089

H0: no ARCH effects   *vs.* H1: ARCH(p) disturbance

The `lags(#)` option specifies the order of the ARCH AR component under the alternative hypothesis. In this example, we have tested the null of no ARCH effects against four separate alternatives: ARCH models of orders 1 through 4. The null is rejected against all four alternatives. Determining lag lengths is always a tricky business, so we are not going to focus yet on the appropriate order of the ARCH process. However, we do interpret these results as indicating the presence of ARCH disturbances of some order.

> **Note:** As we mentioned above, the `estat archlm` command is available only after the `regress` command. This may appear to be a limitation when our original time-series model involves MA parameters, but, as a practical matter, we can still test for ARCH effects. Remember that a model with a finite MA component can be written as an infinite-order AR model. While it's not possible to use `regress` to estimate a regression with an infinite number of terms, for a stationary model, the coefficients on the AR terms die out relatively quickly. So you can estimate a finite-order regression with just-enough AR terms in place of a mixed ARMA model. Determining how many AR terms is "just enough" is a bit of an art, but basically you want to make sure the residual is indistinguishable from white noise.
>
> Another approach, proposed by Bollerslev (1988), is to apply the Box–Jenkins methodology to the $\widehat{\epsilon}_t^2$, the squared residuals from the ARIMA model, for the observable variable. This approach may indicate that a pure AR model for ϵ_t^2 may not provide the best characterization of the variance process. More on that in the next section.

Stata provides the `arch` command to fit ARCH (and many related) models. The syntax is similar to the `arima` syntax:

`arch` *depvar* [`indepvars`] [*if*] [*in*] [*weight*] [`,` `arch(`*numlist*`)` *other_options*]

The `arch` command has many, many options to take account of the many different models of time-varying volatility that have been proposed since Engle's (1982) article. We will cover some of the more useful ones shortly. The `arch` command also allows the `arima()`, `ar()`, and `ma()` options we used in the previous chapter with the `arima` command.

Because the results of the `estat` command suggest the presence of first-order ARCH effects, let's begin by fitting that model.

```
. arch growth, ar(1) arch(1) nolog
ARCH family regression -- AR disturbances
Sample: 1947:2 - 2012:1                    Number of obs   =        260
Distribution: Gaussian                     Wald chi2(1)    =      31.01
Log likelihood = -701.1802                 Prob > chi2     =     0.0000
```

growth	Coef.	OPG Std. Err.	z	P>|z|	[95% Conf. Interval]	
growth						
_cons	3.097168	.3469909	8.93	0.000	2.417078	3.777257
ARMA						
ar						
L1.	.3993747	.0717198	5.57	0.000	.2588064	.539943
ARCH						
arch						
L1.	.2608906	.0750396	3.48	0.001	.1138158	.4079655
_cons	10.23213	.8356407	12.24	0.000	8.594301	11.86995

Compare these estimates with the `arima` estimates above. The coefficient estimates are very similar in the two models. In the ARCH model, the standard error of the constant in the structural equation is slightly smaller than before, while the standard error of ϕ_1 is somewhat larger than before. In the ARCH equation, both ζ and α_1 are significant, and $0 < \alpha_1 < 1$ as is required for stationarity of ϵ_t^2.

We have assumed the errors follow an ARCH(1) process so far, but, of course, they may
follow a higher-order process. In the previous chapter, we reviewed the autocorrelations
and partial autocorrelations of real GDP growth for clues to the order of the ARMA
model. For ARCH models, we have no similar set of statistics to guide us, so we will
resort to the old-fashioned approach of estimating multiple specifications.[6]

```
. arch growth, ar(1) arch(1 2) nolog
ARCH family regression -- AR disturbances
Sample: 1947:2 - 2012:1                       Number of obs    =        260
Distribution: Gaussian                        Wald chi2(1)     =      38.91
Log likelihood = -691.3014                    Prob > chi2      =     0.0000
```

growth	Coef.	OPG Std. Err.	z	P>|z|	[95% Conf. Interval]	
growth						
_cons	3.467476	.3396975	10.21	0.000	2.801681	4.13327
ARMA						
ar						
L1.	.4383648	.0702733	6.24	0.000	.3006316	.576098
ARCH						
arch						
L1.	.3308078	.0839041	3.94	0.000	.1663588	.4952568
L2.	.3870112	.1109962	3.49	0.000	.1694627	.6045598
_cons	5.508447	.852282	6.46	0.000	3.838005	7.178889

---

6. We realize we are committing a cardinal statistical sin. The appropriate approach is to fit an
   encompassing model first, then trim insignificant coefficients. By reversing this order, we complicate
   the interpretation of significance tests, probably beyond hope of untangling. Our aim here is to
   illustrate the **arch** command. Please do not take our example as encouragement of bad research
   habits.

```
. arch growth, ar(1) arch(1 2 3) nolog

ARCH family regression -- AR disturbances
Sample: 1947:2 - 2012:1                      Number of obs   =       260
Distribution: Gaussian                       Wald chi2(1)    =     37.02
Log likelihood = -691.1207                   Prob > chi2     =    0.0000
```

growth	Coef.	OPG Std. Err.	z	P>|z|	[95% Conf. Interval]	
growth						
_cons	3.48945	.3438003	10.15	0.000	2.815614	4.163286
ARMA						
ar						
L1.	.4309567	.0708268	6.08	0.000	.2921388	.5697746
ARCH						
arch						
L1.	.3216774	.0831899	3.87	0.000	.1586282	.4847265
L2.	.4019769	.1126535	3.57	0.000	.1811802	.6227737
L3.	.0245882	.0494727	0.50	0.619	-.0723766	.1215529
_cons	5.216906	.8617869	6.05	0.000	3.527834	6.905977

It appears that a second-order ARCH process may fit the data best. Note that the ARCH(2) model produces slightly different estimates of μ and ϕ_1.

Recall that we fit two time-series models to real GDP growth, an AR(1) model and an MA(2) model. For comparison, here are estimates of the original MA(2) model and the same model with first-, second-, and third-order ARCH effects added.

	MA(2) Coef.	$P > \|z\|$	ARCH(1) Coef.	$P > \|z\|$	ARCH(2) Coef.	$P > \|z\|$	ARCH(3) Coef.	$P > \|z\|$
μ	3.11	0.00	3.18	0.00	3.53	0.00	3.52	0.00
θ_1	0.332	0.00	0.362	0.00	0.376	0.00	0.377	0.00
θ_2	0.219	0.00	0.251	0.00	0.259	0.00	0.261	0.00
ζ			9.70	0.00	5.61	0.00	5.67	0.00
α_1			0.291	0.00	0.338	0.00	0.341	0.00
α_2					0.355	0.00	0.351	0.00
α_3							−0.006	0.90

As before, an ARCH(2) process seems to fit best. Note that the estimate of α_3 in the ARCH(3) model is negative in this case, although it is not significant.

After fitting an ARCH model, Stata's `predict` command can generate an estimate of the conditional variances implied by the model. As before, either dynamic or static predictions are available. Let's use that command to get a look at the conditional variances of the AR(1) model with ARCH(2) disturbances.

```
. arch growth, ar(1) arch(1 2)
  (output omitted)
. predict variance, variance
. format date %dCCYY
. tsline variance
```

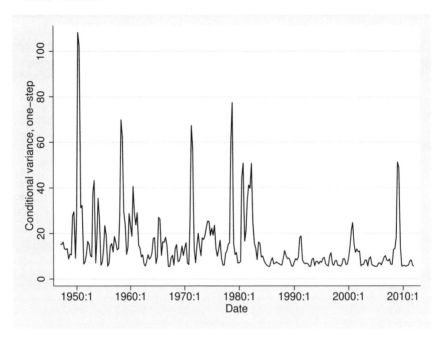

Figure 8.6. The conditional variance of ϵ_t

The conditional variance shown in figure 8.6 exhibits repeated spikes in the years prior to the mid-1980s followed by two decades of relative quiet until the spike in 2008–2009. This pattern echoes the phenomena we observed in figure 8.4 and figure 8.5 above.

It turns out that an alternative formulation of the ARCH model provides a convenient basis for developing extensions to this framework. Suppose that

$$\epsilon_t = \sqrt{h_t} \times \omega_t$$

where ω_t is a white-noise process but now with the additional restriction $E\omega^2 = 1$. In this representation, the variable $\sqrt{h_t}$ scales the white-noise innovations ω_t. Hence, the process for h_t determines the time path of ϵ^2. If that process is

$$h_t = \zeta + \alpha_1 \epsilon_{t-1}^2 + \cdots + \alpha_m \epsilon_{t-1}^2$$

then, as before,

$$E(\epsilon_t^2 | \epsilon_{t-1}^2, \epsilon_{t-2}^2, \ldots) = \zeta + \alpha_1 \epsilon_{t-1}^2 + \cdots + \alpha_m \epsilon_{t-m}^2$$

that is, ϵ_t follows an ARCH(m) process. With a little bit of algebra, we can derive a compact representation for ν_t, the innovation in the ARCH(m) process.[7,8]

$$\nu_t = h_t \times (\omega_t^2 - 1)$$

With this preparation, we can proceed now to some extensions of Engle's original proposal.

8.3   Extensions to the ARCH model

The phenomenon of time-varying volatility had been recognized for decades, but until the publication of Engle's original proposal, researchers lacked a satisfactory technique for modeling this form of heteroskedasticity. Engle's innovation generated a deluge of empirical articles applying this new technique. However, the initial ARCH model addresses only the fundamental observation that volatility often varies over time. Other aspects of time-varying volatility—other long-recognized empirical regularities in financial and economic time series—are not captured by the "vanilla" ARCH model. As a consequence, econometricians developed numerous, often competing extensions to the ARCH model.

Some of the more important empirical regularities are the following:

- The uneven, intermittent, or random arrival of "news" (impactful new information) is commonly cited as cause of time-varying volatility. However, many series appear to react asymmetrically to positive and negative news.

- The conditional mean of y_t, the observable time series, often appears to depend on the current level of volatility. For example, the level of stock prices often declines during periods of unusually high uncertainty.

---

7. Squaring both sides of the equation for ϵ_t produces

$$\epsilon_t^2 = h_t \times \omega_t^2$$

but

$$\epsilon_t^2 = \zeta + \alpha_1 \epsilon_{t-1}^2 + \cdots + \alpha_m \epsilon_{t-m}^2 + \nu_t = h_t + \nu_t$$

hence,

$$h_t + \nu_t = h_t \times \omega_t^2$$

or

$$\nu_t = h_t \times (\omega_t^2 - 1)$$

8. This formulation, although convenient, is more restrictive than the first version of the ARCH model we presented. In particular, somewhat unintuitive restrictions on the α_i are required to guarantee a solution for σ_ν^2, the unconditional variance of the innovation in the AR process for ϵ_t^2. See Hamilton (1994) for details.

- Asset prices, such as stock prices, tend to have distributions with "fat tails". In other words, extreme events (unusually large price increases or decreases) occur more frequently than in a normal distribution. The ARCH model introduces some leptokurtosis[9] in the conditional variance but less than is typically observed in financial time series.

The options to Stata's `arch` command provide these and many other extensions to Engle's original formulation. Covering them all would take us too far afield, but we highlight the most important below.[10]

8.3.1   GARCH: Limiting the order of the model

In many situations, it may appear that a high-order ARCH process is required to provide an adequate description of the time-varying volatility. This possibility does not present any theoretical obstacles. All stationary time series can be represented by an AR process.[11] But high-order processes are unwieldy and difficult to estimate precisely. A better approach is to combine low-order AR and MA components to model the dynamic behavior of the series.

A similar situation can arise when modeling time-varying volatility. The data may suggest that many ARCH terms are required. The GARCH introduced by Bollerslev (1986) can provide a good fit with a lower-order parameterization.

Using the notation we developed at the end of section 8.2, we characterized the time-series properties of the random error in the ARIMA model as an AR process in h_t, the variable that scales the standard normal innovation. We assumed that the ARCH process had some finite order, call it m, but more generally, we can assume that the ARCH process depends on an infinite-order autoregression in h_t,

$$h_t = \zeta + \pi(L)\epsilon_t^2$$

where

$$\pi(L) = \sum_{j=1}^{\infty} \pi_j^* L^j$$

As we discussed in chapter 6, this infinite-order lag polynomial can be written more compactly as the ratio of two finite-order lag polynomials:

$$\pi(L) = \frac{\alpha(L)}{1 - \delta(L)} = \frac{\alpha_1 L + \alpha_2 L^2 + \cdots + \alpha_m L^m}{1 - \delta_1 L - \delta_2 L^2 - \cdots - \delta_r L^r}$$

---

9. The typical definition of kurtosis (or, more precisely, excess kurtosis) is $\mu_4/\sigma^4 - 3$, where μ_4 is the fourth moment about the mean and σ is the standard deviation. A distribution, like the normal, with zero excess kurtosis is called mesokurtic. A distribution with positive excess kurtosis (and, hence, "fat tails") is called leptokurtic.

10. See Bollerslev, Chou, and Kroner (1992) for a useful survey of ARCH and its extensions. They also discuss many of the applications of these extensions to financial time series.

11. An infinite-order autoregression may be required to provide an exact representation.

Multiplying both sides of the equation for h_t by $1 - \delta(L)$ yields

$$\{1 - \delta(L)\} \, h_t = \kappa + \alpha(L)\epsilon_t^2$$

where

$$\kappa = (1 - \delta_1 - \cdots - \delta_r)\zeta$$

This parameterization is the GARCH(r, m) model.[12] As in the basic ARCH model, the requirement that $h_t > 0$ implies that all the coefficients of both lag polynomials are positive.

We will use the time series of inflation in U.S. consumer prices from January 1960 through March 2012 to illustrate how Stata fits a GARCH model. We calculate the inflation rate as the change in the log of the consumer price index (CPI). We will model the monthly rate of inflation, but let's calculate the annual rate of inflation (overlapping 12-month log differences) as a simple, MA smoother to highlight the systematic component of inflation.

```
. use ${ITSUS_DATA}/monthly, clear
(Monthly data for ITSUS)
. tsset
        time variable:  date, Jan 13 to Mar 12
                delta:  1 month
. generate lcpi = log(cpi)
. generate anncpi = 100*S12.lcpi
(12 missing values generated)
. label variable anncpi "Annual CPI inflation"
. generate inflation = 1200*D.lcpi
(1 missing value generated)
. label variable inflation "Monthly CPI inflation"
```

12. It's tempting to interpret the $\delta(L)$ as the AR terms for the variance and the $\alpha(L)$ as the MA terms, but that turns out not to be the case. In fact, if ϵ_t follows a GARCH(r, m) process, then ϵ_t^2 follows an ARMA(p, r) process where p is the larger of r and m. See Hamilton (1994) for details.

```
. twoway (line inflation date) (connected anncpi date, msymbol(+) msize(small)),
> yline(0)
```

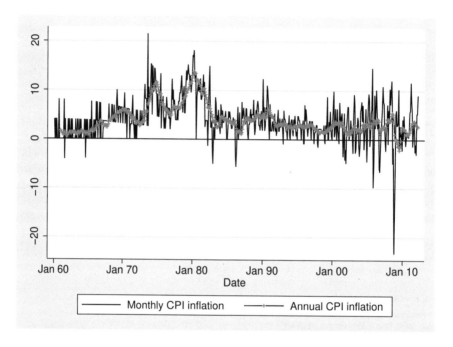

Figure 8.7. Monthly and annual consumer price inflation in the United States

The oscillations in monthly inflation provide an impressionistic sense of the volatility in the series. The level of inflation is highest in the 1970s and early 1980s, during the era of stagflation, but it appears that the volatility may be highest near the end of the series, around the time that deflation occurs.

To save time, we jump straight to the ARIMA and ARCH models for monthly inflation without demonstrating how we landed on this parameterization. (You should work through the Box–Jenkins steps of identifying, fitting, and testing this model to see if you agree with our choices.)

```
. arch inflation, arima(2,0,1) arch(1 2 3) nolog

ARCH family regression -- ARMA disturbances

Sample: Feb 60 - Mar 12                    Number of obs   =      626
Distribution: Gaussian                     Wald chi2(3)    = 65444.26
Log likelihood = -1612.692         ,       Prob > chi2     =   0.0000
```

	Coef.	OPG Std. Err.	z	P>|z|	[95% Conf. Interval]
inflation					
_cons	3.522882	.6551663	5.38	0.000	2.23878 4.806985
ARMA					
ar					
L1.	1.264746	.0568226	22.26	0.000	1.153376 1.376116
L2.	-.2781428	.0536782	-5.18	0.000	-.3833501 -.1729355
ma					
L1.	-.9146949	.0226525	-40.38	0.000	-.959093 -.8702968
ARCH					
arch					
L1.	.3222899	.0724758	4.45	0.000	.1802399 .46434
L2.	.1454913	.0430307	3.38	0.001	.0611527 .2298299
L3.	.0218814	.0388328	0.56	0.573	-.0542295 .0979922
_cons	6.168393	.5877223	10.50	0.000	5.016479 7.320308

```
. predict avar, variance
. label variable avar "ARCH conditional variance"
```

We have fit an ARMA(2,1) model with an ARCH(3) process for the conditional variance. (A fourth-order ARCH term is insignificant.) We have also used Stata's `predict` command to save the ARCH estimate of the conditional variance, in the variable `avar`, for later reference.

As a practical matter, we tend to avoid processes higher than second order, so we are a little suspicious of this ARCH(3) model.[13] It turns out that a GARCH(1,1) model fits these data equally well.

---

13. As we emphasized in chapter 3, modeling and specification choices for time series should be founded as much as possible on an understanding of the process that generates the data. We are modeling monthly inflation estimates based on the CPI published by the United States Bureau of Labor Statistics. With monthly financial data, we often find a difference between end-of-quarter observations and intraquarter observations. End-of-quarter observations often reflect a more rigorous review process because these observations are used in audited quarterly financial statements. While the CPI is published by a government agency rather than a private corporation, there may be some feature of the data-collection process that spreads random increments to information across a quarter. Alternatively, there may be a seasonal adjustment process that affects the dynamic behavior of volatility. Our purpose here is to illustrate Stata's implementation of GARCH, so we will not explore these possibilities further. However, you should review these possibilities carefully in any real research project.

```
. arch inflation, arima(2,0,1) arch(1) garch(1) nolog

ARCH family regression -- ARMA disturbances

Sample: Feb 60 - Mar 12                       Number of obs    =        626
Distribution: Gaussian                        Wald chi2(3)     =   51266.27
Log likelihood = -1603.898                    Prob > chi2      =     0.0000
```

inflation	Coef.	OPG Std. Err.	z	P>|z|	[95% Conf. Interval]	
inflation						
_cons	3.441357	.6031881	5.71	0.000	2.25913	4.623584
ARMA						
ar						
L1.	1.225343	.0491122	24.95	0.000	1.129085	1.321601
L2.	-.240727	.0462019	-5.21	0.000	-.331281	-.1501731
ma						
L1.	-.9060132	.0242611	-37.34	0.000	-.9535641	-.8584623
ARCH						
arch						
L1.	.1976492	.0391789	5.04	0.000	.12086	.2744384
garch						
L1.	.7181568	.051498	13.95	0.000	.6172227	.819091
_cons	1.065144	.2871044	3.71	0.000	.5024296	1.627858

```
. predict gvar, variance

. label variable gvar "GARCH conditional variance"

. correlate gvar avar
(obs=627)
```

	gvar	avar
gvar	1.0000	
avar	0.8674	1.0000

The ARCH and GARCH specifications produce similar estimates of the conditional variance. The GARCH model just uses a more parsimonious specification to obtain its estimate.[14]

---

14. Admittedly, in this example, we reduce the number of parameters by only one. The point is the GARCH model can provide significant reductions in the model degrees of freedom with little or no reduction in model fit in many cases.

Figure 8.8 displays a scatterplot of the ARCH and GARCH conditional variance estimates.[15] This figure provides a graphical confirmation of the high degree of association between the two estimates, although the relationship is not precisely linear.[16]

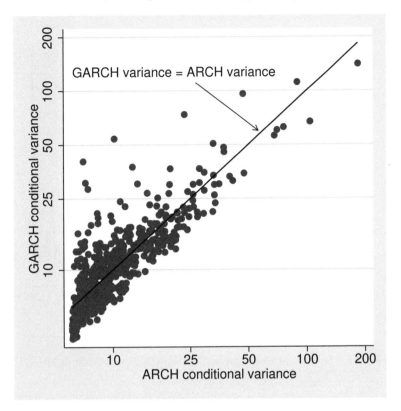

Figure 8.8. Comparison of GARCH and ARCH estimates of conditional variance

15. Both series are plotted on a log scale to improve the readability of the graph.
16. If you take a closer look at subsets of the time line, you will see that the GARCH estimates of the conditional variance have more "inertia" than the ARCH estimates; that is, they take longer to fall back from the peak levels of conditional variance.

Figure 8.9 displays the GARCH estimate of the conditional variance of monthly inflation in the U.S. CPI. As we suspected, the volatility increased dramatically during the
global financial crisis in late 2008 and early 2009.

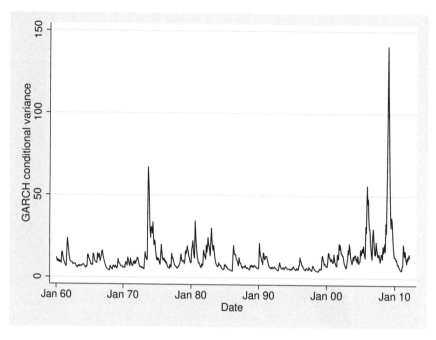

Figure 8.9. Conditional variance of monthly consumer inflation

It turns out that a GARCH(1,1) specification is flexible enough to handle a wide
variety of processes. In this example, higher-order ARCH and GARCH terms are negative
(which violates our assumptions), statistically insignificant, or both.

8.3.2   Other extensions

The GARCH model provides a remarkably useful technical solution (a low-order flexible
functional form) to a technical problem (the need for too many terms in a pure ARCH
model). While many, many other extensions to the ARCH model have been proposed,
most of them are designed to capture one or another empirical regularity of financial
time series that the pure ARCH model does not handle.

Asymmetric responses to "news"

In financial markets, bad news (for instance, the unexpected failure of a major financial institution) sometimes has a larger impact on volatility than good news. The models we have discussed so far cannot capture this feature. The arrival of news is represented by a nonzero value of ν_t, the innovation in the ARCH process. The magnitude of the response to news is proportional to $|\nu_t|$; that is, it depends on the size but not the sign of ν_t.

The exponential GARCH (EGARCH) model introduced by Nelson (1991) provides a specification that allows an asymmetric response to positive and negative news. In the EGARCH model, h_t, the time-varying scale factor for the conditional variance, is modeled as

$$\log h_t = \zeta + \sum_{j=1}^{\infty} \pi_j \left(|\nu_{t-j}| - E|\nu_{t-j}| + \aleph \nu_{t-j} \right)$$

The first two terms in the parentheses—$(|\nu_{t-j}| - E|\nu_{t-j}|)$—indicate that unexpected events ("news") tend to increase volatility temporarily. If $\pi_j > 0$, a larger-than-expected $|\nu_{t-j}|$ increases the variance of ϵ_t. The third term—$\aleph \nu_{t-j}$—allows positive and negative news to affect volatility differently. If $\aleph = 0$, there is no asymmetry—positive shocks have the same impact as negative shocks of the same magnitude. If $-1 < \aleph < 0$, positive shocks have a smaller impact than negative shocks. If $\aleph < -1$, the effect is reversed.

This parameterization of the EGARCH model follows Hamilton (1994). The notation and implementation in Stata is a bit different. Stata denotes the π's as γ's. And Stata expands $\aleph$ into a series of α's, one for each lag of ν_t. I know it's confusing. There is a staggering variety of ARCH-like models, and each author has proposed a slightly different way of summarizing and contrasting them.

Note that the EGARCH model specifies a process for $\log h_t$ rather than h_t. As a consequence, the variance h_t is always positive regardless of the signs of the π_j coefficients; thus no restrictions need to be imposed on the estimation of these parameters. ϵ_t, h_t, and $\log h_t$ are all stationary if $\sum_{j=1}^{\infty} \pi_j^2 < \infty$. Moreover, just as in the GARCH model, a high-order $\pi(L)$ can be reparameterized as the ratio of two lower-order lag polynomials, which results in the model

$$\{1 - \delta(L)\} \log h_t = \zeta + \alpha(L)(|\nu_{t-j}| - E|\nu_{t-j}| + \aleph \nu_{t-j})$$

Stata's implementation of EGARCH incorporates this extension. The `earch()` option specifies the terms in $\alpha(L)$, while the `egarch()` option specifies the terms in $\delta(L)$.

Let's test for asymmetric response in our model of monthly consumer price inflation. We will start with a model that includes only the α and $\aleph$ terms. As in our first ARCH specification, a third-order process is required to characterize the time-varying volatility.

```
. arch inflation, arima(2,0,1) earch(1 2 3) nolog

ARCH family regression -- ARMA disturbances

Sample: Feb 60 - Mar 12                    Number of obs   =        626
Distribution: Gaussian                     Wald chi2(3)    =   72162.84
Log likelihood = -1609.768                 Prob > chi2     =     0.0000
```

inflation	Coef.	OPG Std. Err.	z	P>|z|	[95% Conf. Interval]	
inflation						
_cons	3.503142	.6382883	5.49	0.000	2.25212	4.754164
ARMA						
ar						
L1.	1.267455	.0532998	23.78	0.000	1.162989	1.371921
L2.	-.2804079	.0500075	-5.61	0.000	-.3784208	-.1823951
ma						
L1.	-.9209139	.0218328	-42.18	0.000	-.9637054	-.8781223
ARCH						
earch						
L1.	-.0264705	.0584797	-0.45	0.651	-.1410886	.0881476
L2.	.0998423	.0426103	2.34	0.019	.0163276	.1833571
L3.	-.0431957	.055159	-0.78	0.434	-.1513053	.064914
earch_a						
L1.	.5837493	.0947076	6.16	0.000	.3981258	.7693727
L2.	.356895	.0714128	5.00	0.000	.2169285	.4968615
L3.	.1469453	.0890959	1.65	0.099	-.0276795	.32157
_cons	2.33561	.0695702	33.57	0.000	2.199255	2.471965

The earch_a rows display the estimated coefficients of $\alpha(L)$, that is, the symmetric impacts of the $|\nu_{t-j}| - E|\nu_{t-j}|$. The earch rows display the estimates of the asymmetric impacts. In this example, there is mixed evidence of asymmetric responses to volatility shocks—only the second lag is statistically significant. Just to be sure, we test them jointly.

```
. test [ARCH]:L.earch L2.earch L3.earch

 ( 1)  [ARCH]L.earch = 0
 ( 2)  [ARCH]L2.earch = 0
 ( 3)  [ARCH]L3.earch = 0

           chi2( 3) =     7.29
         Prob > chi2 =   0.0631
```

Unfortunately, not definitive.

As we saw above, in our discussion of the GARCH model, we can replace the third-order ARCH specification with a GARCH(1,1) model. We use the `egarch()` option to effect a similar change here.[17] With this specification, there clearly is no evidence of an asymmetric response to shocks.

```
. arch inflation, arima(2,0,1) earch(1) egarch(1) nolog

ARCH family regression -- ARMA disturbances
Sample: Feb 60 - Mar 12                      Number of obs   =        626
Distribution: Gaussian                       Wald chi2(3)    =   54461.66
Log likelihood = -1603.061                   Prob > chi2     =     0.0000
```

inflation	Coef.	OPG Std. Err.	z	P>|z|	[95% Conf. Interval]	
inflation						
_cons	3.642093	.6782566	5.37	0.000	2.312735	4.971452
ARMA						
ar						
L1.	1.255734	.0495535	25.34	0.000	1.158611	1.352857
L2.	-.2690062	.0462554	-5.82	0.000	-.3596652	-.1783472
ma						
L1.	-.9124448	.0247753	-36.83	0.000	-.9610036	-.863886
ARCH						
earch						
L1.	.0343094	.028904	1.19	0.235	-.0223415	.0909602
earch_a						
L1.	.317298	.0453013	7.00	0.000	.2285091	.406087
egarch						
L1.	.9040266	.0275575	32.81	0.000	.8500149	.9580384
_cons	.2277388	.0629286	3.62	0.000	.104401	.3510767

Variations in volatility affect the mean of the observable series

In some cases, there may be reason to believe the mean of the observable variable, y_t, is associated with the conditional variance of the random innovation. For instance, some economists have speculated that high inflation tends to be accompanied by high levels of price variability.[18] In finance, most theories suggest that investors require a

---

17. In preliminary estimates of this EGARCH model, with a slightly different sample, the `arch` command failed to converge. Models in the ARCH family frequently are difficult to estimate. It may be difficult or impossible to maximize the likelihood function. If you encounter these difficulties, Stata offers some help in the form of options that can control the maximization process. These tools cannot guarantee a solution, and they require a fair degree of sophistication (and luck) to use them successfully. Fortunately, we did not require these options with our final sample.

18. Interestingly, our estimates in this chapter appear to suggest the opposite—the conditional variance is highest when deflation is a real possibility. Even so, there seems to be a relationship between the level of inflation and the conditional variance.

higher return when the perceived risk (that is, variance) of an asset is higher. These theories motivated Engle, Lilien, and Robins (1987) to propose the ARCH-in-mean, or ARCH-M, model. In this model, the conditional variance, h_t, appears as a regressor in the equation for y_t.

To fit an ARCH-M model in Stata, add the `archm` option. For instance,

```
arch inflation, arima(2,0,1) arch(1) garch(1) archm
```

would add an ARCH-M term to our GARCH(1,1) model of monthly inflation. (Try it. The ARCH-M term is insignificant.) Alternatively, Stata allows you to specify an arbitrary lag polynomial in ARCH-M terms with the `archmlags()` option. Finally, you can add an arbitrary function of the conditional variance to the equation for y_t with the `archmexp()` option.

Nonnormal errors

We noted above that the distributions of asset prices possess "fat tails"; that is, extreme outcomes occur more frequently than is predicted by the normal distribution. You can use the `distribution()` option to specify that the random errors follow a t distribution or a generalized error distribution instead of a normal distribution. Stata estimates the degrees of freedom for the t distribution and the shape parameter for the generalized error distribution, but you can override those estimates by adding a legal value to the `distribution` option; for instance, `distribution(ged 3)` forces Stata to set the shape parameter to 3.

Odds and ends

We have just scratched the surface of proposed models of time-varying volatility. We have not even covered all the models available in Stata. The table below is a concise attempt to fill that gap. It lists issues that you might be facing, such as asymmetric responses, and lists the related models Stata offers through the `arch` command. The number of choices can be overwhelming. As always, you are on solid ground if your choice of model is tied to knowledge of the dynamic process generating your data.

Table 8.1. Extensions to the ARCH model and `arch` command

Reason	Approach	Models	Options
ARCH specification alone requires too many lags	Combine ARCH terms (lags of ϵ_t^2) with GARCH terms (lags of σ_t^2), the conditional variance	GARCH	garch(*numlist*)
Volatility responds asymmetrically to positive and negative "news"	Add terms that differentiate the impact of positive and negative "news"	simple asymmetric ARCH	saarch(*numlist*)
		EGARCH	earch(*numlist*) egarch(*numlist*)
		threshold ARCH	abarch(*numlist*) atarch(*numlist*) sdgarch(*numlist*)
		Glosten, Jagannathan, and Runkle	tarch(*numlist*)
Linear combination of lagged ϵ_t^2 terms not adequate	Introduce nonlinearity by shifting the minimum impact of the ith term from $\epsilon_{t-i} = 0$ to $\epsilon_{t-i} = \kappa_i$	nonlinear ARCH	narch(*numlist*)
		nonlinear power ARCH with one shift	narchk(*numlist*)
The conditional mean of y_t depends on σ_t^2, the conditional variance	Add current and lagged values of σ_t^2 to the state equation	ARCH in mean	archm(*numlist*) or archmlags(*numlist*)
Model in a power function of ϵ_t rather than ϵ_t^2 seems more appropriate	Power function versions of most of the previous specifications	power ARCH	parch(*numlist*)
		asymmetric power ARCH	aparch(*numlist*)
		nonlinear power ARCH or nonlinear power ARCH with one shift	nparch(*numlist*) or nparchk(*numlist*)
Innovations may not follow a normal distribution	Specify a different error distribution		distribution(*dist* [*#*])
Observable variables influence the conditional variance	Add observable variables to the conditional variance equation		het(*varlist*)

8.4   Points to remember

- Researchers have long noted what appear to be episodes of relatively high and relatively low volatility (that is, conditional variance), especially in economic and financial time series. The standard ARIMA model assumes constant residual variance; hence, it does not capture this phenomenon.

- The ARCH model of Engle proposed an AR structure for the conditional variance of a time series while retaining the constant unconditional variance. This model provided the first practical method for modeling time-varying volatility, and its appearance triggered an explosion in both empirical and theoretical research on time-varying volatility.

- Many extensions to Engle's ARCH specification have been proposed and used successfully. Bollerslev's GARCH model provides a flexible way of fitting the conditional variance with relatively few parameters. Other proposals (exponential GARCH, ARCH-M, etc.) extend the model to capture additional empirical regularities.

- Models in the ARCH family often are difficult to estimate, and you may need to experiment with some of the parameters that control Stata's maximization algorithm to obtain results.

Stata commands and features discussed

arch ([TS] **arch**): ARCH family of estimators; sections 8.2 and 8.3

egen ([D] **egen**): Extensions to generate; section 8.1

estat archlm ([R] **regress postestimation time series**): Test for ARCH effects in the residuals; section 8.2

predict ([TS] **arch postestimation**): Postestimation tools for arch; section 8.3

table ([R] **table**): Tables of summary statistics; section 8.1

test ([TS] **arch postestimation**): Postestimation tools for arch; section 8.3

9 Models of multiple time series

Chapter map

9.1 Vector autoregressions. The vector autoregression (VAR) as the extension of the AR model to multiple time series. The three types of VARs: reduced-form, recursive, and structural.

9.2 A VAR of the U.S. macroeconomy. A three-equation VAR for inflation, unemployment, and interest rates. How to estimate the reduced-form version of this VAR, check it for stationarity, and forecast the dependent variables.

9.3 Who's on first? Perspectives on the temporal sequence of variable movements and its relationship to causation. Cross correlations, Granger causality, impulse–response functions (IRFs), and forecast-error variance decompositions (FEVDs).

9.4 SVARs. VARs and structural econometric models (SEMs). The return of identifying restrictions. Short-run and long-run structural vector autoregression (SVARs).

9.5 Points to remember. VARs, SVARs, orderings, and restrictions.

9.6 Looking ahead. Modeling nonstationary variables. The role of trends.

So far, we have developed models for a single time series. These models provide practical methods for forecasting a time series on the basis of its past values and (optionally) the past and present values of additional, exogenous factors. In the previous chapter, we extended these models in ways that allow us to track and forecast the dynamic behavior of the variance of a time series in addition to its level.

In many cases, we are interested in the interaction of several endogenous time series. In the macroeconomy, for instance, we understand that inflation, output, and unemployment (and, perhaps, additional variables) interact and influence the future paths of each other. In this chapter, we explore techniques for modeling these interactions.

The logical first step in modeling multiple time series is the VAR—a straightforward rewrite of the univariate AR model as a vector process. The next section introduces VARs and identifies three types of VARs: reduced-form VARs, recursive VARs, and SVARs. As its name suggests, the reduced-form VAR incorporates the fewest maintained hypotheses and represents the VAR in its purest form. Recursive VARs add assumptions about the order of causation among the dependent variables, and SVARs add even more assumptions of varied types. When these assumptions are well founded, recursive VARs and SVARs provide useful insights into the dynamic behavior of the dependent variables.

To illustrate these three types of VARs, the subsequent section introduces a three-equation VAR of the U.S. macroeconomy. Before the section is finished, we show how Stata can estimate the reduced-form version of this VAR, check it for stationarity, and forecast future values of the dependent variables. We also discuss one approach to assessing the quality of the VAR forecasts. In the next section, we approach the thorny questions of temporal precedence and causation that underlie the recursive VAR. Starting with simple cross correlations, we proceed to the confusingly named concept of Granger causation, and, finally, to IRFs and FEVDs. These latter statistics represent attempts to characterize the channels of causation among the dependent variables in a VAR. The next section discusses some of the ways SVARs incorporate additional assumptions about the relationships among the dependent variables.

9.1   Vector autoregressions

A simple univariate AR(p) model (without exogenous variables) is written as

$$y_t = \mu + \phi_1 y_{t-1} + \cdots + \phi_p y_{t-p} + \epsilon_t$$

or, more compactly, as

$$\phi(L) y_t = \mu + \epsilon_t$$

y_t is a function of a constant (μ), p prior values of y_t, and a random disturbance, ϵ_t.

Consider now a vector of n jointly endogenous variables

$$\mathbf{y}_t = \begin{bmatrix} y_{1,t} \\ y_{2,t} \\ \vdots \\ y_{n,t} \end{bmatrix}$$

We can model this n-element vector as function of n constants, p prior values of the vector $\mathbf{y}_t$, and a vector of n random disturbances, $\boldsymbol{\epsilon}_t$,

$$\mathbf{y}_t = \boldsymbol{\mu} + \boldsymbol{\Phi}_1 \mathbf{y}_{t-1} + \cdots + \boldsymbol{\Phi}_p \mathbf{y}_{t-p} + \boldsymbol{\epsilon}_t$$

In this equation, $\boldsymbol{\mu}$ is the n-element vector of constants,

$$\boldsymbol{\mu} = \begin{bmatrix} \mu_1 \\ \mu_2 \\ \vdots \\ \mu_n \end{bmatrix}$$

the $\boldsymbol{\Phi}_i$ are matrices of coefficients,

$$\boldsymbol{\Phi}_i = \begin{bmatrix} \phi_{i,11} & \phi_{i,12} & \cdots & \phi_{i,1n} \\ \phi_{i,21} & \phi_{i,22} & \cdots & \phi_{i,2n} \\ \vdots & \vdots & \ddots & \vdots \\ \phi_{i,n1} & \phi_{i,n2} & \cdots & \phi_{i,nn} \end{bmatrix}$$

and $\boldsymbol{\epsilon}_t$ is the n-element vector of random disturbances

$$\boldsymbol{\epsilon}_t = \begin{bmatrix} \epsilon_{1,t} \\ \epsilon_{2,t} \\ \vdots \\ \epsilon_{n,t} \end{bmatrix}$$

where

$$E\boldsymbol{\epsilon}_t = 0$$

and

$$E\boldsymbol{\epsilon}_t\boldsymbol{\epsilon}_s' = \begin{cases} \boldsymbol{\Sigma}, & t = s \\ \mathbf{0}, & t \neq s \end{cases}$$

Notice that the elements of $\boldsymbol{\epsilon}_t$ can be contemporaneously correlated.

We can write this pth-order VAR more compactly as

$$\boldsymbol{\Phi}(L)\mathbf{y}_t = \boldsymbol{\mu} + \boldsymbol{\epsilon}_t$$

where $\boldsymbol{\Phi}(L)$ is a matrix polynomial in the lag operator

$$\boldsymbol{\Phi}(L) \equiv \mathbf{I} - \boldsymbol{\Phi}_1(L) - \cdots - \boldsymbol{\Phi}_p(L)$$

To make the interactions between the elements of $\mathbf{y}_t$ clear, we extract the equation for the ith endogenous time series from the basic VAR

$$\begin{aligned} y_{i,t} = {}& \mu_i + \phi_{1,i1}y_{1,t-1} + \cdots + \phi_{1,in}y_{n,t-1} \\ & + \phi_{2,i1}y_{1,t-2} \cdots + \phi_{2,in}y_{n,t-2} + \cdots \\ & + \phi_{p,i1}y_{1,t-p} + \cdots + \phi_{p,in}y_{n,t-p} + \epsilon_{i,t} \end{aligned}$$

Each of the n endogenous time series—the $y_{i,t}$—depends on its own lagged values and the lagged values of all the other endogenous time series. However, as a consequence of the dependence on lagged values of all the elements of $\mathbf{y}_t$, $y_{i,t}$ is influenced not only by the current and lagged values of its own random disturbance, $\epsilon_{i,t}$, but also by the lagged values of all the other elements of ϵ_t.[1] This plethora of dependencies makes it difficult to trace the multiple channels of influence between the $y_{i,t}$ and the $\epsilon_{i,t}$. The methods developed for analyzing VARs are designed to help us make sense of these interactions.[2]

9.1.1   Three types of VARs

Christopher Sims (1980) proposed VARs in part to counteract what he saw as the "incredible" a priori identifying assumptions used in the estimation of large-scale macroeconometric models. He argued that VARs retain the essential value of multiple-equation models (that is, the information about the interactions of the jointly endogenous variables) without relying on overly strong identifying assumptions. Moreover, the forecasting accuracy of VARs compares favorably with, and frequently exceeds, that of structural, simultaneous-equation models. Sims's argument was persuasive, and VARs have become an essential part of the time-series toolkit.

The VAR, as we described it above, is a reduced-form model. Each equation can be estimated consistently by least squares.[3] The VAR provides a convenient way to characterize the distribution of $\mathbf{y}_t$ and to forecast it while taking account of the interactions among its elements. To go much further though, some additional assumptions are required.

---

1. Even if the elements of ϵ_t are uncorrelated, that is,

$$\boldsymbol{\Sigma} = \begin{bmatrix} \sigma_{11} & 0 & 0 & \cdots & 0 \\ 0 & \sigma_{22} & 0 & \cdots & 0 \\ 0 & 0 & \sigma_{33} & \cdots & 0 \\ \vdots & \vdots & \vdots & \ddots & \vdots \\ 0 & 0 & 0 & \cdots & \sigma_{nn} \end{bmatrix}$$

2. We can complicate things further by including a k-element vector of exogenous variables.

$$\boldsymbol{\Phi}(L)\mathbf{y}_t = \boldsymbol{\mu} + \mathbf{B}\mathbf{x}_t + \epsilon_t$$

We even can add lagged values (a so-called distributed lag) of the exogenous variables.

$$\boldsymbol{\Phi}(L)\mathbf{y}_t = \boldsymbol{\mu} + \mathbf{B}(L)\mathbf{x}_t + \epsilon_t$$

While these extensions are useful, their treatment is somewhat tangential to the essential features of VARs. Accordingly, we put them aside in the discussion that follows.

3. The correlations among the elements of ϵ suggests that some type of generalized least squares (GLS) estimation may be appropriate. However, in this basic VAR, each equation contains the same regressors and there are no constraints on the parameters. GLS reduces to ordinary least-squares in this case. When we start to add complexity to the VAR model, GLS will provide greater efficiency, and Stata will use an iterated seemingly unrelated regression method to fit the model.

A recursive VAR specifies a priority of influence among the elements of $\mathbf{y}_t$. The equations in the system are ordered so that the error term in each equation is uncorrelated with the errors in the preceding equations. (We will discuss below how to impose this condition.) This recursive form allows us to make definite statements about the sequence of causation. For instance, a random shock to, say, the third equation in a five-equation recursive VAR will have a current-period impact on the third, fourth, and fifth elements of $\mathbf{y}_t$, but will not affect the first and second elements until period $t + 1$. The ordering imposed in a recursive VAR makes a significant impact on the inferences drawn from the model. Unfortunately, it typically is difficult to make a compelling argument for one ordering over another, and there are $n!$ possible orderings for the n equations in a VAR. Despite this difficulty, the analysis of recursive VARs has remained popular and can, in some cases, be very useful.

While VARs are powerful tools for forecasting, inevitably, macroeconomists want to make stronger statements about the relationships between the elements of $\mathbf{y}_t$ than can be produced from a reduced-form or recursive VAR. They would like to predict the dynamic impact of policies in a way that recognizes the wealth of interactions among macroeconomic variables. SVARs incorporate additional restrictions on the model that permit this type of analysis. However, just as with recursive VARs, these additional constraints can be a way for the dubious identifying assumptions highlighted by Sims to creep back into the model.

In the next section, we put a reduced-form VAR—the simplest form of VAR—through its paces: identification, estimation, testing for stationarity, and forecasting. All of these steps will take some time to illustrate and will require a number of new Stata commands, but these steps prepare the way for recursive and SVARs in subsequent sections.

9.2   A VAR of the U.S. macroeconomy

To illustrate the methods for estimating and analyzing VARs, we borrow an example from the excellent overview article by Stock and Watson (2001).[4,5] They fit a three-equation model for U.S. inflation, unemployment, and interest rates to demonstrate aspects of VARs. Although this illustrative model is small—the entire macroeconomy is summarized by three variables—it does provide a reasonable example of a complete model. The model traces the dynamic interactions of real output (proxied by the unemployment rate), prices (the inflation rate), and monetary policy (the Federal funds rate) and allows us to consider the impact of alternative policy rules.

---

4. In their article, Stock and Watson divide VARs into the three types listed above: reduced-form, recursive, and structural.

5. I encourage you to compare the estimates reported in this chapter with the estimates published in Stock and Watson. While the results are qualitatively similar, you will find differences. Some of the differences can be attributed to the difference in the sample period and, perhaps, to differences in the numerical properties of the algorithms used by Stata and the algorithms used by the programs used by Stock and Watson. Also there have been revisions to the inflation and unemployment series since the time Stock and Watson calculated their estimates.

Stock and Watson fit their model by using quarterly U.S. data from 1960:1 through 2000:4. Inflation is measured by the annualized quarterly percentage change in the GDP price deflator. The unemployment rate is measured by the civilian unemployment rate. The interest rate is measured by the Federal funds rate. Both the unemployment and the interest rates are quarterly averages of their monthly values. For this example, we update the sample to 2012:1.

Let's take a look at the data.

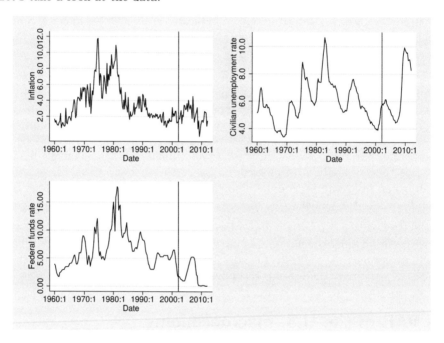

Figure 9.1. U.S. inflation, unemployment, and the Federal funds rate, 1960:1–2012:1

We are going to reserve the most recent 10 years—2002:2 through 2012:1—for out-of-sample forecasts. The vertical lines in figure 9.1 indicate the separation between in-sample and out-of-sample data.

There are a couple of items to note about the data in figure 9.1. The in-sample data on inflation and the Federal funds rate contain episodes of extremely high values in the mid-1970s and early 1980s. Conversely, the out-of-sample data include extremely low values of these two measures reflecting the post-9/11 Federal Reserve easing and, more recently, the fallout from the global financial crisis. In contrast, the unemployment rate covers similar ranges in the in-sample and out-of-sample periods.

```
. use ${ITSUS_DATA}/varexample
(Quarterly macroeconomic statistics for VAR examples)
. summarize fedfunds inflation unrate if date<tq(2001q3)

    Variable |       Obs        Mean    Std. Dev.       Min        Max
-------------+--------------------------------------------------------
    fedfunds |       166    6.570402    3.170146   1.683333      17.78
   inflation |       166    3.837343    2.494568   .5674297   11.79167
      unrate |       166    5.938353    1.513248        3.4   10.66667

. summarize fedfunds inflation unrate if date>=tq(2001q3)

    Variable |       Obs        Mean    Std. Dev.       Min        Max
-------------+--------------------------------------------------------
    fedfunds |        43    1.966279    1.800846   .0733333   5.256667
   inflation |        43    2.171986    1.053403   -.448682   4.558042
      unrate |        43    6.504651    1.890404   4.433333   9.933333
```

9.2.1   Using Stata to estimate a reduced-form VAR

A reduced-form VAR places few restrictions on the data, provides a useful characterization of the distribution of $\mathbf{y}_t$, and generates joint forecasts of the elements of $\mathbf{y}_t$. Let's see how to carry out those steps in Stata.

Recall that when we fitted a univariate ARIMA model for GDP in chapter 7, we began by examining the autocorrelations of the log real GDP for stationarity. We ended up by taking the first difference of log real GDP to induce stationarity and fitting a model for the rate of growth in real GDP.

VARs must obey more restrictive conditions. A VAR is stationary if its first and second moments [$E(\mathbf{y}_t)$ and $E(\mathbf{y}_t\mathbf{y}'_{t-j})$, respectively] are independent of t. These conditions imply that each of the elements of $\mathbf{y}_t$ is stationary. Additionally, stationarity of a VAR requires a stability condition that we test only *after* we estimate the VAR. Consequently, we will defer the check for stationarity.

The next step in constructing our model is the identification of the number of lags of $\mathbf{y}_t$ to include in the VAR. In the univariate model, we examined the pattern of autocorrelations and partial autocorrelations for clues as to the appropriate lag length, although we discovered that the sampling error of these statistics makes any identification a tentative, trial-and-error process. We rely on a different set of statistics to identify a VAR, but the tentative nature of identification is similar to the univariate case. To identify the appropriate lag length of a VAR, we estimate multiple VARs of varying lag lengths and compute a variety of test statistics. While each statistic provides a definite choice of lag length, the statistics frequently disagree.

Stata's `varsoc` command saves us the trouble of estimating each of the candidate VARs. It estimates the entire set silently and reports a table of test statistics. The syntax of the `varsoc` command is

`varsoc` *depvarlist* [*if*] [*in*] [, <u>max</u>lag(*#*) *other_options*]

For illustration, we will test up to a maximum of 12 lags on our estimation sample.

```
. varsoc inflation unrate fedfunds if date<=tq(2002q1), maxlag(12)
   Selection-order criteria
   Sample:  1963:1 - 2002:1                        Number of obs      =      157
```

lag	LL	LR	df	p	FPE	AIC	HQIC	SBIC
0	-1006.3				76.995	12.8574	12.8811	12.9158
1	-464.619	1083.4	9	0.000	.086985	6.07158	6.16645	6.30518
2	-417.517	94.204	9	0.000	.053544	5.5862	5.75223	5.995*
3	-399.775	35.485	9	0.000	.047917	5.47484	5.71202*	6.05883
4	-390.923	17.703	9	0.039	.048038	5.47673	5.78506	6.23592
5	-381.039	19.768	9	0.019	.047547	5.46546	5.84495	6.39986
6	-364.239	33.599	9	0.000	.043112	5.36611	5.81675	6.4757
7	-359.33	9.8187	9	0.365	.045509	5.41822	5.94002	6.70301
8	-348.539	21.582	9	0.010	.0446	5.3954	5.98836	6.85539
9	-334.836	27.406	9	0.001	.042148	5.33549	5.9996	6.97068
10	-326.368	16.936	9	0.050	.042613	5.34226	6.07753	7.15265
11	-315.755	21.226	9	0.012	.041961*	5.32172*	6.12814	7.30731
12	-306.787	17.937*	9	0.036	.042239	5.32212	6.19969	7.48291

```
Endogenous:  inflation unrate fedfunds
Exogenous:   _cons
```

For each lag length, the table reports (from left to right), the log of the likelihood function (LL); a likelihood-ratio test statistic followed by its degrees of freedom and p-value (LR, df, and p); and four information criteria: Akaike's final prediction error (FPE); Akaike's information criterion (AIC), Hannan and Quinn's information criterion (HQIC), and Schwarz's Bayesian information criterion (SBIC).[6] For each statistic, the preferred value is marked with a *.

For each lag length, the likelihood-ratio tests the hypothesis that the coefficients of the last lag are jointly equal to 0, that is, one less lag is appropriate. Starting from the maximum lag considered, the first rejection is marked as the appropriate lag length. As a consequence, the outcome of this test is sensitive to the maximum number of lags considered. In this example, the likelihood-ratio test selects 12 lags of $\mathbf{y}_t$, the maximum tested (likelihood ratio = 17.937, p-value = 0.036).

The four information criteria are based on information theory and are supposed to indicate the relative information lost when the data are fit using different specifications.[7] The lag length that produces the minimum value of the information statistic is the preferred specification. In this example, the FPE and AIC prefer 11 lags, the HQIC prefers 3 lags, and the SBIC chooses 2 lags.

---

6. Formally, the FPE is not an information criterion. However, like the other criteria, we seek to minimize it.

7. Each of these information criteria provides a slightly different trade-off between model complexity (measured by the number of parameters) and goodness of fit (measured by the log of the likelihood function). Additional parameters are penalized (increase the value of the criterion) at the same time that they improve the goodness of fit (decrease the sum of squared residuals, which decreases the value of the criterion). Stock and Watson (2019, 540–546) provide an intuitive description of the AIC and SBIC.

As is frequently the case, this assortment of test statistics leaves us without clear direction. According to Lütkepohl (2005), the SBIC and HQIC provide consistent estimates of the true lag order, while the FPE and AIC overestimate the lag order with positive probability. In any event, we are going to use four lags to match the specification chosen by Stock and Watson.

Stata's `varbasic` command estimates a simple reduced form VAR (no exogenous variables) and provides the most commonly used features of VAR estimation and analysis. The syntax is

`varbasic` *depvarlist* [*if*] [*in*] [, `lags`(*numlist*) `irf` `fevd` `nograph` `step`(#)]

The `lags()` option is self-explanatory. If you omit it, `varbasic` includes the first and second lags of each of the variables. The remaining options specify which statistic to display (or not if the `nograph` option is specified) and for how many steps ahead (the `steps()` option, with a default of eight steps). By default, `varbasic` displays the orthogonalized impulse–response function (OIRF)—we will explain that in a second— but it can replace that with a display of the nonorthogonalized IRF or the FEVD. Brace yourself. We are going to introduce a lot of new terms.

So let's see what happens when we use `varbasic` to estimate our three-equation VAR. We will continue to restrict the estimation sample to reserve 10 years of out-of-sample data for comparison with our VAR forecast.

```
. varbasic inflation unrate fedfunds if date<=tq(2002q1), lags(1 2 3 4) step(12)
> nograph

Vector autoregression

Sample:  1961:1 - 2002:1                     Number of obs    =        165
Log likelihood = -407.6184                   AIC              =   5.413556
FPE            =  .0450901                    HQIC             =   5.711566
Det(Sigma_ml)  =  .0280771                    SBIC             =   6.147689

Equation          Parms      RMSE     R-sq      chi2     P>chi2

inflation            13    1.05454   0.8346   832.4144   0.0000
unrate               13    .231832   0.9784   7467.331   0.0000
fedfunds             13    .886177   0.9283   2135.223   0.0000
```

	Coef.	Std. Err.	z	P>|z|	[95% Conf.	Interval]
inflation						
inflation						
L1.	.51057	.0764711	6.68	0.000	.3606895	.6604505
L2.	.1167881	.0853305	1.37	0.171	-.0504567	.2840329
L3.	.1097551	.0867862	1.26	0.206	-.0603428	.279853
L4.	.2537744	.0802771	3.16	0.002	.096434	.4111147
unrate						
L1.	-.7837757	.3845296	-2.04	0.042	-1.53744	-.0301115
L2.	1.231028	.6542283	1.88	0.060	-.0512355	2.513292
L3.	-.9411817	.6568343	-1.43	0.152	-2.228553	.3461898
L4.	.3300222	.366324	0.90	0.368	-.3879596	1.048004
fedfunds						
L1.	.1812778	.1025155	1.77	0.077	-.0196489	.3822045
L2.	-.1190267	.1373793	-0.87	0.386	-.3882852	.1502318
L3.	-.026651	.1359343	-0.20	0.845	-.2930774	.2397754
L4.	-.0432113	.1038741	-0.42	0.677	-.2468007	.1603781
_cons	1.062324	.3661876	2.90	0.004	.3446093	1.780038
unrate						
inflation						
L1.	.0112462	.0168116	0.67	0.504	-.0217039	.0441962
L2.	-.011175	.0187592	-0.60	0.551	-.0479424	.0255925
L3.	.0168998	.0190793	0.89	0.376	-.0204948	.0542945

(output omitted)

Even this modestly sized VAR (three variables, four lags) contains an enormous number of coefficients. We cut off the `varbasic` output after the coefficients from the inflation equation and part of the unemployment rate equation. How can we make sense of all this output? Let's start by looking at the statistics reported at the top of the output, before the endless list of coefficients.

After reminding us that we are estimating a VAR, `varbasic` reports the time period covered by the sample and the number of observations in the sample. Note that there are 169 quarters in the period from the first quarter of 1960 through the first quarter of 2002, but `varbasic` reports only 165 observations. Where are the missing four observations? As it happens, VARs are estimated based on the conditional likelihood function, that is, the likelihood function conditioned on the first p observations (Hamilton 1994, 291). In our example, there are four lags (p is equal to four), so the first four observations comprise the conditioning information. For simplicity, the term "conditional" generally is dropped, and we refer to the conditional likelihood function simply as the likelihood function and the estimate based on this function as the maximum likelihood estimate.

The next three rows report the log of the (conditional) likelihood function, the four information criteria we saw above in the output of the `varsoc` command, and the determinant of the maximum likelihood estimate of the covariance matrix of the residuals (`Det(Sigma_ml)`). None of these statistics are directly useful in assessing a single estimated specification. Rather they are used to compare alternative specifications. The log of the likelihood and the determinant of $\widehat{\Sigma}$ are used to construct likelihood-ratio tests of potential restrictions. While the information criteria do not lead to a formal hypothesis test with a p-value, they can be used to compare any two specifications, while the likelihood-ratio test can only be used for nested models. For both the likelihood-ratio test and comparisons of information criteria, the models should be estimated on the same sample.

The next few rows display summary statistics for each of the three equations in this VAR. `varbasic` indicates there are 13 parameters in each equation: a constant plus four lag coefficients for each of the three endogenous variables (not counting any of the elements in Σ). `varbasic` also reports the root mean squared error (RMSE) and R^2 (`R-sq`) for each equation. In this example, these statistics indicate a noticeably poorer fit for the inflation equation than for the other two equations. The last two statistics are the test statistic (`chi2`) for the Wald test of the regression coefficients (aside from the constant) and the p-value for the test.

By default, `varbasic` produces a graph of the OIRF (say that 10 times fast!). IRFs lead directly into the topic of recursive VARs, and we want to finish understanding our example as a reduced form VAR before we start puzzling out the priority of influence among our three macro variables. As a result, we suppressed the graph with the `nograph` option.

Now that we have estimated a candidate VAR, let's see whether it obeys the stability condition—the multiple variable test of stationarity—that we mentioned above.

9.2.2 Testing a VAR for stationarity

Recall that a univariate autoregressive process is stationary if all the roots of $\phi(z) = 0$ lie outside the unit circle. An analogous condition applies to VARs. A VAR is stationary if all the roots of

$$|\Phi(z)| = 0$$

lie outside the unit circle.[8] Equivalently, all the eigenvalues of the companion matrix must lie inside the unit circle. Stata's `varstable` command tests this latter condition.

---

8. See Hamilton (1994, 259) for an explanation and pp. 285–286 for a proof of this statement.

So what the heck is the companion matrix? It turns out that a pth-order VAR can always be rewritten as a first-order VAR, and the matrix of lag coefficients in the first-order representation is called the companion matrix.[9]

Let's demonstrate. We write a pth-order VAR as

$$\mathbf{y}_t = \boldsymbol{\mu} + \boldsymbol{\Phi}_1 \mathbf{y}_{t-1} + \cdots + \boldsymbol{\Phi}_p \mathbf{y}_{t-p} + \boldsymbol{\epsilon}_t$$

where $\mathbf{y}_t$ is the n-element vector of endogenous variables. We begin the transformation to a first-order VAR by subtracting $\boldsymbol{\mu}$, the vector of constants, from $\mathbf{y}_t$, then stacking the current and $p-1$ lags of this vector difference into an np-element vector, like this:

$$\widetilde{\mathbf{y}}_t \equiv \begin{bmatrix} \mathbf{y}_t - \boldsymbol{\mu} \\ \mathbf{y}_{t-1} - \boldsymbol{\mu} \\ \vdots \\ \mathbf{y}_{t-p+1} - \boldsymbol{\mu} \end{bmatrix}$$

Then we build the np-by-np companion matrix like this

$$\widetilde{\boldsymbol{\Phi}} \equiv \begin{bmatrix} \boldsymbol{\Phi}_1 & \boldsymbol{\Phi}_2 & \cdots & \boldsymbol{\Phi}_{p-1} & \boldsymbol{\Phi}_p \\ \mathbf{I}_n & \mathbf{0} & \cdots & \mathbf{0} & \mathbf{0} \\ \mathbf{0} & \mathbf{I}_n & \cdots & \mathbf{0} & \mathbf{0} \\ \vdots & \vdots & \ddots & \vdots & \vdots \\ \mathbf{0} & \mathbf{0} & \cdots & \mathbf{I}_n & \mathbf{0} \end{bmatrix}$$

and stack the n-element vector of random disturbances on top of $n(p-1)$ zeroes like this:

$$\widetilde{\boldsymbol{\epsilon}}_t \equiv \begin{bmatrix} \boldsymbol{\epsilon}_t \\ \mathbf{0} \\ \vdots \\ \mathbf{0} \end{bmatrix}$$

Now we have the first-order VAR

$$\widetilde{\mathbf{y}}_t = \widetilde{\boldsymbol{\Phi}} \widetilde{\mathbf{y}}_{t-1} + \widetilde{\boldsymbol{\epsilon}}_t$$

Now that you understand what the companion matrix is, you never have to think about it again. Just remember that a VAR is stationary if all the eigenvalues of the companion matrix lie inside the unit circle, and the `varstable` command will calculate the eigenvalues for you.

The syntax of the `varstable` command is

`varstable [ , estimates(`*estname*`) graph` *other_options* `]`

---

9. The companion matrix is a convenient way to express any higher-order polynomial—scalar or matrix, with lag operators or not—as a first-order polynomial. Many proofs are more convenient in terms of the companion matrix than in the original, higher-order form.

By default, `varstable` displays a table of the eigenvalues. The `graph` option adds an attractive plot of the eigenvalues in polar coordinates; that is, the real part of the eigenvalue is measured along the horizontal axis and the imaginary part along the vertical axis of the plot.

Like all Stata estimation commands, the `varbasic` command (and the more general `var` command, which we have not demonstrated yet) stores a host of results that later commands can use. By default, `varstable` uses results stored by the most recent `var` or `varbasic` or `var svar` (another command we have not gotten to yet) command. However, estimation results can be saved to a file by using the `estimates` command, and the `estimates()` option can be used to point `varstable` at one of these files in place of the most recent estimates.

Let's check on the stationarity of our example model. We will include the graph.

```
. varstable, graph
   Eigenvalue stability condition
```

Eigenvalue	Modulus
.9729189 + .05537318*i*	.974493
.9729189 − .05537318*i*	.974493
.7945809 + .2277735*i*	.826583
.7945809 − .2277735*i*	.826583
−.00758325 + .693766*i*	.693807
−.00758325 − .693766*i*	.693807
.05002827 + .5616101*i*	.563834
.05002827 − .5616101*i*	.563834
−.5356091	.535609
−.4003269	.400327
.1295862 + .274046*i*	.30314
.1295862 − .274046*i*	.30314

```
All the eigenvalues lie inside the unit circle.
VAR satisfies stability condition.
```

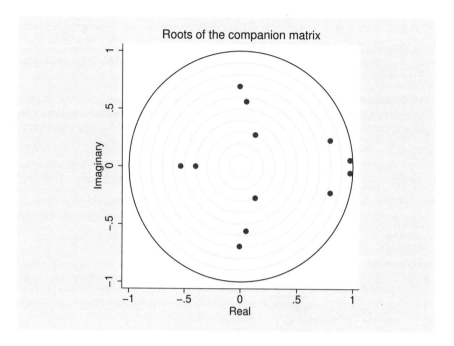

Figure 9.2. Eigenvalues of the companion matrix

The companion matrix is np by np, so there are np eigenvalues. In our VAR, there are three variables ($n = 3$) and four lags ($p = 4$), so there are twelve eigenvalues. (Actually six pairs because complex roots always come as a pair of complex complements.) While all the eigenvalues lie inside the unit circle, note that one pair is close to the limit, indicating that some shocks may not die out quickly.

Other tests

Before we move on to forecasting, let's mention some other tests you may want to run on your estimated VAR.

We typically assume that the order of the VAR (p) is sufficiently high to guarantee that the residuals are white noise, and there is no residual autocorrelation.[10] Stata's `varlmar` command performs a test of the null hypothesis of no residual autocorrelation.[11] The syntax is

varlmar $\left[\right.$, m̲lag(#) e̲s̲t̲imates(*estname*) s̲e̲p̲arator(#) $\left.\right]$

The `mlag()` option specifies the maximum order of autocorrelation to consider. By default, `varlmar` considers up to two lags.

```
. varlmar
  Lagrange-multiplier test
```

lag	chi2	df	Prob > chi2
1	21.3601	9	0.01114
2	7.1795	9	0.61844

```
  H0: no autocorrelation at lag order
```

The Lagrange multiplier indicates evidence of first-order residual autocorrelation at the 5% but not the 1% level.[12]

For some purposes, we may want to assume the VAR disturbances are normally distributed. A variety of tests for normality are provided by the **varnorm** command. The syntax of **varnorm** is

varnorm $\left[\right.$, j̲bera s̲kewness k̲urtosis e̲s̲t̲imates(*estname*) *other_options* $\left.\right]$

---

10. The assumption of stationarity implies that a pure AR process can eliminate any measurable serial correlation; however, the order of the AR model may have to be inconveniently high. In the univariate case, we used mixed models—models that combine both autoregressive and moving-average components—to guarantee white-noise residuals with a low-order model. While there are vector models that incorporate moving-average components, they are not used frequently.
11. `varlmar` performs a Lagrange multiplier test. The test statistic for the null of no residual autocorrelation at s lags is

$$\text{Lagrange−multiplier}_s \equiv (T-d-0.5)\ln\left(\frac{|\widehat{\Sigma}|}{|\widehat{\Sigma}_s|}\right)$$

$\widehat{\Sigma}$ is the estimated covariance matrix of the residuals from the VAR. $\widehat{\Sigma}_s$ is the estimated covariance matrix of the residuals from an augmented VAR where s lags of the estimated VAR residuals have been added to the system of equations.
12. In an actual research project, this evidence of residual autocorrelation would lead us to further analysis and, perhaps, respecification of our VAR. Because our purpose here is to explain and illustrate the methods of VAR analysis—and because we want to maintain comparability with the original example by Stock and Watson (2001)—we will retain our current specification.

. varnorm

Jarque-Bera test

Equation	chi2	df	Prob > chi2
inflation	6.079	2	0.04786
unrate	29.569	2	0.00000
fedfunds	246.787	2	0.00000
ALL	282.435	6	0.00000

Skewness test

Equation	Skewness	chi2	df	Prob > chi2
inflation	.32874	2.972	1	0.08472
unrate	.67241	12.434	1	0.00042
fedfunds	1.145	36.052	1	0.00000
ALL		51.458	3	0.00000

Kurtosis test

Equation	Kurtosis	chi2	df	Prob > chi2
inflation	3.6723	3.107	1	0.07794
unrate	4.5787	17.135	1	0.00003
fedfunds	8.5365	210.735	1	0.00000
ALL		230.977	3	0.00000

By default, the Jarque–Bera, skewness, and kurtosis tests all are presented, but an individual test can be requested with the appropriate option. In our example model, the null hypothesis of normal disturbances is clearly rejected by the Jarque–Bera test for the unemployment and Federal funds equations but only marginally rejected for the inflation equation ($p = 0.048$).

Finally, we used the `varsoc` command to select an appropriate order ex ante for our VAR. The `varwle` command provides a formal test ex post of the significance of the lags in each equation of the VAR. The syntax of the `varwle` command is

varwle $\big[$, <u>est</u>imates(*estname*) <u>sepa</u>rator(*#*) $\big]$

. varwle

Equation: inflation

lag	chi2	df	Prob > chi2
1	65.63708	3	0.000
2	7.541783	3	0.056
3	4.200774	3	0.241
4	10.33434	3	0.016

Equation: unrate

lag	chi2	df	Prob > chi2
1	375.257	3	0.000
2	21.64073	3	0.000
3	1.497849	3	0.683
4	.6964667	3	0.874

Equation: fedfunds

lag	chi2	df	Prob > chi2
1	251.803	3	0.000
2	33.38775	3	0.000
3	14.83628	3	0.002
4	6.102115	3	0.107

Equation: All

lag	chi2	df	Prob > chi2
1	600.3899	9	0.000
2	44.87808	9	0.000
3	20.28046	9	0.016
4	19.51237	9	0.021

The third lag is not significant in the inflation and unemployment rate equations, and the fourth lag is not significant in the unemployment rate and Federal funds equations. However, both of these lags are significant when the VAR as a whole is considered.

9.2.3   Forecasting

Calculating forecasts from a VAR is straightforward. As with any time-series model, the key decision is whether you want to produce one-step-ahead forecasts or dynamic forecasts. In a one-step-ahead forecast, only $\mathbf{y}_t$ is unknown (and $\boldsymbol{\epsilon}_t$, of course). All lagged variables—anything dated earlier than t—are observed. Dynamic forecasts occur whenever we want to predict more than one period ahead, and we are forced to include predicted values of at least some of the lagged, right-hand-side variables.[13]

Stata's workhorse `predict` command generated both one-step-ahead and dynamic forecasts from univariate ARIMA models. For VARs, however, `predict` produces only the one-step-ahead forecasts, while the `fcast` command calculates and displays dynamic forecasts. While dynamic forecasts are more useful in practice, both types of forecasts have their place in time-series analysis.

Let's start by calculating one-step-ahead forecasts. The allowed options for the `predict` command are context sensitive; they depend on the estimation command that preceded it. After VAR estimation, the syntax is

`predict` $\begin{bmatrix} type \end{bmatrix}$ *newvar* $\begin{bmatrix} if \end{bmatrix}$ $\begin{bmatrix} in \end{bmatrix}$ $\begin{bmatrix} , \underline{\texttt{equation}}(eqno\begin{bmatrix} , eqno \end{bmatrix}) \ statistic \end{bmatrix}$

Because a VAR is a multiple-equation model, you have to specify the equation for which you want predictions. You can specify the equation by a number that indicates its order in the model. For example, to predict y_{it}, you can specify `equation(#i)`. Alternatively, you can specify the equation name, which in the case of a VAR is the name of the dependent variable you wish to predict. If you omit the `equation()` option, Stata assumes you want predictions of the first variable in the VAR.

The optional statistic is `xb` (the predicted value), `stdp` (the standard error of the predicted value), or `residuals`. The default is `xb`.

---

13. A partial exception arises when there are gaps in the lag polynomial. For example, consider a variable that follows the fourth-order univariate autoregression

$$y_t = \phi_4 y_{t-4} + \epsilon_t$$

In period $t-1$, the first four predictions $(\widehat{y}_t, \widehat{y}_{t+1}, \widehat{y}_{t+2}, \widehat{y}_{t+3})$ are based on observed values of the lagged dependent variable, but $\widehat{y}_{t+4}$ is based on $\widehat{y}_t$. Another complication arises when exogenous variables are included in the time-series model. To predict y_t more than one step ahead, predictions of the exogenous variables must be generated somehow. One common practice is to specify a separate time-series model (either a VAR or a set of univariate autoregressions) for the exogenous variables.

Here is an example of calculating one-step-ahead forecasts from our three-variable VAR:

```
. use ${ITSUS_DATA}/varexample, clear
(Quarterly economic statistics)
. quietly var inflation unrate fedfunds if date<=tq(2001q2), lags(1 2 3 4)
. predict f_inflation
(option xb assumed; fitted values)
(4 missing values generated)
. predict test1, equation(#1)
(option xb assumed; fitted values)
(4 missing values generated)
. predict test2, equation(inflation)
(option xb assumed; fitted values)
(4 missing values generated)
. label variable f_inflation "forecast"
. compare f_inflation test1
```

	count	minimum	difference average	maximum
f_infla~n=test1	205			
jointly defined	205	0	0	0
jointly missing	4			
total	209			

```
. compare f_inflation test2
```

	count	minimum	difference average	maximum
f_infla~n=test2	205			
jointly defined	205	0	0	0
jointly missing	4			
total	209			

```
. drop test1 test2
```

We snuck in a new Stata command, var, which you can think of as varbasic without training wheels. var offers some additional flexibility, but it does not include commonly used graphs of summary information. Instead, var stores information internally that can be used by later commands to produce a wide variety of statistics and graphs. We will come back to the var command and its options later as we explore other features of VARs, but for now, let's get back to forecasting.

Notice the adverb quietly in front of the var command. quietly tells Stata to process the following command normally but to suppress the display of any output. We have already seen these lengthy VAR estimates—they are identical with the estimates produced by the varbasic command in section 9.2.1—and we just want to load them into Stata's memory so that predict can use them. As before, we ended the estimation sample at 2001q2, reserving the last ten years of data for out-of-sample forecasts.

We calculated the one-step-ahead forecast of inflation three times to illustrate the three different ways of getting to the same place. (In each case, `predict` told us that it could not generate predictions in the first four observations, because it needs four lags of each of the variables to calculate a prediction.) The first calculation is the lazy version—we take advantage of the fact that inflation is the first variable in the VAR. In the second calculation, we specify predictions from the first equation, which happens to be the equation for inflation. The third calculation is the recommended version. We indicated by name the variable we wanted to predict. The `compare` commands that follow verify that all three calculations produce identical results.

Now that we have calculated one-step-ahead forecasts of inflation, let us do something with them. We can compare them with actual inflation rates in the out-of-sample period.

```
. tsline inflation f_inflation if date>tq(2001q2), lpattern(solid dash)
```

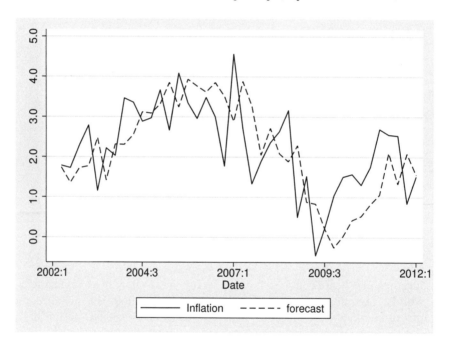

Figure 9.3. U.S. inflation, forecasts and actuals, 2002:2–2012:1

Or we can examine the model misses.

```
. predict r_inflation, equation(inflation) residuals
(4 missing values generated)
. tsline r_inflation, yline(0)
```

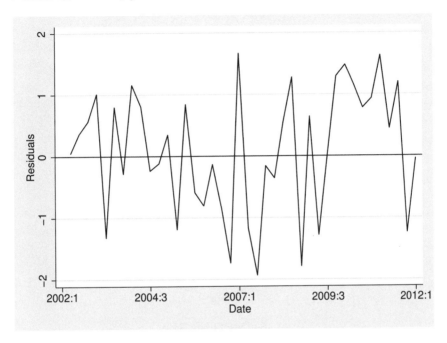

Figure 9.4. Forecast errors of U.S. inflation, 2001:2–2012:1

Or we can get really fancy and add a confidence band for the forecasts.

```
. predict s_inflation, equation(inflation) stdp
(4 missing values generated)
. generate l_inflation = f_inflation - 2*s_inflation
(4 missing values generated)
. generate u_inflation = f_inflation + 2*s_inflation
(4 missing values generated)
. twoway (rarea u_inflation l_inflation date, fintensity(inten10))
> (line inflation f_inflation date, lpattern(solid dash)) in -40/1,
> yline(0) legend(order(2 3 1) label(1 "95% CI")) name(inflation, replace)
```

I will not even try to explain that last graph command—it's a doozy![14] Figure 9.5 displays this graph along with corresponding ones for the unemployment rate and the Federal funds rate.

---

14. And I could not have puzzled it out myself without consulting the outstanding "cookbook" for Stata graphics, *A Visual Guide to Stata Graphics* by Michael N. Mitchell (2012). I strongly recommend you order the latest edition from the Stata Press website; see https://www.stata-press.com/books/visual-guide-to-stata-graphics/.

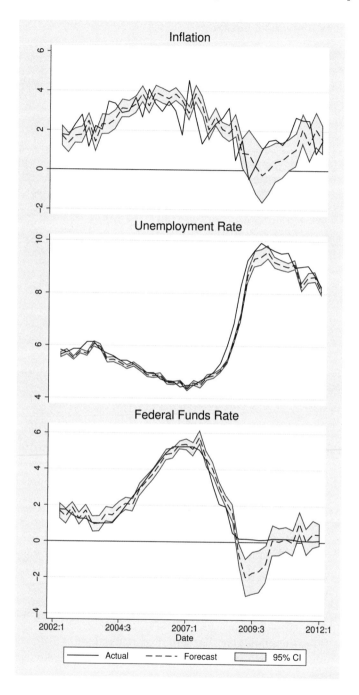

Figure 9.5. Forecasts of U.S. inflation, unemployment, and the Federal funds rate, 2001:3–2011:2

One-step-ahead forecasts rarely drift too far away from actuals because they "true up" to the most recent actuals every period. Nonetheless, some interesting features of these forecasts are apparent in figure 9.5. For all three variables, the uncertainty around the forecasts increased (the confidence interval widened) at the end of 2008 as the global financial crisis peaked. The impact is most noticeable for inflation, which has a wider confidence band throughout, but all three variables exhibit this pattern. Quarterly inflation appears particularly difficult to predict. Its oscillations frequently push it outside the confidence interval.

The unemployment rate is more predictable, but the actuals appear to come from separate regimes—a low-unemployment regime (average 5.3%) prior to 2008, a sudden increase in unemployment over the course of 2008, and a high-unemployment regime thereafter (average 9.2%). Notice that the forecasts systematically underpredict the unemployment rate, with the worst performance in 2009. The forecasts underpredict the unemployment rate in 32 of the 40 forecast quarters. In contrast, the forecasts of inflation and the Federal funds rate are nearly equally balanced between overprediction and underprediction.

Over the course of 2008, the Federal Reserve lowered the Federal funds rate sharply in response to the deteriorating financial environment. The rate averaged 4.5% in the fourth quarter of 2007, but by the fourth quarter of 2008, it had dropped to 0.5%. In 2009, the Fed pushed the rate close to 0, and the rate averaged just 0.1% in the first quarter of 2012. Our VAR model projected negative Federal funds rate for all of 2009 and for several subsequent quarters. Our linear model does not include a nonnegativity constraint, but it is unlikely we would report out this type of prediction.

We did a fair amount of work to produce these one-step-ahead forecasts and graphs. In practice, we are usually more interested in dynamic forecasts, and Stata makes it much easier to generate dynamic VAR forecasts. How easy? Take a look.

```
. fcast compute F_, step(40)

. fcast graph F_inflation F_unrate F_fedfunds, observed lpattern(dash)
```

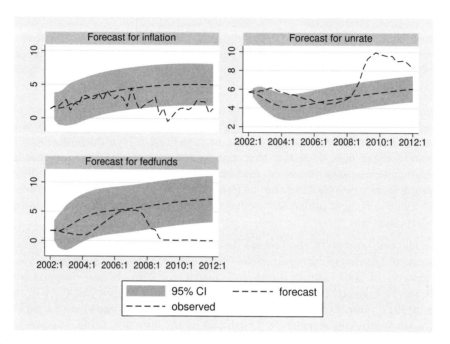

Figure 9.6. Dynamic forecasts of U.S. inflation, unemployment, and the Federal funds rate, 2002:2–2012:1

That is pretty easy. The `fcast compute` command calculates all the forecasts and confidence intervals and saves the information in variables with a prefix chosen by you (F_ in this example). `fcast graph` displays the graphs for all the dependent variables in the VAR. The syntax of `fcast compute` is[15]

`fcast compute` *prefix* [, <u>step</u>(*#*) <u>d</u>ynamic(*time_constant*) <u>esti</u>mate(*estname*)

  `replace` *other_options*]

`step()` indicates the number of periods to forecast. We specified `step(40)` to cover the entire 10-year out-of-sample period. By default, the dynamic forecasts begin in the period following the end of the estimation sample. `fcast compute` calculates dynamic forecasts, upper and lower bounds of the 95% confidence interval, and the standard errors of the forecast.

---

15. In this section, we use the `fcast` command. Another alternative is to use the suite of `forecast` commands. This suite of commands can be used with **var**, **vec**, and other estimation commands. See [TS] **forecast**.

```
. describe F_infl*

               storage  display    value
variable name  type     format     label    variable label
```

```
F_inflation     double %10.0g               F_inflation, dyn(2002:2)
F_inflation_LB  double %10.0g               95% LB for F_inflation
F_inflation_UB  double %10.0g               95% UB for F_inflation
F_inflation_SE  double %10.0g               SE for F_inflation

. list date inflation F_inflation if date>tq(1999q4) & date<tq(2003q1)
```

	date	inflat~n	F_infla~n
161.	2000:1	3.2	.
162.	2000:2	2.0	.
163.	2000:3	2.4	.
164.	2000:4	2.1	.
165.	2001:1	2.7	.
166.	2001:2	2.7	.
167.	2001:3	1.3	.
168.	2001:4	1.2	.
169.	2002:1	1.4	1.4136602
170.	2002:2	1.8	1.7398057
171.	2002:3	1.7	1.4791789
172.	2002:4	2.3	1.5471971

You can start the dynamic forecasts earlier (but not later) by specifying the dynamic option. In this case, no standard errors are calculated.[16]

```
. fcast compute Z_, step(40) dynamic(tq(2000q3))
since tq(2000q3) is in the estimation sample, nose is implicitly specified

. describe Z_*

               storage  display    value
variable name  type     format     label    variable label
```

```
Z_inflation  double %10.0g                 Z_inflation, dyn(2000:3)
Z_unrate     double %10.0g                 Z_unrate, dyn(2000:3)
Z_fedfunds   double %10.0g                 Z_fedfunds, dyn(2000:3)
```

---

16. You can also suppress the standard errors with the **nose** option. Further, you can ask Stata to calculate bootstrap estimates of the standard errors rather than asymptotic estimates.

```
. list date inflation F_inflation Z_inflation if date>tq(2000q4) &
> date<tq(2003q1)
```

	date	inflat~n	F_infla~n	Z_infla~n
165.	2001:1	2.7	.	2.958903
166.	2001:2	2.7	.	2.8533951
167.	2001:3	1.3	.	2.9386774
168.	2001:4	1.2	.	3.0119638
169.	2002:1	1.4	1.4136602	3.0785473
170.	2002:2	1.8	1.7398057	3.0999383
171.	2002:3	1.7	1.4791789	3.158313
172.	2002:4	2.3	1.5471971	3.2303194

VAR estimates can be given a name and saved for later use. The `estimate()` option allows you to specify a saved estimate to use in place of the most recent one.

The syntax of `fcast graph` is

fcast graph *varlist* [*if*] [*in*] [, noci o̲bserved o̲bopts(*cline_option*)

   other_graph_options]

The *varlist* contains the names of one or more of the forecasted variables. `noci` suppresses the display of the confidence bands. `observed` directs Stata to display the actual values along with the predictions. The appearance of the actuals can be altered via the `obopts()` option. And you can add some other graph options as well. In our example, we specified that forecasts be connected by a dashed line by including the `lpattern(dash)` option.

Figure 9.7 overlays the one-step-ahead forecasts on the graph of the dynamic forecasts.

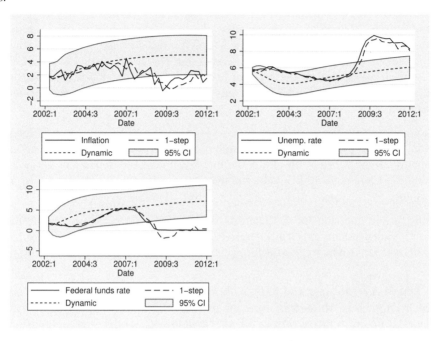

Figure 9.7. Dynamic and one-step-ahead forecasts

As you would expect, the one-step-ahead forecasts track the actuals more closely than the dynamic forecasts, even for short forecast horizons. The dynamic forecasts are very smooth relative to either the actuals or the one-step-ahead forecasts. The dynamic forecasts of inflation and the Federal funds rate systematically overpredict the actuals. And, of course, the dynamic forecasts have no way of reacting to the financial crisis in late 2008. The "regime" shift in unemployment is missed completely.

Nonetheless, relative to the very wide—note the vertical scales—standard errors of the forecasts, the dynamic forecasts perform reasonably. Despite the extremely long, 10-year forecast horizon, the actuals remain within the confidence bands for an impressively long time.

Evaluating a VAR forecast

One of the appeals of VARs is the ease of calculating dynamic forecasts that take account of the interdependence of the elements in $\mathbf{y}_t$. As a reduced-form representation, a VAR places few restrictions on the estimated relationships among the elements of $\mathbf{y}_t$—essentially just the selection of p, the order of the VAR, and the linearity of each equation; hence, VAR forecasts may be less vulnerable to the problems that can arise in structural models when one or more of the relationships are misspecified.

Just how good is the forecast from a VAR? The previous section provided a qualitative impression of performance for our three-equation VAR. We can be more formal though. We can compare the performance of the VAR forecasts with predictions generated by simpler techniques. For example, when we evaluate a static regression, we measure the improvement in model fit over the fit from simply predicting the dependent variable by its sample mean. (The F test of the regression and the R^2 statistic are based on this comparison.) If we cannot do better than just using the mean as a predictor, then the model does not provide much insight. We can calculate this comparison for a time-series model as well.

A more interesting benchmark for time-series models is the comparison of the model prediction with the prediction from a random walk. The random walk model asserts that the only useful information about the future value of the dependent variable is its current value—there is no additional dynamic behavior that can be leveraged to predict future values.

Finally, a key question for a VAR forecast is whether it improves on the forecast provided by independent univariate time-series models for each element of $\mathbf{y}_t$. If the univariate models do just as good a job, there is no advantage to the extra complexity of the VAR.

The table below displays the RMSEs obtained from pseudo out-of-sample forecasts obtained from each of these four forecast methods. Pseudo out-of-sample forecasts are calculated by an iterative process. First, a candidate model is estimated by using data for periods 1 through t, and forecasts are calculated for periods $t + 1$ through T. Next the estimation sample is extended by one period—the new sample covers periods 1 through $t + 1$—and new forecasts are calculated for periods $t + 2$ through T. This process of extending the sample by one period, reestimating, and reforecasting continues until the data are exhausted and the RMSE of the forecasts is calculated. Pseudo out-of-sample forecasts are intended to replicate the type of forecast that occurs frequently in practice. Each period, as a new observation becomes available, a forecaster can update his or her model estimates with this new information.[17]

With a bit of work, Stata can calculate all of these comparisons, but the process is a bit messy and tedious. I wrote a Stata command called `varbench` to perform these calculations together and display them neatly. See chapter 1 for details on how to obtain this command.[18]

---

17. These forecasts are called "pseudo" because they use the most currently published version of the data series. In reality, data are revised frequently, many times significantly. A real forecaster would incorporate both new (but not settled) observations and updates and revisions to previous observations in each reestimate.

18. But be careful! `varbench` is a crude command. It uses brute force in place of intelligence. Instead of using Stata's matrix language, Mata, to update the univariate autoregression and VAR estimates efficiently, `varbench` simply reestimates each regression. This approach is extremely slow but still more convenient than trying to calculate the RMSEs manually. On occasion, I have written a Stata command that the folks at Stata feel might be useful to a wider audience than just me. In that case, they replace my clumsy programming with some really beautiful (and efficient) code. Who knows—we might get lucky.

The syntax of `varbench` is

varbench *depvarlist* $\big[$, <u>l</u>ags(*numlist*) <u>s</u>teps(*numlist*) <u>start</u>(*string*) <u>end</u>(*string*)
  <u>pref</u>ix(*string*) $\big]$

The only thing you have to type is `varbench` and the list of dependent variables. `varbench` will make guesses for all the options if you omit them.

The `lags()` option specifies the lags to include in the univariate AR and VAR models. You must specify each of the lags you want included. For example, `lags(1 2 3 4)` will include the first four lags of each of the dependent variables, but `lags(4)` will include only the fourth lag.

The `steps()` option specifies the horizon of the forecasts you would like to compare. For example, `steps(2 4 8)` directs `varbench` to compare the RMSEs of the 2-step-ahead, 4-step-ahead, and 8-step-ahead forecasts from each method.

The `start()` and `end()` options specify the sample periods to use for updating estimates. For example, `start(tq(1985q1)` indicates that the initial estimates will include observations from the beginning of the dataset through the first quarter of 1985, and the indicated step-ahead forecasts will be calculated from that point and saved. Similarly, `end(tq(2000q4))` indicates that the final estimates will be based on observations from the beginning of the dataset through the fourth quarter of 2000.[19]

You should never have to specify the `prefix()` option. We saw above that `fcast compute` stores VAR forecasts in new variables whose names combine a user-specified prefix with the forecasted variable's name. `varbench` will try several prefixes unlikely to occur in your dataset, but it may fail to find a name that does not already exist. In that event, you can specify a prefix that you know will work.[20]

---

19. Use the `start()` and `end()` options wisely. They determine the time it takes for `varbench` to run.
20. In addition to general inefficiency, `varbench` suffers from some obvious gaps. It does not save the various forecasts, it does not allow for including exogenous variables, it does not allow for comparisons to ARIMA alternatives, and so on. Feel free to add those features.

Let's compare the RMSEs of these alternative pseudo out-of-sample forecasts for our reduced-form VAR.[21]

```
. varbench inflation unrate fedfunds, lags(1 2 3 4) start(tq(2002q1))
RMSE of simulated out-of-sample forecasts
End of sample: 2002:1 to 2010:1
Inflation
                    Method                        % Improvement
         ---------------------------        --------------------
Horizon  Mean     RW     AR    VAR          Mean     RW     AR
   2     1.77    1.15   1.02   1.09           38      5     -7
   4     1.76    1.33   1.15   1.40           20     -6    -22
   8     1.78    1.51   1.36   1.87           -5    -24    -38

Civilian unemployment rate
                    Method                        % Improvement
         ---------------------------        --------------------
Horizon  Mean     RW     AR    VAR          Mean     RW     AR
   2     1.82    0.75   0.51   0.67           63     11    -31
   4     2.01    1.37   1.10   1.49           26     -9    -35
   8     2.26    2.27   1.96   2.66          -18    -17    -35

Federal funds rate
                    Method                        % Improvement
         ---------------------------        --------------------
Horizon  Mean     RW     AR    VAR          Mean     RW     AR
   2     4.29    0.88   0.91   1.16           73    -32    -27
   4     4.39    1.62   1.69   1.68           62     -4      0
   8     4.53    2.71   2.89   2.84           37     -5      2
```

For each of the three dependent variables, **varbench** displays a table of RMSEs for each forecast horizon and the percentage improvement of the VAR forecast compared with the forecasts using the mean, random walk, and a univariate autoregression.

In this example, the VAR forecast is an improvement on simpler forecasts in less than half the comparisons. For inflation, the relative performance of the VAR deteriorates as the forecast horizon increases. For the 2-step-ahead forecast, the RMSE of the VAR forecast is 38% smaller than the RMSE of the mean forecast, 5% smaller than the RMSE of the random-walk forecast, and 7% larger than the RMSE of the univariate AR forecast. At 8 steps ahead, the VAR forecast is less accurate than all three alternatives. A similar pattern holds for forecasts of the unemployment rate. For the Federal funds rate, the VAR forecast is more accurate at all three horizons than the mean forecast but less accurate than the random-walk forecasts. The 2-step-ahead VAR forecast underperforms the AR forecast, but at longer horizons, it's a dead heat.

These daunting results serve as a reminder that there is no such thing as a forecasting method that is "best" in all circumstances. A VAR is convenient and holds the promise of incorporating extra information from the additional dependent variables, but there is no guarantee that it will perform well for your data.[22]

---

21. This **varbench** command took minutes to finish on my MacBook Pro. I told you **varbench** was inefficient.

22. Stock and Watson (2001) calculate RMSEs for the period 1985 Q1 through 2000 Q4 and find the VAR performs better, although it does not always outperform the alternatives.

When statistical results are puzzling, it's worth looking at the data more closely. Look back at figure 9.1. There appear to be two inflation regimes, a regime with high and variable inflation prior to 1985 and a more quiescent period from 1985 and later. Perhaps there is some shift in the dynamic relationships among the variables from one regime to the next. To follow up on this notion, we calculate the relative performance of the alternative forecast methods using only data from 1985 and later.

```
. drop if year(dofq(date))<1985
(100 observations deleted)
. varbench inflation unrate fedfunds, lags(1 2 3 4) start(tq(2002q1))
RMSE of simulated out-of-sample forecasts
End of sample: 2002:1 to 2010:1
Inflation
                      Method                    % Improvement
              -----------------------        --------------------
Horizon   Mean    RW     AR    VAR       Mean    RW     AR
    2     1.13   1.15   1.10   1.01       11     13      8
    4     1.14   1.33   1.11   1.15       -1     14     -3
    8     1.17   1.51   1.17   1.49      -28      1    -27

Civilian unemployment rate
                      Method                    % Improvement
              -----------------------        --------------------
Horizon   Mean    RW     AR    VAR       Mean    RW     AR
    2     1.85   0.75   1.75   0.57       69     23     67
    4     2.07   1.37   1.97   1.22       41     11     38
    8     2.35   2.27   2.27   2.38       -1     -5     -5

Federal funds rate
                      Method                    % Improvement
              -----------------------        --------------------
Horizon   Mean    RW     AR    VAR       Mean    RW     AR
    2     3.47   0.88   3.49   0.86       75      2     75
    4     3.57   1.62   3.59   1.63       54     -1     55
    8     3.69   2.71   3.76   3.12       15    -15     17
```

The results have changed—sometimes for the better, sometimes for the worse.

These results are merely suggestive. A more serious investigation would be required to sort out the sources of the instability of relative forecast performance.[23] But this example does demonstrate the importance of looking at the data from several different angles before coming to a conclusion.

9.3 Who's on first?

Up to this point, we have used our three-variable VAR as a reduced-form model with no restrictions. If our only interest were in forecasting and characterizing the joint distribution of inflation, unemployment, and the Federal funds rate, that would be enough. Invariably, though, we would like to be able to say more. At a minimum, we would like

---

23. There are lots of possibilities. Setting the order of the VAR shorter or longer than four lags may improve performance. Perhaps we have left out important exogenous variables that are shifting performance. Perhaps there is a regime change that we have missed. Lots of work to do here.

to be able to sharpen our description of the temporal relationships among the variables. We would like to be able to say, for instance, that decreases in the Federal funds rate (which represent an expansionary monetary policy) trigger a temporary reduction in unemployment two quarters later, then increase inflation in later quarters as the unemployment rate returns to its natural level. This sequence of events is roughly consistent with a simple version of the Phillips curve/natural rate of unemployment theory of the macroeconomy.[24] The essence of this type of analysis is the ability to place events in order—event A happens first, followed by event B, then event C. This is the realm of recursive VARs.

It is important to recognize at the start that this goal may be too ambitious. The data may not support a definitive ordering of events. Or more precisely, the data may be consistent with multiple orderings. As a result, we have to select among possible orderings on the basis of additional assumptions, prior beliefs, or theories about the causal relationships among the variables in our VAR. The data alone will not decide for us.

This situation is no different for VARs than for any other reduced-form model. To make stronger statements about the relationships among variables, we must be willing to make identifying assumptions. However, because VARs were promoted, in part, to eliminate what were seen as unrealistic or overly strong identifying assumptions in macroeconomic models, it is generally desirable to limit the use of identifying assumptions as much as possible.

We are going to transition from our reduced-form VAR to a recursive VAR in easy steps. First, we will look at the pairwise cross correlations among our variables—a technique for characterizing temporal relationships that does not require a VAR. Then we will introduce methods for summarizing the temporal information embedded in the estimated parameters of the VAR. We will start with Granger causality tests. These tests characterize temporal relationships in terms of predictability—does variable A significantly help predict variable B after we have already taken account of lagged values of variable B? Granger causality tests fall somewhere between the reduced-form and recursive approaches to VARs. They make statements about the order of events but do not require additional identifying assumptions. Finally, we will introduce the two signature techniques of the recursive VAR—FEVDs and IRFs—and demonstrate how they can be used.

9.3.1   Cross correlations

When there are many jointly dependent variables in a VAR, it can be difficult to apprehend all the interrelationships among the variables. Often it's useful to ground our intuition by reviewing the joint distributions of pairs of variables.[25] Recall that the

---

24. This being macroeconomics, this sequence also is consistent with alternative theories.

25. Indeed Box and Jenkins (1976) devote substantial attention to transfer function models, that is, models of the relationships between one input variable and one output variable. This emphasis reflects their focus on time-series models of industrial control processes.

distribution of a stationary, normally distributed time series is characterized completely by its mean and autocovariance function. Similarly, the bivariate distribution of a pair of stationary, normally distributed time series, y_t and x_t, is characterized completely by the means of y_t and x_t and by the cross-covariance function

$$\gamma_{yx}(k) \equiv E\left\{(y_t - \mu_y)(x_{t+k} - \mu_x)\right\}$$

The cross-correlation function is

$$\rho_{yx}(k) = \frac{\gamma_{yx}(k)}{\sigma_y \sigma_x}$$

The autocorrelation function, ρ_k, is symmetric about 0, that is, $\rho_k = \rho_{-k}$. The same is not true of the cross-correlation function. In general, $\rho_{yx}(k) \neq \rho_{yx}(-k)$. However, $\rho_{yx}(k) \equiv \rho_{xy}(-k)$.[26]

The **xcorr** command calculates and displays cross-correlation functions. The syntax is

xcorr *varname1 varname2* $\left[\,if\,\right]$ $\left[\,in\,\right]$ $\left[\,,\; \underline{\text{lags}}(\#)\; \underline{\text{table}}\; other\_options\,\right]$

The **lags()** option indicates the maximum lag and lead to display. If you omit this option, Stata will choose a maximum based on the size of your sample. By default, Stata displays the cross-correlation function as a graph. If you prefer to see the values of the cross correlations, specify the **table** option.

Let's see what the cross-correlation functions can tell us about our three macroeconomic variables. We start with the inflation and unemployment rates, and we calculate the cross correlations over the same sample we used to estimate our VAR.

```
. xcorr unrate inflation if date<tq(2002q1), name(left)
. xcorr inflation unrate if date<tq(2002q1), name(right)
. graph combine left right, name(leftright)
```

---

26. To convince yourself of these statements, consider the two-equation system

$$
\begin{aligned}
y_t &= \alpha_{11} y_{t-1} + \alpha_{12} x_{t-1} + \epsilon_{1,t} \\
x_t &= \alpha_{22} x_{t-1} + \epsilon_{2,t}
\end{aligned}
$$

It's straightforward to show

$$
\begin{aligned}
\gamma_{yx}(1) &= \frac{\alpha_{11}\alpha_{22}}{1 - \alpha_{11}\alpha_{22}}\gamma_{xx}(1) \\
&\neq \alpha_{11}\alpha_{22}\gamma_{yx}(0) + \alpha_{12}\alpha_{22}\gamma_{xx}(0) \\
&= \gamma_{yx}(-1)
\end{aligned}
$$

and

$$\gamma_{yx}(1) = \gamma_{xy}(-1)$$

We have used the `name()` option of the graph command so that we can refer to the individual cross-correlation graphs later.[27] We display only the final, combined graph in figure 9.8.

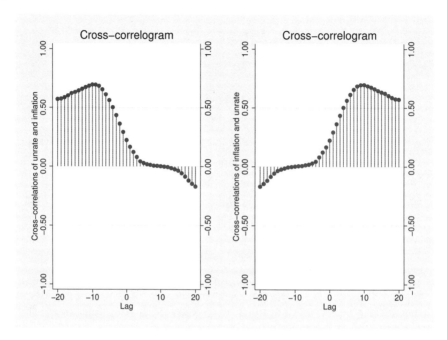

Figure 9.8. Cross correlations of inflation and unemployment rates, 1960:2–2001:2

How do we read these graphs?[28] Let's look at the left-hand graph. The correlation for $k = -10$ is around 0.7. Is that the correlation of `L10.inflation` and `unrate` or the correlation of `L10.unrate` and `inflation`? You can tell by reading the *Methods and formulas* section of the `xcorr` entry in the *Time-Series Reference Manual* ([TS]), or, like me, you can use the lazy man's method.

---

27. Among the unspecified *other_options* in the syntax diagram of the `xcorr` command are many of the options to refine the appearance of the graphs.
28. To start, they confirm that the cross-correlation function is not symmetric around 0; that is, $\rho_{yx}(k) \neq \rho_{yx}(-k)$. Also switching the order of the two variables produces a mirror image of the original function; that is, $\rho_{yx}(k) = \rho_{xy}(-k)$.

```
. correlate inflation L10.inflation unrate L10.unrate if date<=tq(2002q1)
(obs=159)
```

	inflat~n	L10. inflat~n	unrate	L10. unrate
inflation				
--.	1.0000			
L10.	0.5551	1.0000		
unrate				
--.	0.2398	0.7136	1.0000	
L10.	-0.0020	0.1962	0.4167	1.0000

The positive correlation of 0.71 represents the correlation between the inflation rate lagged 10 quarters and the current quarter's unemployment rate. Similarly, the correlation of roughly 0 (-0.002) represents the correlation between the unemployment rate lagged 10 quarters and the current quarter's inflation rate or, equivalently, the correlation between the current quarter's unemployment rate and the inflation rate 10 quarters ahead.[29] So when you type

```
xcorr y x
```

you can read the resulting graph as the correlations between different leads and lags of the variable x and the current value of the variable y.

Now that we have cleared that up, what do these graphs tell us? Increases in the inflation rate today are correlated with future increases in the unemployment rate, but increases in the unemployment rate appear to be uncorrelated with future inflation. This pattern might occur if the Federal Reserve believes it can control inflation but not unemployment. In this situation, the Fed will tighten monetary policy (raise the Federal funds rate) when inflation rises above levels the Fed deems acceptable. This monetary tightening generates what appears to be a gradual but persistent increase in the unemployment rate.

Of course, this is only one possible explanation of these cross correlations. Let's take a look at all the pairwise cross correlations to see if we can glean any more information. Figure 9.9 combines nine different cross-correlation function estimates. Each column displays the cross correlations for one pair of variables. For example, the column labeled "Unemployment/inflation" displays graphs produced by the command

```
xcorr inflation unrate
```

for different sample periods. The first row displays cross correlations for the period prior to 1985. The second row focuses on the period from 1985 through the end of our estimation sample. And the third row displays graphs for the 10-year out-of-sample period, that is, the second quarter of 2002 through the first quarter of 2012. To make the graphs more readable, we specified the `lags(12)` option to restrict the display to three years of lags and leads.

---

29. Assuming bivariate stationarity.

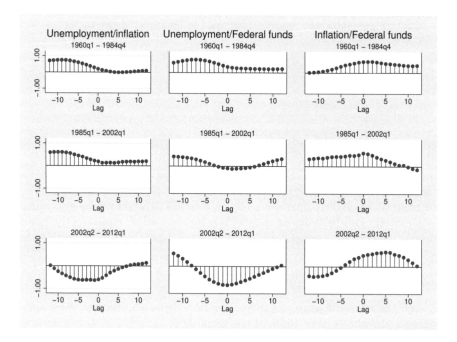

Figure 9.9. Cross correlations of inflation, unemployment, and the Federal funds rate

The division of our data into an early period (pre-1985), a late period (1985–2002q1), and an out-of-sample period (2002q2–2012q1) is somewhat arbitrary, but it highlights some of the challenges of interpreting dynamic macroeconomic relationships. Start with the column labeled "Unemployment/inflation". The cross-correlation functions are very similar in the early and late periods. Lagged inflation rates are positively correlated with current unemployment rates, but there is little correlation between current unemployment and future inflation. However, the cross correlations in the out-of-sample period are strikingly different from either the early- or the late-period estimates. Do these differences represent a change in monetary policy, fiscal policy, or the influences of novel global economic shocks? Difficult to say, but something clearly has changed.

We see different patterns of results in the next two columns. In the middle column, labeled "Unemployment/Federal funds", the cross-correlation function seems to evolve. In the early period, lagged increases in the Federal funds rate (tighter monetary policy) are associated with higher unemployment in the current quarter. In fact, the cross correlations are at least modestly positive at all lags and leads. In the late period, the correlations between current unemployment and the future funds rate turn modestly negative, consistent with a Federal Reserve policy of easing monetary policy in reaction to increases in the unemployment rate. That pattern becomes more pronounced in the out-of-sample period. In fact, the correlations on both sides of a 0 lag (contemporaneous correlation) become negative.

The third column, labeled "Inflation/Federal funds", presents even more instability. In the early period, the Federal funds rate from two to three years earlier is uncorrelated with current inflation, but there are positive correlations at all other lags and leads. The late period looks like a mirror image of the early period, with positive correlations at all but the longest leads.[30] And the pattern in the out-of-sample period does not resemble either of the other two cross-correlation function estimates.

What are we to make of this apparent instability? First, remember that we have not looked at any measures of statistical significance, so it's unclear which of these differences represent ordinary sampling error and which represent a systematic change in the bivariate relationships.[31] Second, if the three-variable VAR represents a complete characterization of the joint probability distribution of inflation, unemployment, and the Federal funds rate, shifts in the bivariate relationships may reflect a shift in the forces determining one of the three variables. The obvious candidate is the Federal funds rate because it is determined by the monetary policy of the Federal Reserve. The funds rate is determined by the Federal Open Market Committee of the Fed at its regular meetings. The composition of the Federal Open Market Committee—the chairman of the Fed, the other members of the Board of Governors, and a shifting subset of the presidents of the 12 Federal Reserve Banks—changes over time, and different points of view and attitudes may prevail over time. Third, just as these bivariate relationships represent a partial view of the relationships between all three variables, our three-variable VAR may also be "too small". In other words, we may have left out one or more jointly endogenous variables—for instance, a variable representing the influence of foreign demand—and the behavior of those left-out variables may be driving the apparent instability of the relationships among inflation, unemployment, and the funds rate. Finally, we cannot ignore the possibility that structural macroeconomic relationships have shifted over time, although this last possibility is perhaps the hardest to verify and estimate.

This review of the cross-correlation functions has emphasized qualitative patterns among the variables. In the next sections, we introduce more formal statistical techniques for characterizing the timing patterns of variable movements.

9.3.2 Summarizing temporal relationships in a VAR

Even a small VAR contains a lot of parameters. While the relationships among the elements of $\mathbf{y}_t$ are completely determined by $\mathbf{\Phi}(L)$ and $\mathbf{\Sigma}$, it is nearly impossible to interpret those relationships by inspecting the estimated parameters. Instead we need ways to summarize the information contained in these parameters. Three measures commonly used to provide these summaries are Granger causality tests, FEVDs, and IRFs.

---

30. This reversal is so striking that I double checked my Stata commands to make sure I had not mistyped something.
31. But recall from our discussion of the autocorrelation function in chapter 7 that the estimated standard errors of these correlations can be misleading at times.

Granger causality

In a VAR, each random disturbance influences all the endogenous variables. Indeed the whole point of estimating a VAR is to characterize the joint distribution of the elements of $\mathbf{y}_t$. Nonetheless, random disturbances may exhibit their influence in some of the endogenous variables earlier and others later. Granger causality tests check a VAR for evidence of this type of temporal ordering by testing whether lagged values of one variable, say, $y_{j,t}$, improve the forecast of another variable, say, $y_{i,t}$ after the lagged values of $y_{i,t}$ are taken into account.[32]

Consider a single equation from a simple VAR:

$$
\begin{aligned}
y_{i,t} \;=\; & \mu_i + \phi_{1,i1}y_{1,t-1} + \cdots + \phi_{1,in}y_{n,t-1} \\
& + \phi_{2,i1}y_{1,t-2} \cdots + \phi_{2,in}y_{n,t-2} + \cdots \\
& + \phi_{p,i1}y_{1,t-p} + \cdots + \phi_{p,in}y_{n,t-p} + \epsilon_{i,t}
\end{aligned}
$$

The Granger causality test that $y_{j,t}$ does not Granger cause $y_{i,t}$ is a test of the null hypothesis

$$ H0 : \phi_{1,ij} = \phi_{2,ij} = \cdots = \phi_{p,ij} = 0 $$

If $y_{j,t}$ does not Granger-cause $y_{i,t}$, then $y_{j,t}$ can safely be ignored when calculating forecasts of $y_{i,t}$.

It may be easier to interpret this test if we write it in matrix form. Ignoring the vector of constant terms for the moment, a VAR is written in matrix form as

$$ \boldsymbol{\Phi}(L)\mathbf{y}_t = \boldsymbol{\epsilon}_t $$

We can partition the vector of endogenous variables into two parts, $\mathbf{y}_1$ and $\mathbf{y}_2$, and write the system as

$$
\begin{bmatrix} \boldsymbol{\Phi}_{11}(L) & \boldsymbol{\Phi}_{12}(L) \\ \boldsymbol{\Phi}_{21}(L) & \boldsymbol{\Phi}_{22}(L) \end{bmatrix}
\begin{bmatrix} \mathbf{y}_{1t} \\ \mathbf{y}_{2t} \end{bmatrix}
= \begin{bmatrix} \boldsymbol{\epsilon}_{1t} \\ \boldsymbol{\epsilon}_{2t} \end{bmatrix}
$$

The statement that $\mathbf{y}_{2t}$ does not Granger-cause $\mathbf{y}_{1t}$ is equivalent to the statement that $\boldsymbol{\Phi}_{12}(L) = 0$, and the system of equations can be written as the block-triangular system

$$
\begin{bmatrix} \boldsymbol{\Phi}_{11}(L) & 0 \\ \boldsymbol{\Phi}_{21}(L) & \boldsymbol{\Phi}_{22}(L) \end{bmatrix}
\begin{bmatrix} \mathbf{y}_{1t} \\ \mathbf{y}_{2t} \end{bmatrix}
= \begin{bmatrix} \boldsymbol{\epsilon}_{1t} \\ \boldsymbol{\epsilon}_{2t} \end{bmatrix}
$$

The term "Granger causality" makes some statisticians uncomfortable. One of the earliest lessons in statistics is "correlation does not imply causation". Yet the Granger causality test is based on the correlation (or lack of correlation) of lagged values of one variable with current values of another variable. The time-series nature of this test

---

32. Well, sort of. $y_{j,t}$ may influence another variable, say, $y_{k,t}$, which does improve the forecast of $y_{i,t}$. In other words, $y_{k,t}$ may mask the influence of $y_{j,t}$. Does this matter? If $y_{j,t}$ is under the control of policymakers and $y_{k,t}$ is not, it might.

makes it tempting—and sometimes appropriate—to interpret the results as a measure of a chain of causality among variables. However, an assertion of causality requires some nonstatistical basis or explanation in addition to the statistical evidence provided by the Granger causality test.

We could use the Stata `test` command to construct a Granger causality test. For example, in our example VAR, we could test whether inflation Granger-causes the unemployment rate.

```
. test [unrate] : L1.inflation L2.inflation L3.inflation L4.inflation

 ( 1)   [unrate]L.inflation = 0
 ( 2)   [unrate]L2.inflation = 0
 ( 3)   [unrate]L3.inflation = 0
 ( 4)   [unrate]L4.inflation = 0

           chi2(  4) =      1.49
         Prob > chi2 =    0.8282
```

Note that this test excludes only the four lags of inflation. The lagged values of the Federal funds rate remain in the equation along with the lagged values of the unemployment rate.[33]

Apparently, inflation does not Granger-cause the unemployment rate, at least during our estimation period.[34] This result may seem surprising because our review of the cross-correlation functions above indicated that lagged inflation is positively correlated with the current unemployment rate. Remember, though, that the coefficients tested here apply only to the first four lags of inflation, while the largest positive cross correlations occurred at lags of about 2 1/2 years. More importantly, the cross-correlation functions measure simple correlations, while the Granger causality tests reports on the incremental predictiveness of inflation after the influence of prior values of the unemployment rate already are taken into account.

---

33. In some contexts, authors use "Granger causality" to denote the result of exclusion tests in a bivariate VAR, for instance, in a VAR containing only the unemployment and inflation rate. In our examples, Granger causality—or the lack of it—denotes the results of the exclusion test within our full VAR, that is, taking account of all other variables in the system.
34. None of the four lag coefficients on inflation in the unemployment rate equation are individually significant—the smallest p-value is 0.376—but the insignificance of the individual coefficients does not guarantee that the coefficients are jointly insignificant.

The `vargranger` command[35] provides a convenient way of running all the possible Granger causality tests for a VAR.

```
. vargranger
```

Granger causality Wald tests

Equation	Excluded	chi2	df	Prob > chi2
inflation	unrate	8.5437	4	0.074
inflation	fedfunds	3.6304	4	0.458
inflation	ALL	25.259	8	0.001
unrate	inflation	1.4914	4	0.828
unrate	fedfunds	17.372	4	0.002
unrate	ALL	37.94	8	0.000
fedfunds	inflation	25.304	4	0.000
fedfunds	unrate	31.53	4	0.000
fedfunds	ALL	45.747	8	0.000

`vargranger` displays the results of all the pairwise Granger causality tests and tests that none of the other elements of $\mathbf{y}_t$ Granger-cause $y_{i,t}$. Note that the χ^2 test statistic and *p*-value of the test that inflation Granger-causes the unemployment rate match the results of the `test` command above.

It appears that neither the Federal funds rate nor the unemployment rate Granger-causes inflation, although the unemployment rate is significant at the 10% level. However, these variables are highly significant when tested jointly.[36,37] As we previously noted, inflation does not Granger-cause the unemployment rate, but the Federal funds rate does. And both inflation and the unemployment rate Granger-cause the Federal funds rate.

So what can we make of these test results? Do these results give us any clues about the paths of influence and causality in the U.S. economy? Perhaps. The significance of inflation and unemployment in the equation for the Federal funds rate makes sense because the Federal Reserve relies on the Federal funds rate as one of its key policy instruments as it tries to promote both full employment and price stability. Note, however, that the significance of the coefficients on lagged inflation and unemployment does not imply that the Federal Reserve is "backward looking". The Federal Reserve may base its policy purely on its view of future inflation and unemployment. Because inflation and unemployment are strongly associated with their lagged values,[38] a "forward-looking" policy rule also implies a significant relationship between today's Federal funds rate and lagged values of inflation and unemployment. The Granger causality of the Federal

---

35. By now, you have noticed that Stata has a lot of commands that start with the characters `var`. When we get to vector error-correction models, you will see a lot of commands that start with `vec`.
36. Use caution in interpreting the *p*-values in this table. When multiple tests are conducted, as they are here, nominal *p*-values overstate the significance of test statistics.
37. In contrast to our results, Stock and Watson (2001) found that unemployment Granger-causes inflation at the 5% level. Their other Granger causality results are qualitatively similar to ours.
38. You can use the `test` command to verify this association.

funds rate in the unemployment rate equation is consistent with the view that monetary policy can impact the real economy in the short run. And the lack of association of either unemployment or the funds rate with inflation may reflect extremely long lags between monetary expansion and ensuing inflation. Or not. These results are suggestive, but not definitive.[39]

How to impose order

We have arrived at the decisive break between reduced-form and recursive VARs. Both cross-correlation functions and Granger causality tests make statements about temporal ordering without imposing prior restrictions on the sequence of impacts among the variables. To go further, though, we have to assume a recursive order on the random shocks to the variables. The method for imposing this order was popularized by Sims (1980). However, this type of assumption is not innocuous. The interpretation of the relationships among the variables is affected significantly by the ordering chosen. As a result, this approach is most appropriate when there are sound theoretical reasons for choosing one ordering over another.

To ground our intuition, we start with a univariate autoregressive model.

$$y_t = \mu + \phi_1 y_{t-1} + \cdots + \phi_p y_{t-p} + \epsilon_t$$

If y_t is stationary, we can write this model in moving-average form:

$$y_t = \nu + \sum_{i=0}^{\infty} \psi_i \epsilon_{t-i}$$

Consider the following experiment. Set all the ϵ_t to 0 except for ϵ_{t-k}, which is set to 1. The ψ_i trace the impact over time of this random impulse.

$$
\begin{aligned}
y_{t-k} - \nu &= \psi_0 \\
y_{t-k+1} - \nu &= \psi_1 \\
y_{t-k+2} - \nu &= \psi_2 \\
\vdots \quad \vdots \quad &\vdots \\
y_t &= \psi_k
\end{aligned}
$$

We call the sequence of ψ_i's the IRF.

A similar analysis can be applied to a VAR. Ignoring for the moment any exogenous variables, we write a VAR as a pth-order autoregression:

$$\mathbf{y}_t = \boldsymbol{\mu} + \boldsymbol{\Phi}_1 \mathbf{y}_{t-1} + \cdots + \boldsymbol{\Phi}_p \mathbf{y}_{t-p} + \boldsymbol{\epsilon}_t$$

---

39. See Sargent (1976) for a clear explanation of the substantial difficulties in using these types of correlations as evidence for a particular economic viewpoint.

Assuming stability, we can rewrite this model in moving-average form

$$\mathbf{y}_t = \boldsymbol{\nu} + \sum_{i=0}^{\infty} \boldsymbol{\Psi}_i \boldsymbol{\epsilon}_{t-i}$$

where $\boldsymbol{\Psi}_0 = \mathbf{I}_n$. The $\boldsymbol{\Psi}_i$ are the simple IRFs for this VAR. As in the univariate case, the IRF traces out the impact of a single random shock. In particular, the j, kth element of $\boldsymbol{\Psi}_i$ measures the impact of a one-unit shock to the kth element of $\boldsymbol{\epsilon}_t$ on the jth element of $\mathbf{y}_t$ after i periods, holding all else constant.

Unfortunately, the experiment of disturbing one element of $\boldsymbol{\epsilon}_t$ and holding all others constant is not very interesting. The elements of $\boldsymbol{\epsilon}_t$ are correlated with correlation matrix $\boldsymbol{\Sigma}$; thus we can place no causal interpretation on our impulse experiment.

This limitation is significant. We very much want to be able to predict the impact of a single disturbance on the evolution of $\mathbf{y}_t$. In our three-variable VAR of the macroeconomy, for instance, if some random shock (say, an unexpected transitory drop in foreign demand) affects the unemployment rate, we would like to be able to make predictions about the monetary policy response and the subsequent sequence of inflation and unemployment rates and consequent further adjustments to monetary policy. The simple IRFs cannot help us answer that question.

If we are willing, however, to impose a recursive order on the impact of random shocks, we can answer questions of this type. Here is how.

We can decompose the correlation matrix $\boldsymbol{\Sigma}$ into the product of a lower triangular matrix, $\mathbf{P}$ and its transpose[40]

$$\boldsymbol{\Sigma} = \mathbf{P}\mathbf{P}'$$

Thus

$$\mathbf{P}^{-1}\boldsymbol{\Sigma}\mathbf{P}'^{-1} = \mathbf{I}_n$$

As a result, we can use $\mathbf{P}^{-1}$ to convert $\boldsymbol{\epsilon}_t$ to a vector of uncorrelated random disturbances:

$$E\{\mathbf{P}^{-1}\boldsymbol{\epsilon}_t(\mathbf{P}^{-1}\boldsymbol{\epsilon}_t)'\} = \mathbf{P}^{-1}E(\boldsymbol{\epsilon}_t\boldsymbol{\epsilon}_t')\mathbf{P}'^{-1} = \mathbf{P}^{-1}\boldsymbol{\Sigma}\mathbf{P}'^{-1} = \mathbf{I}_n$$

So with complete generality, we can rewrite our moving-average representation as

$$\mathbf{y}_t = \boldsymbol{\nu} + \sum_{i=0}^{\infty} \boldsymbol{\Psi}_i \mathbf{P}\mathbf{P}^{-1}\boldsymbol{\epsilon}_{t-i} = \boldsymbol{\nu} + \sum_{i=0}^{\infty} \boldsymbol{\Xi}_i \mathbf{v}_{t-i}$$

where

$$\boldsymbol{\Xi}_i \equiv \boldsymbol{\Psi}_i \mathbf{P}$$

---

40. Because correlation matrices are real, symmetric, and positive definite, the Cholesky decomposition can be used to calculate $\mathbf{P}$. Heuristically, $\mathbf{P}$ is often referred to as the square root of $\boldsymbol{\Sigma}$. The choice of the Cholesky decomposition was popularized by Sims (1980), and it has become the standard approach. There is an element of arbitrariness in this technique, which we discuss further in what follows.

and

$$\mathbf{v}_t = \mathbf{P}^{-1} \boldsymbol{\epsilon}_t$$

By construction, the elements of $\mathbf{v}_t$ are mutually orthogonal—$\mathbf{v}_t \mathbf{v}_t' = \mathbf{I}$—so we can make causal statements about shocks to one element of $\mathbf{v}_t$ without considering correlations with the other elements. The $\boldsymbol{\Xi}_i$ are called the OIRFs. They describe how these orthogonalized shocks work their way through the system over time. It can be challenging to grasp exactly how the OIRFs work, so we will take our time explaining them.

The essential feature of OIRFs is the order they impose on the impacts of the orthogonalized shocks. We can see this most easily if we premultiply both sides of $\mathbf{v}_t = \mathbf{P}^{-1} \boldsymbol{\epsilon}_t$ by $\mathbf{P}$ to obtain $\mathbf{P} \mathbf{v}_t = \boldsymbol{\epsilon}_t$ or

$$
\begin{bmatrix}
p_{11} & 0 & \cdots & 0 \\
p_{21} & p_{22} & \cdots & 0 \\
\vdots & \vdots & \ddots & \vdots \\
p_{n1} & p_{n2} & \cdots & p_{nn}
\end{bmatrix}
\begin{bmatrix}
v_{1t} \\
v_{2t} \\
\vdots \\
v_{nt}
\end{bmatrix}
=
\begin{bmatrix}
\epsilon_{1t} \\
\epsilon_{2t} \\
\vdots \\
\epsilon_{nt}
\end{bmatrix}
$$

Now, as a thought experiment, let's consider an orthogonalized, one-unit shock to y_{2t}, the second variable in our VAR. In other words, we set $v_{2t} = 1$ and all the other elements of $\mathbf{v}_t$ to 0.[41] Because the first row in $\mathbf{P}$ contains all zeroes except for p_{11}, this shock to v_{2t} has no impact on ϵ_{1t} and, hence, no impact on y_{1t}. Moving on to the second equation above, we have

$$\epsilon_{2t} = p_{21} v_{1t} + p_{22} v_{2t} = p_{21} \times 0 + p_{22} \times 1 = p_{22}$$

so the impact on y_{2t} is equal to p_{22}.

Things get more interesting when we move to the third and later elements of $\boldsymbol{\epsilon}_t$ and $\mathbf{y}_t$. Using the same logic as above, we have $\epsilon_{3t} = p_{32}$, $\epsilon_{4t} = p_{42}$, and so on. In other words, a one-unit shock to v_{2t} affects $\epsilon_{2t}, \epsilon_{3t}, \ldots$ and, hence, $y_{2t}, y_{3t}, \ldots$ but not ϵ_{1t} or y_{1t}. More generally, a one-unit shock to v_{jt} affects variables in period t in the jth and higher equations but has no impact in period t on variables in prior equations.

This property is of the essence of a recursive VAR. Orthogonalized shocks to the first equation have a contemporaneous impact on all the variables in the system. Shocks to the second equation have a contemporaneous impact on all subsequent equations but not the first. And shocks to the last equation do not have an immediate impact on any other equation. It is important to remember that this property of the OIRF is a consequence of the order we have chosen for the decomposition of $\boldsymbol{\Sigma}$ into $\mathbf{PP}'$. If we reordered the equations in the VAR, we would impose a different sequence of impacts.

The whole point of constructing a recursive VAR is to simplify the interpretation of shocks to the system. If we shock a single element of $\boldsymbol{\epsilon}_t$, the correlations among the

---

41. There is nothing magical about the convention of setting $v_{2t} = 1$. This choice just makes it easier to interpret the sequence of impacts on the other variables in the system. To calculate the impacts of an arbitrary-sized shock, say, v^* to v_{2t}, just multiply the entire OIRF by v^*.

elements of $\boldsymbol{\epsilon}_t$ confound our attempts to make crisp inferences. But the cost of clarifying the channels of influence among the variables in our VAR is the requirement that we impose an ordering on the system a priori.

So far, we have considered only the *initial* impact of an orthogonalized shock, that is, the period t impact. But we also are interested in how these impacts play out over time. To understand these dynamic impacts, we write the moving-average representation of the OIRFs in more detail.

$$
\begin{bmatrix} y_{1t} \\ y_{2t} \\ \vdots \\ y_{nt} \end{bmatrix} = \begin{bmatrix} \nu_1 \\ \nu_2 \\ \vdots \\ \nu_n \end{bmatrix} + \begin{bmatrix} p_{11} & 0 & \cdots & 0 \\ p_{21} & p_{22} & \cdots & 0 \\ \vdots & \vdots & \ddots & \vdots \\ p_{n1} & p_{n2} & \cdots & p_{nn} \end{bmatrix} \begin{bmatrix} v_{1t} \\ v_{2t} \\ \vdots \\ v_{nt} \end{bmatrix}
$$

$$
+ \begin{bmatrix} \xi_{1,11} & \xi_{1,12} & \cdots & \xi_{1,1n} \\ \xi_{1,21} & \xi_{1,22} & \cdots & \xi_{1,2n} \\ \vdots & \vdots & \ddots & \vdots \\ \xi_{1,n1} & \xi_{1,n2} & \cdots & \xi_{1,nn} \end{bmatrix} \begin{bmatrix} v_{1,t-1} \\ v_{2,t-1} \\ \vdots \\ v_{n,t-1} \end{bmatrix}
$$

$$
+ \begin{bmatrix} \xi_{2,11} & \xi_{2,12} & \cdots & \xi_{2,1n} \\ \xi_{2,21} & \xi_{2,22} & \cdots & \xi_{2,2n} \\ \vdots & \vdots & \ddots & \vdots \\ \xi_{2,n1} & \xi_{2,n2} & \cdots & \xi_{2,nn} \end{bmatrix} \begin{bmatrix} v_{1,t-2} \\ v_{2,t-2} \\ \vdots \\ v_{n,t-2} \end{bmatrix} + \cdots
$$

Remember that $\boldsymbol{\Xi}_i \equiv \boldsymbol{\Psi}_i \mathbf{P}$ and $\boldsymbol{\Psi}_0 \equiv \mathbf{I}_n$. This equality accounts for the appearance of $\mathbf{P}$ as the first matrix on the right-hand side of the moving-average representation.

Now let's trace the effects over time of our hypothetical orthogonalized shock to v_{2t}. We have already calculated the period t impacts above—no impact on y_{1t}, y_{2t} increases by p_{22} units, y_{3t} increases by p_{23} units, and so on. In period $t+1$, $y_{1,t+1}$ increases by $\xi_{1,12}$ units, $y_{2,t+1}$ increases by $\xi_{1,22}$ units, $y_{3,t+1}$ increases by $\xi_{1,32}$ units, and so on. In general, then, the impact on variable i of an orthogonalized shock to equation j that occurred k periods in the past is given by the (i, j)th element of $\boldsymbol{\Xi}_k$.

The transformation we have chosen has imposed a recursive system of impacts on the equations in the VAR. Shocks to one variable initially will affect only variables later in the ordering. Does this make sense? In some cases, yes. Consider our three-equation VAR. The Granger causality tests suggest (but do not mandate) an ordering. Inflation and unemployment appear to precede the Federal funds rate; that is, they predict future levels of the funds rate even after accounting for prior values of the funds rate. In contrast, the inflation rate is not Granger-caused by either of the other two variables. So, perhaps, an ordering of inflation $\rightarrow$ unemployment $\rightarrow$ Federal funds rate may make sense. However, it's a stretch to tie this particular ordering to any traditional macroeconomic theory or model. And other orderings may make as much sense. Therein lies the challenge of OIRFs. It can be difficult to defend a particular ordering, and if the ordering is suspect, any causal interpretations are equally suspect. Despite these challenges, OIRFs commonly are used in analyzing VARs.

Before we demonstrate how to use Stata to calculate IRFs and OIRF, we will discuss FEVDs, another set of statistics that require us to select a recursive ordering.

FEVDs

If there were no random disturbances—if ϵ_t were always 0—a VAR would fit the data perfectly, and we could forecast $\mathbf{y}_t$ without error. Real-world VARs cannot predict perfectly. However, they can help us attribute forecast errors to their source.

We begin by writing the one-step-ahead forecast in terms of the moving-average representation of the VAR

$$\widehat{\mathbf{y}}_t(1) = \boldsymbol{\nu} + \sum_{i=1}^{\infty} \boldsymbol{\Psi}_i \epsilon_{t-i}$$

where $\widehat{\mathbf{y}}_t(1)$ denotes the forecast of $\mathbf{y}_{t+1}$ formed at period t. Subtracting this expression from the realized value of $\mathbf{y}_{t+1}$ yields

$$\mathbf{y}_{t+1} - \widehat{\mathbf{y}}_t(1) = \boldsymbol{\Psi}_0 \epsilon_{t+1} = \epsilon_{t+1}$$

because $\boldsymbol{\Psi}_0 \equiv I_n$.

The one-step-ahead forecast error in, say, the ith variable in the VAR depends only on $\epsilon_{i,t+1}$, the random disturbance in the ith equation.

This independence of forecast errors no longer applies for dynamic forecasts. The forecast error for the h-step-ahead forecast is[42]

$$\mathbf{y}_{t+h} - \widehat{\mathbf{y}}_t(h) = \sum_{i=0}^{h-1} \boldsymbol{\Psi}_i \epsilon_{t+h-i}$$

From this equation, it is clear that the error in forecasting $y_{i,t+h}$ depends on the random disturbances to all n equations.

As with IRFs, we are limited in the types of statements we can make about these forecast errors. If we are willing, however, to impose a recursive order, we can attribute the forecast errors for each variable to the random disturbances in individual equations. Using the Cholesky decomposition we described above, we can rewrite the forecast errors as

$$\mathbf{y}_{t+h} - \widehat{\mathbf{y}}_t(h) = \sum_{i=0}^{h-1} \boldsymbol{\Xi}_i \mathbf{v}_{t+h-i}$$

This parameterization enables us to measure the fraction of the total forecast-error variance that is attributable to each orthogonalized shock. As before, this attribution is more reliable if the recursive ordering is supported by sound theoretical arguments.

---

42. See Lütkepohl (2005, sec. 2.2.2) for details.

Using Stata to calculate IRFs and FEVDs

At this point, you might be guessing that Stata provides one or more commands that start with the letters `var` to calculate IRFs and FEVDs. If so, you have been fooled. Instead Stata provides an entire suite of compound commands that start with `irf`. The `irf create` command calculates IRFs and FEVDs, the `irf graph` command displays them, and so on.

There is a logic to this explosion of additional commands. Stata organizes the workflow of all statistical analyses similarly. You begin by bringing your data into Stata and performing any necessary variable transformations. (Likely commands include `use`, `merge`, `generate`, and `replace`.) Next you typically "look" at your data—you calculate summary statistics, examine simple graphs, and review some preliminary calculations. (In the case of VARs, an example of preliminary calculations is provided by the `varsoc` command.) Next you fit one or more statistical models (`varbasic` and `var`). Finally, you choose among the many postestimation commands to perform tests on the model estimates (`test`, `varstable`, `vargranger`), calculate forecasts (`predict` and `fcast`), and perform additional calculations based on the model estimates. (Here is where the `irf` commands come in.)

This process works smoothly because Stata stores results from virtually every command for later use. For simple commands, Stata may store only one or two pieces of information. As an example, the `count` command, which counts the number of observations that satisfy user-specified conditions, stores only the count that is displayed. For more complex commands such as estimation commands, Stata may store extensive results. It takes two pages in the *Time-Series Reference Manual* ([TS]) to list all the items stored by the `var` command. These results remain in memory until you execute a subsequent `var` command.

Most of the time, all this storing and reusing of results is invisible to you. Stata goes about its business and does not bother you with the details. However, if you program your own Stata commands (like the `varbench` command we used to compare forecasts), these stored results can be very useful.

The suite of `irf` commands exposes a bit more of the Stata machinery than usual. The main reason is the amount of "housekeeping" associated with IRFs. In our simple, three-equation VAR, for instance, there are nine OIRFs (the dynamic response of each of the three variables to impulses from each of the three equations) for each ordering of the equations, and there are six potential orderings.[43] Typically, you will want to compare OIRFs from alternative orderings to determine how sensitive the results are to the choice of ordering. You also will probably want to compare results from different specifications of the VAR. This type of comparison can only be done if you can save OIRFs (and FEVDs, and ...) from each specification. The various `irf` commands allow you to save, drop, rename, and review an unlimited number of IRF-related statistics.

---

43. For an n-equation VAR, there are n^2 OIRFs for each ordering and $n!$ orderings.

A secondary reason for this approach is convenience. The `irf` commands provide a wide variety of the most useful graphs and tables of results. In principle, Stata could simply calculate these results and leave it to you to graph and list them as you see fit, but the tools provided by the various `irf` commands make these tasks much easier.

Enough preliminaries. Let's calculate. We will start by refreshing our VAR estimates to make sure that Stata grabs the right results when it creates IRFs and FEVDs.

```
. use ${ITSUS_DATA}/varexample, clear
(Quarterly macroeconomic statistics for VAR examples)
. quietly var inflation unrate fedfunds if date<tq(2002q1), lags(1 2 3 4)
. irf create iuf, set(macrovar) step(20)
(file macrovar.irf created)
(file macrovar.irf now active)
(file macrovar.irf updated)
```

There is that `var` command again. For completeness, let's take a look at its syntax now.

var *depvarlist* [*if*] [*in*] [, <u>ex</u>og(*varlist*) <u>lags</u>(*numlist*) <u>noc</u>onstant

other_options]

In contrast to `varbasic`, the `var` command allows you to specify exogenous variables and to suppress the constant term in the model. In addition, the *other_options* include ways to add linear constraints to the estimated parameters, control the estimation method, and make other advanced adjustments. However, `var` does not automatically produce a graph of OIRFs (or IRFs or FEVDs), which `varbasic` does as a convenience feature. With the `var` command, you will have to use the commands described below to generate the same graph. In exchange, though, the commands below provide you much more flexibility in calculating and displaying results.[44]

The `irf create` command calculates all the IRF-related results you will ever want to see. The syntax is

irf create *irfname* [, set(*filename*[, replace]) <u>step</u>(#) <u>order</u>(*varlist*)

replace *other_options*]

You do not specify any statistics in the `irf create` command—all of them are calculated whether you need them or not. By default, the statistics are calculated for eight steps. In our example, we asked for 20 steps (five years of quarterly data).

Although you do not specify the statistics, you do have to choose a name for them and to indicate a place to store them. The *irfname* is the name you choose for the results. We chose `iuf` for the *irfname*. (More on that in a moment.) If you have used `irf create` previously in your Stata session, `irf create` assumes you want these results stored in the currently active IRF dataset. In our example, we used the `set()` option to create a new dataset called `macrovar` to store our results. The messages below

---

44. These IRF-related commands also can be used after a `varbasic` command.

the command tell us that Stata created a file called `macrovar.irf`, made it active (which means Stata will assume we are referring to `macrovar.irf` unless we specify otherwise), and updated it (that is, added all the IRF-related results associated with the *irfname* `iuf` to `macrovar.irf`).

To calculate the IRF-related statistics, `irf create` needs to know what ordering to use. If you do not specify the `order()` option, Stata uses the order in which you listed the variables in the var, in this case, `inflation` → `unrate` → `fedfunds`. Hence, the *irfname* `iuf`: `i` for `inflation`, `u` for `unrate`, and `f` for `fedfunds`.

The file `macrovar.irf` is an ordinary Stata dataset. You can `use` it and do anything you want with the results. Of course, if you change the contents, you run the risk of confusing Stata, but it's worthwhile knowing you can grab the detailed calculations if for some reason the suite of `irf` commands does not meet all of your needs.

Let's see what is in `macrovar.irf`.

```
. describe using macrovar.irf
Contains data
  obs:           189                          1 Jan 2020 15:14
  vars:           23
```

variable name	storage type	display format	value label	variable label
irf	double	%10.0g		impulse response function (irf)
step	int	%10.0g		step
cirf	double	%10.0g		cumulative irf
oirf	double	%10.0g		orthogonalized irf
coirf	double	%10.0g		cumulative orthogonalized irf
sirf	double	%10.0g		structural irf
dm	double	%10.0g		dynamic multipliers
cdm	double	%10.0g		cumulative dynamic multipliers
stdirf	double	%10.0g		std error of irf
stdcirf	double	%10.0g		std error of cirf
stdoirf	double	%10.0g		std error of oirf
stdcoirf	double	%10.0g		std error of coirf
stdsirf	double	%10.0g		std error of sirf
stddm	double	%10.0g		std error of dm
stdcdm	double	%10.0g		std error of cdm
fevd	double	%10.0g		fraction of mse due to impulse
sfevd	double	%10.0g		(structural) fraction of mse due to impulse
mse	double	%10.0g		SE of forecast of response variable
stdfevd	double	%10.0g		std error of fevd
stdsfevd	double	%10.0g		std error of sfevd
irfname	str15	%15s		name of results
response	str9	%9s		response variable
impulse	str9	%9s		impulse variable

```
Sorted by:
```

If you are really curious, you can

```
use macrovar.irf, clear
```

and poke around, but you can puzzle out a lot of the contents just from the results of this `describe`. First, note that there are 189 observations. There are nine possible OIRFs in a three-equation VAR (combinations of one impulse variable with one response variable), and we requested calculations for 20 steps (actually 21 steps if you count the 0th step). Twenty-one steps times nine OIRF combinations gives you 189 observations. The last three variables contain names—the *irfname* `iuf` in this example and the impulse–response pairs for each observation. Combined with the `step` number, there is enough information to lay out results in tables and graphs. Everything else is some sort of statistic. Some of them are familiar (IRFs, OIRFs, and FEVDs), while others are still awaiting an explanation (but later).

Let's get started reviewing the results. We will begin with a table of the OIRFs.

```
. irf table oirf
  (output omitted)
```

Trust me, you did not want to see the results of that command. The `irf` commands can overwhelm you with output if you are not careful. In this case, the `irf table` printed nine tables combined into three stacked panels. In each table, 21 steps of OIRFs were displayed for one impulse–response pair along with the lower and upper bounds of the 95% confidence interval for the OIRF. Too much.

Let's start more simply. We will look at just one OIRF. We will trace the response of the unemployment rate to a shock to the inflation rate equation.

```
. irf table oirf, impulse(inflation) response(unrate)
            Results from iuf
```

step	(1) oirf	(1) Lower	(1) Upper
0	-.016029	-.049936	.017879
1	-.011948	-.071699	.047802
2	-.006998	-.08421	.070214
3	.021008	-.064855	.106871
4	.049584	-.028713	.127882
5	.078124	.004966	.151283
6	.101808	.029918	.173697
7	.127087	.054326	.199848
8	.150455	.074532	.226378
9	.169473	.089312	.249634
10	.183933	.100116	.267749
11	.195821	.109096	.282545
12	.20517	.115922	.294418
13	.211682	.120305	.30306
14	.215802	.12239	.309215
15	.218352	.122636	.314069
16	.219652	.121202	.318102
17	.219812	.118218	.321406
18	.219068	.114009	.324127
19	.217679	.108972	.326387
20	.215748	.10334	.328155

```
95% lower and upper bounds reported
(1) irfname = iuf, impulse = inflation, and response = unrate
```

Much better. According to this table, a random one-percentage-point increase in the inflation rate does not have an immediate impact on the unemployment rate. (The 95% confidence interval spans 0 for the first four quarters.) After a year, the unemployment rate creeps up modestly, perhaps as a result of Federal Reserve actions to suppress the rise in inflation.

If you want to see more of the OIRFs at once, you can get a compact presentation by omitting the upper and lower confidence bounds.

```
. irf table oirf, noci impulse(inflation)
          Results from iuf
```

step	(1) oirf	(2) oirf	(3) oirf
0	1.01215	-.016029	.14597
1	.555796	-.011948	.276469
2	.424355	-.006998	.438761
3	.441264	.021008	.448045
4	.58891	.049584	.351073
5	.51615	.078124	.362236
6	.459134	.101808	.419734
7	.452217	.127087	.414354
8	.459495	.150455	.37176
9	.428517	.169473	.363001
10	.397348	.183933	.37333
11	.379571	.195821	.368875
12	.363632	.20517	.355987
13	.340407	.211682	.352288
14	.317351	.215802	.353935
15	.297792	.218352	.351859
16	.278632	.219652	.347233
17	.258057	.219812	.343959
18	.237693	.219068	.340946
19	.21828	.217679	.335964
20	.199065	.215748	.329317

```
(1) irfname = iuf, impulse = inflation, and response = inflation
(2) irfname = iuf, impulse = inflation, and response = unrate
(3) irfname = iuf, impulse = inflation, and response = fedfunds
```

The legend at the bottom identifies which OIRF is in each column.

The syntax of `irf table` is

irf table *stat* [, set(*filename*) irf(*irfnames*) step(*#*) impulse(*impulsevar*)

    response(*endogvars*) individual title("*text*") *other_options*]

As before, `set()` and `irf()` specify the `.irf` dataset and named set of IRF results to use. If they are omitted, Stata uses the active `.irf` dataset and all the *irfnames* in that dataset (which can be a lot, so be careful). The `step()` can limit the number of steps displayed, but it cannot extend them beyond the number specified in the `irf create` command. Use the `impulse()` and `response()` options to limit the impulse–response pairs displayed. `individual` tells Stata to display results in individual tables rather than paste the tables together horizontally. And `title` allows you to replace the default title (`Results from iuf` in our example) with a title of your choosing.

You can choose from nine different statistics to tabulate. We have already discussed three of them (IRF, OIRF, and FEVD). There are also cumulative versions of the IRF (`cirf`) and OIRF (`coirf`). These statistics do the obvious—they calculate the cumulative impact of an impulse.

Stata also calculates dynamic multipliers (`dm`) and cumulative dynamic multipliers (`cdm`). These statistics measure the impact of a one-period change in an exogenous variable. Just as a random impulse to a single equation is propagated to all the equations after the first period, a change to an exogenous variable works its way through the entire VAR. Dynamic multipliers keep track of all of these effects.

The last two statistics are the SVAR versions of IRFs (`sirf`) and FEVDs (`sfevd`).

> **Useful tip:** I am old fashioned and stubborn, so I typically ignore the existence of Stata's excellent menu system, but there is no reason you have to be as pig headed as me. The `var`, `vec`, and `irf` families of commands offer so much flexibility that it can be difficult to remember exactly how to summon the results you are after. The menu system and associated dialog boxes can make your life a lot easier. For example, if you type `db irf table`, Stata will present a (reasonably) intuitive dialog box that will walk you through the steps of specifying exactly the table you want. And if `irf table` does not push you over the edge into using dialog boxes, wait until we get to `irf cgraph`.

To make the results a little more readable, we have reformatted the results of

```
irf table fevd
```

in table 9.1.

Table 9.1. FEVD

Inflation rate

Horizon	Inflation	Unemployment	Funds
1	100	0	0
4	91	8	2
8	87	11	1
12	87	11	2
16	87	10	3
20	86	9	5

Unemployment rate

Horizon	Inflation	Unemployment	Funds
1	1	99	0
4	0	98	2
8	6	82	12
12	19	62	18
16	30	53	17
20	38	47	15

Federal funds rate

Horizon	Inflation	Unemployment	Funds
1	3	20	77
4	14	46	39
8	18	53	29
12	24	51	26
16	29	47	24
20	33	45	22

Let's use the top panel of the table—the panel that reports the FEVDs for the inflation rate—to see how to read this table. Because we have placed inflation first in our ordering, 100% of the forecast-error variance in the first step is attributed to the error in the inflation equation. Four steps ahead, 91% of the variance is still attributed to the error in the inflation equation, 8% is attributed to the error in the unemployment equation, and 2% is attributed to the Federal funds disturbance.[45] The message of this top panel is clear. The error variance in inflation forecasts are almost exclusively due to uncertainty in the inflation equation (at least in this ordering). These results echo the Granger causality tests above, where inflation is only weakly related to unemployment and the funds rate after the impact of prior inflation rates is taken into account.

The lower two panels show a very different picture. The errors in predicting the unemployment rate are sensitive to disturbances in the inflation equation. After 12-quarters, almost 40% of the error variance in unemployment forecasts is split between contributions from shocks to the inflation and funds equations.[46] And the disturbances to all three equations have significant shares in the error variance of funds rate forecasts.

The `inflation` → `unrate` → `fedfunds` ordering we have chosen is apparent in this table. Inflation is ordered first, so at 1-step-ahead unemployment and the funds rate have no impact on the uncertainty in the inflation equation. The unemployment rate is ordered second, so only inflation and the unemployment rate can influence the 1-step-ahead FEVD for unemployment. And because the Federal funds is last in order, all three variables contribute to the 1-step-ahead FEVD.

The OIRF and FEVD tables can be difficult to read—there are a lot of data to absorb. Often it is easier to absorb this information in graphical form. The `irf graph` command makes it easy to see all the OIRFs or FEVDs at once.

The syntax of the `irf graph` command is

`irf graph` *stat* [, `set(`*filename*`)` <u>`irf`</u>`(`*irfnames*`)` <u>`imp`</u>`ulse(`*impulsevar*`)`

  <u>`r`</u>`esponse(`*endogvars*`)` `noci` <u>`lstep`</u>`(#)` <u>`ustep`</u>`(#)` *other_options*]

The *stat* is one of the nine statistics we described above (that is, `irf`, `oirf`, ...), and the options are similar to the options for the `irf table` command. The *other_options* include not only the `title()` option but also most of the other options available for `twoway` graphs (see [G-3] ***twoway_options***).

---

45. Rows may not sum to 100% because of rounding.
46. However, this many steps ahead, the confidence interval around these estimates has expanded to the point that we cannot distinguish most of these differences statistically.

Figure 9.10 displays all the OIRFs for the `inflation` → `unrate` → `fedfunds` ordering.[47] The title of each graph in figure 9.10 indicates the *irfname* (`iuf` in this case), the impulse variable, and the response variable. All three graphs in the top row display the impacts of Federal funds impulses. The graphs in the middle row display the impacts of inflation impulses, and the graphs in the bottom row display the impacts of unemployment impulses. From left to right, the columns display the responses of the Federal funds rate, inflation, and unemployment, respectively.

```
. irf graph oirf, yline(0)
```

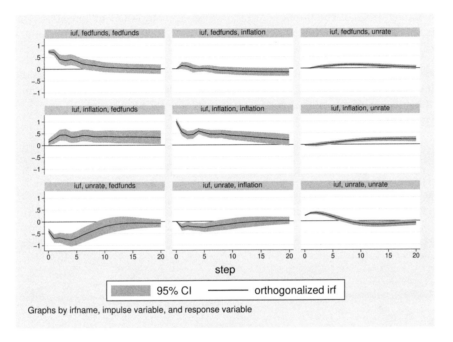

Figure 9.10. Orthogonalized impulse–response functions for the `iuf` ordering

These impulse–response functions present some puzzles. The Federal funds rate—the primary instrument of Federal Reserve monetary policy—appears to have little influence on either inflation or unemployment. However, the impacts of inflation impulses make more sense. An increase in the inflation rate induces a long-lived increase in the Federal funds rate—a plausible reaction if the Fed is determined to control inflation. Increased inflation also generates an increase in the unemployment rate about a year-and-a-half later, again a plausible result of Fed reactions. Note also that the response of inflation to the initial inflation impulse is very long lived, suggesting a high degree of inertia in inflation. Increases in inflation are difficult to eradicate. The impacts of

---

47. You may recall that we did not display the graph produced by the `varbasic` command we used in section 9.2.1 because we had not discussed recursive VARs yet. That graph would have been layed out just like figure 9.10.

unemployment impulses also appear reasonable. An increase in unemployment elicits an easing of monetary policy (a decrease in the Federal funds rate) that persists for about three years. In addition, inflation is reduced modestly.

Does this interpretation hang together? Not entirely. Remember that the Federal funds rate appears last in our Cholesky ordering. As a result, inflation and unemployment "soak up" a lot of the explanatory power, leading to the impression that independent random shocks to the funds rate have only anemic impacts on the economy. Another ordering might produce a different impression of the effectiveness of monetary policy. Another logical hurdle: these graphs attribute little power to policy, but the explanations of the impacts of inflation and unemployment impulses rest on an assumption that monetary policy reactions to disturbances in the macroeconomy trigger measurable follow-on effects.

This example highlights some of the challenges of recursive VARs. Orderings often are arbitrary, and the data alone rarely permit of a single interpretation. The temptation to see a confirmation of your preferred theory in the murky patterns of the OIRFs can be overwhelming.

We can demonstrate the fragility of these relationships by reordering our equations.

```
. irf create fui, order(fedfunds unrate inflation) step(20)
(file macrovar.irf updated)

. irf cgraph (iuf fedfunds inflation oirf) (fui fedfunds inflation oirf)
> (iuf fedfunds unrate oirf) (fui fedfunds unrate oirf), noci yline(0)
> ylabel(-.2 -.1 0 .1 .2) legend(off)
```

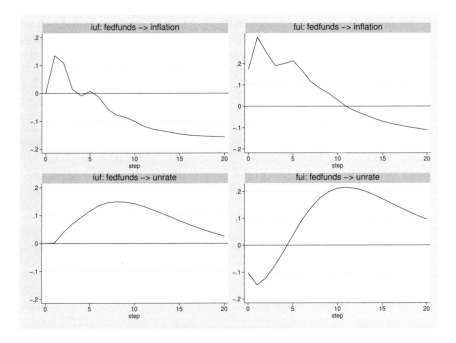

Figure 9.11. Comparison of OIRFs for two different orderings

We have created a second ordering named `fui` (for `fedfunds` → `unrate` → `inflation`), the reverse of the order we have been analyzing. Because `macrovar` is still the active `.irf` file, these results are stored in `macrovar.irf` along with the previously created `iuf` results.[48] We use a new command, `irf cgraph`, to combine graphs depicting the selected results of both orderings and display them side by side in figure 9.11. The graphs in the top row display the response of inflation in each ordering to Federal funds impulses. The graphs in the bottom row display the response of unemployment to Federal funds impulses. To make the graphs more readable, we have suppressed the confidence intervals and the legends, which identify which statistic is graphed.

These two orderings produce noticeable differences. In the `iuf` ordering, an increase in the funds rate has very little impact on inflation. (Recall from figure 9.10 that these estimated responses are not significantly different from 0.) In contrast, in the `fui` ordering an increase in the funds rate increases the inflation rate for over two years before decreasing it.[49]

---

48. Same variables in `macrovar.irf` as before, but double the number of observations.

49. If you type `irf graph oirf, irf(fui) impulse(fedfunds) response(inflation)`, you will see that only the initial increase in inflation in the first five quarters is significant.

The difference in the response of unemployment is even greater. In the `iuf` ordering, monetary tightening (increase in the funds rate) increases the unemployment rate almost immediately. In the reverse ordering, an increase in the funds rate initially decreases unemployment, then increases it after a year.

Sometimes it can be difficult to assess the essential differences between OIRFs. In figure 9.11, for example, the general shapes of the pair of OIRFs for the impact of Fed funds shocks on inflation are not that dissimilar, but the levels of the responses differ. Sometimes a better way to assess the overall differences generated by these two orderings is to examine the cumulative impact of each impulse. The cumulative OIRFs provide that view. COIRFs represent the sum of the individual period impulse–responses. Recall that the sequence of responses of y_{it} to a shock to the jth equation is given by

$$\xi_{0,ij}(=p_{ij}), \xi_{1,ij}, \xi_{2,ij}, \dots$$

The Kth value of the COIRF is given by[50]

$$\text{COIRF}(i, j, K) = \sum_{k=0}^{K} \xi_{k,ij}$$

---

50. For some variables, it may not make sense to attach an interpretation to the sum of the impulse–responses, which represent period-by-period deviations from a baseline time path. Nonetheless, the COIRF presents a visually more coherent image of these impacts, which can make it easier to identify meaningful differences in the more volatile OIRFs.

Let's take a look at the COIRFs for our VAR.

```
. irf cgraph (iuf fedfunds inflation coirf) (fui fedfunds inflation coirf)
> (iuf fedfunds unrate coirf) (fui fedfunds unrate coirf), noci yline(0)
> ylabel(-2 -1 0 1 2) legend(off)
```

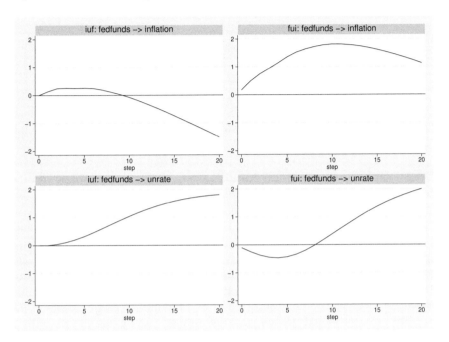

Figure 9.12. Cumulative OIRFs for two different orderings

Figure 9.12 makes clear the complete disagreement between the two orderings on the cumulative impact of Federal funds impulses on inflation. In contrast, the response of the unemployment differs mainly in timing. Clearly, our view of the dynamics of the macroeconomy depend on the recursive ordering we impose on the VAR.

The `cgraph` we snuck in above creates and combines IRF related graphs, including graphs based on different orderings. The syntax is

irf cgraph (*spec_1*) $\big[$ (*spec_2*) ... $\big[$ (*spec_N*) $\big]\big]$ $\big[$, *options* $\big]$

where the graph *spec*s have the form

(*irfname impulsevar responsevar stat* $\big[$, *spec_options* $\big]$)

The available statistics are the same nine statistics as in the `irf graph` command. Most of the options are related to refining the appearance of the graphs, although you can also `set()` the `.irf` file and suppress the confidence intervals. Stata's `irf ctable` command provides essentially the same functionality for IRF-related tables instead of graphs.

Somewhat confusingly, both the `irf graph` and `irf cgraph` commands can combine IRF-related graphs. The `irf graph` command creates and combines them automatically. For instance,

```
irf graph oirf
```

displays all the OIRFs stored in the current `.irf` file. In contrast, the `irf cgraph` command forces you to specify each graph individually but allows you to mix and match graphs, orderings, and statistics however you wish.

> **Useful tip:** In chapter 7, we calculated the IRF for a univariate autoregressive moving-average model the hard way—we did the calculations by hand. However, Stata is perfectly happy to calculate a VAR for one variable (a univariate VAR!). Then we can use the `irf` commands to calculate and display the IRF for this univariate model. The solution is exact for a pure AR model but only approximate for a model that includes a moving-average component. In this latter case, you may want to add additional lags to the AR specification until the residual is indistinguishable from white noise. Alternatively, you can write a Stata program (ado-file) to calculate the exact IRF from the estimated AR and moving-average parameters. Personally, I prefer the lazy man's method.

9.4 SVARs

VARs became a mainstay of macroeconomic time-series analysis in large part because of the arguments by Christopher Sims (1980) and others against the prevailing practice of analyzing macroeconomic policy questions in the context of SEMs. The issue that concerned Sims and other economists of the rational expectations and equilibrium business cycle school was the unavoidable use of "incredible" (in Sims's words) identifying assumptions to estimate SEMs.[51] VARs, as reduced-form models, avoided these types of assumptions. Instead they focused on the dynamic properties of the VARs as clues to the workings of the macroeconomy.

---

51. An important aspect of this issue for rational expectations advocates was the Lucas critique (Lucas 1976), an attack on the conventional methods used for econometric policy analysis. According to the Lucas critique, rational agents anticipate systematic policy behavior. As a result, policy interventions—that is, systematic changes in these policy rules—will change expectations and, hence, behavior. In other words, among the "incredible" assumptions criticized by Sims was the assumption that consumer responses to changes in the value of a policy variable (such as the Federal funds rate) are invariant to predictable changes in the behavior of the policy authority. While this critique was directed at traditional SEMs, it complicates the interpretation of VARs as well.

As we have seen in the sections above, reduced-form dynamics alone fall short of answering the questions we wish to pose. As a result, time-series analysts look for the minimal-identifying assumptions that can narrow the range of possible interpretations of the data. Specifying a recursive ordering for the orthogonalized-error terms is one method that has gained adherents. However, economic theory rarely implies a simple ordering, so recursive analyses frequently are unpersuasive. To make further progress, we usually need to make stronger identifying assumptions in a form that is more directly tied to theory. SVARs are the result. They are the result of overlaying a reduced-form VAR with identifying assumptions.

In some respects, we have come full circle from Sims's path-breaking article. To make strong statements about the present and future impacts of economic policy actions and random shocks to the economy, we have to make the type of assumptions that originally led Sims to champion VARs as an alternative to SEMs. However, we are not complete backsliders. The use of a reduced-form VAR as the starting point of econometric analysis is a departure from prior practice. This change of perspective has been most fully articulated in the general-to-specific approach developed at the London School of Economics by David Hendry and others.

Let's see how to specify a SVAR. Ignoring the vector of constants for simplicity, recall that we can write a generic, reduced-form VAR as

$$\boldsymbol{\Phi}(L)\mathbf{y}_t = \boldsymbol{\epsilon}_t$$

where

$$\boldsymbol{\Phi}(L) \equiv I - \boldsymbol{\Phi}_1 L - \boldsymbol{\Phi}_2 L^2 \cdots - \boldsymbol{\Phi}_p L^p$$

In other words, $\boldsymbol{\Phi}_0 = I$. Assuming white-noise disturbances, the covariance matrix of the disturbances is

$$E\boldsymbol{\epsilon}_t\boldsymbol{\epsilon}'_s = \begin{cases} \boldsymbol{\Sigma}, & t = s \\ \mathbf{0}, & t \neq s \end{cases}$$

Thus, in the reduced-form VAR, there are no contemporaneous, endogenous variables on the right-hand side of the equation;[52] hence, all the coefficients in $\boldsymbol{\Phi}(L)$ can be estimated consistently by classical least squares.

While this may not be the most familiar representation, we can write an SEM as

$$\mathbf{A}\left(\boldsymbol{\Phi}(L)\mathbf{y}_t\right) = \mathbf{A}\boldsymbol{\epsilon}_t \equiv \mathbf{B}\boldsymbol{\nu}_t$$

where

$$E\left(\boldsymbol{\nu}_t\boldsymbol{\nu}'_s\right) = \begin{cases} \mathbf{I}, & t = s \\ \mathbf{0}, & t \neq s \end{cases}$$

and $\mathbf{A}$ is full rank.

---

52. While we have omitted them for convenience, contemporaneous exogenous (or predetermined) variables may appear in the VAR equations. If the same list of exogenous variables appears in each equation, they add no identifying information.

The introduction of the **B** matrix simplifies the error structure: it transforms the disturbance vector into $\boldsymbol{\nu}_t$, whose elements are uncorrelated. However, the introduction of the **A** matrix introduces additional contemporaneous endogenous variables to each equation:

$$\mathbf{A}\mathbf{y}_t = \mathbf{A}(\boldsymbol{\Phi}_1 \mathbf{y}_{t-1} + \cdots + \boldsymbol{\Phi}_p \mathbf{y}_{t-p}) + \mathbf{B}\boldsymbol{\nu}_t$$

To estimate the structural parameters **A** and **B**, we need to find identifying assumptions. To motivate the SVAR approach to identification, we rewrite this SEM as a reduced-form VAR again.

$$\mathbf{y}_t = \boldsymbol{\Phi}_1 \mathbf{y}_{t-1} + \cdots + \boldsymbol{\Phi}_p \mathbf{y}_{t-p} + \mathbf{A}^{-1}\mathbf{B}\boldsymbol{\nu}_t$$

To simplify this equation, we constrain $\mathbf{B} = \mathbf{I}$.[53] Then the structural parameters of interest, the **A**, are isolated in the covariance matrix of $\mathbf{A}^{-1}\boldsymbol{\nu}_t$, which is $\boldsymbol{\Sigma}$. Our identification problem is to solve the p equations of

$$\mathbf{A}^{-1}\mathbf{A}^{-1'} = \boldsymbol{\Sigma}$$

for the p^2 elements of **A**. Because $\boldsymbol{\Sigma}$ is symmetric, it contains only $(p+1)p/2$ free elements. Consequently, we need $(p^2 - p)/2$ restrictions to exactly identify all the elements of **A**. If we have additional restrictions, we have an overidentified model, and we can test the identifying assumptions in the usual way.

This analysis should sound familiar to anyone who has studied the textbook approach to simultaneous equation models. What we have done so far is state the order condition for identification of the parameters in our SEM.[54] In this approach, we would search for zero restrictions; that is, assumptions that some endogenous or predetermined variables do not appear in some of the equations (in other words, their parameters in those equations were equal to 0) and cross-equation restrictions on parameter values (for example, the assumption that two parameters are equal). Ideally, these restrictions are grounded in economic theory, but in practice, the connections sometimes can be tenuous (hence, Sims's description of these assumptions as "incredible").

The approach described above defines a short-run SVAR. Restrictions on the elements of **A** are statements about the contemporaneous relationships among the endogenous variables.[55] We can also place restrictions on the long-run behavior of a VAR. If we define

$$\overline{\mathbf{A}} = \left(\mathbf{I} - \boldsymbol{\Phi}_1 L - \boldsymbol{\Phi}_2 L^2 - \cdots - \boldsymbol{\Phi}_p L^p\right)$$

then we can write our reduced-form VAR as

$$\overline{\mathbf{A}}\mathbf{y}_t = \boldsymbol{\epsilon}_t$$

---

53. We can always normalize this system so that $\mathbf{B} = \mathbf{I}$. Simply premultiply both sides of the structural equation by $\mathbf{B}^{-1}$. Then our structural parameters of interest become $\mathbf{A}^* \equiv \mathbf{B}^{-1}\mathbf{A}$.

54. As in a conventional SEM, the order condition is necessary but not sufficient for identification. An associated rank condition also is necessary.

55. Restrictions can also be placed on the elements of **B**. These restrictions are statements about the contemporaneous relationships between the random shocks in $\boldsymbol{\epsilon}$ and the endogenous variables.

The inverse of $\overline{\mathbf{A}}$ is the matrix of long-run effects of the reduced-form VAR shocks[56]

$$\mathbf{y}_t = \overline{\mathbf{A}}^{-1}\boldsymbol{\epsilon}_t$$

For instance, the long-run impact on $\mathbf{y}_i$ of a one-unit change in ϵ_{jt} is given by the ijth element of $\overline{\mathbf{A}}^{-1}$.

Rewriting our SVAR in terms of $\overline{\mathbf{A}}$, we have

$$\mathbf{A}\overline{\mathbf{A}}\mathbf{y}_t = \mathbf{B}\boldsymbol{\epsilon}_t$$

If we normalize this equation so that $\mathbf{A} = \mathbf{I}$,[57] we have

$$\mathbf{y}_t = \overline{\mathbf{A}}^{-1}\mathbf{B}\boldsymbol{\epsilon}_t = \mathbf{C}\boldsymbol{\nu}_t$$

Thus $\mathbf{C}$ is the matrix of long-run responses to the orthogonalized shocks. As in the short-run SVAR, the identification problem revolves around finding appropriate restrictions on the $(p^2 - p)/2$ free elements of $\mathbf{CC}'$.

9.4.1   Examples of a short-run SVAR

A recursive VAR provides an example of an exactly identified SVAR, although we might want to argue against calling it an SVAR in the absence of a well-founded theoretical justification for the particular ordering chosen. Nonetheless, it provides a useful example of how to use Stata's `svar` command. We will use the `svar` command to impose the same restrictions implied by the `fedfunds` $\rightarrow$ `unrate` $\rightarrow$ `inflation` ordering.

First, let's estimate the recursive VAR and store the IRFs in a new `.irf` file.

```
. use ${ITSUS_DATA}/varexample, clear
(Quarterly economic statistics)
. quietly var fedfunds unrate inflation if date<tq(2001q3), lags(1 2 3 4)
. irf create fui, set(svar) step(20)
(file svar.irf created)
(file svar.irf now active)
(file svar.irf updated)
```

Now let's estimate this same system as an SVAR. The syntax of the `svar` command for short-run SVARs is

`svar` *depvarlist* $\big[$*if*$\big]$ $\big[$*in*$\big]$, { <u>acon</u>straints(*constraints_a*) <u>aeq</u>(*matrix_aeq*)

<u>acns</u>(*matrix_acns*) <u>bcon</u>straints(*constraints_b*) <u>beq</u>(*matrix_beq*)

<u>bcns</u>(*matrix_bcns*)} $\big[$*short_run_options*$\big]$

The first three options provide alternative ways to specify restrictions on the $\mathbf{A}$ matrix. The second three options provide the same ways of applying restrictions to

---

56. The assumption of stability guarantees the existence of $\overline{\mathbf{A}}^{-1}$.
57. Multiplying both sides by $\mathbf{A}^{-1}$ simply redefines the elements of $\mathbf{B}$.

the **B** matrix. At least one of the six options must be specified. If one or more of the **A** options is specified and none of the **B** options are specified, **B** will be normalized to the identity matrix. Similarly, if one or more of the **B** options is specified and none of the **A** options are specified, **A** will be normalized to the identity matrix.

aconstraints() applies restrictions that were previously defined by using Stata's constraint command. We will see an example of this option in the subsequent example. The next two options, aeq() and acns(), take as arguments constraint matrices that were previously defined by using Stata's matrix command. In the aeq() command, each element of the matrix must be either a missing value, which indicates a free (that is, unconstrained) parameter, or a real number, which indicates a specific value to assign to the parameter (possibly 0, if the parameter is to be dropped). In the acns() command, each element of the matrix must be a missing value (free parameter), 0 (dropped parameter), or a positive integer. Each positive integer must appear at least twice, and the positive integers indicate parameters that are equal. Any elements coded with a '1' indicate a set of parameters with the same value; any elements coded with a '2' indicate another set of parameters with the same value (but not necessarily the same value as the '1' parameters); and so on.

In this case, it's easier to demonstrate the command than to explain it. To match the fedfunds → unrate → inflation recursive model, we want to impose the specification

$$\mathbf{A}\epsilon_t = \mathbf{B}\nu_t$$

or, in more detail,

$$
\begin{bmatrix}
1 & 0 & 0 \\
a_{21} & 1 & 0 \\
a_{31} & a_{32} & 1
\end{bmatrix}
\begin{bmatrix}
\epsilon_{\text{fedfunds},t} \\
\epsilon_{\text{unrate},t} \\
\epsilon_{\text{inflation},t}
\end{bmatrix}
= \nu_t
$$

The first row of **A** specifies that the current Federal funds rate is unaffected by contemporaneous shocks to the unemployment and inflation rate equations. The second row specifies that the unemployment rate is unaffected by contemporaneous shocks to the inflation rate but affected by contemporaneous shocks to the funds rate. The third row indicates that the inflation rate is affected by all contemporaneous shocks. Finally, we have constrained **B** to be the identify matrix.[58]

---

58. This specification also imposes the assumption that shocks to the Federal funds rate do not affect unemployment and inflation in the current quarter. While economists of all stripes have long accepted Milton Friedman's claim that monetary policy changes affect the economy only after long and variable lags, there remains a great deal of debate about exactly how long and how variable those lags are. Friedman favored a six-month average lag between policy changes and economic impacts, but it is conceivable that some of the impacts are observable within a single quarter. Our aim here is not to take a stand on this macroeconomic question, but to match the recursive estimates—which incorporate the same assumption.

Let's translate this equation to Stata terminology. First, we define the **A** matrix.

```
. matrix A = [1,0,0\.,1,0\.,.,1]
. matrix list A
A[3,3]
     c1  c2  c3
r1    1   0   0
r2    .   1   0
r3    .   .   1
```

Remember that the missing values $(.)$ represent the free parameters a_{21}, a_{31}, a_{32}.

Next we define the **B** matrix so the errors are orthogonal.

```
. matrix B = [.,0,0\0,.,0\0,0,.]
. matrix list B
symmetric B[3,3]
     c1   c2   c3
r1    .
r2    0    .
r3    0    0    .
```

Now we can estimate the SVAR.

```
. svar fedfunds unrate inflation if date<=tq(2002q1), lags(1 2 3 4) aeq(A)
> beq(B)
Estimating short-run parameters

Iteration 0:     log likelihood = -722.70627
Iteration 1:     log likelihood = -579.49558
Iteration 2:     log likelihood = -437.55373
Iteration 3:     log likelihood =  -409.6353
Iteration 4:     log likelihood = -407.62757
Iteration 5:     log likelihood = -407.61837
Iteration 6:     log likelihood = -407.61837

Structural vector autoregression

 ( 1)   [/A]1_1 = 1
 ( 2)   [/A]1_2 = 0
 ( 3)   [/A]1_3 = 0
 ( 4)   [/A]2_2 = 1
 ( 5)   [/A]2_3 = 0
 ( 6)   [/A]3_3 = 1
 ( 7)   [/B]1_2 = 0
 ( 8)   [/B]1_3 = 0
 ( 9)   [/B]2_1 = 0
 (10)   [/B]2_3 = 0
 (11)   [/B]3_1 = 0
 (12)   [/B]3_2 = 0

Sample:  1961:1 - 2002:1           Number of obs     =         165
Exactly identified model           Log likelihood    =   -407.6184
```

		Coef.	Std. Err.	z	P>|z|	[95% Conf.	Interval]
/A							
	1_1	1	(constrained)				
	2_1	.1203299	.018084	6.65	0.000	.0848858	.155774
	3_1	-.2090148	.1027809	-2.03	0.042	-.4104617	-.007568
	1_2	0	(constrained)				
	2_2	1	(constrained)				
	3_2	-.0398154	.3928788	-0.10	0.919	-.8098437	.7302128
	1_3	0	(constrained)				
	2_3	0	(constrained)				
	3_3	1	(constrained)				
/B							
	1_1	.8505507	.0468213	18.17	0.000	.7587827	.9423187
	2_1	0	(constrained)				
	3_1	0	(constrained)				
	1_2	0	(constrained)				
	2_2	.1975775	.0108763	18.17	0.000	.1762604	.2188946
	3_2	0	(constrained)				
	1_3	0	(constrained)				
	2_3	0	(constrained)				
	3_3	.9970985	.0548885	18.17	0.000	.8895191	1.104678

An SVAR is estimated by maximizing the log-likelihood function after applying the constraints, hence, the six iterations at the top of the output that report the progress of the maximization. The next block of output echoes the constraints we applied. The third block of output reports the estimates of the unconstrained elements of **A** and **B**. Note that Stata recognizes this specification as an exactly identified model.

How can we be sure that we have fit the model implied by the `fui`-ordered recursive VAR? Let's check the IRFs. For the SVAR, we request the structural IRF; however, in this example, they are identical to the OIRFs by construction.

```
. irf create svar, set(svar) step(20)
(file svar.irf now active)
(file svar.irf updated)

. irf cgraph (fui fedfunds inflation oirf) (svar fedfunds inflation sirf),
> legend(off)
```

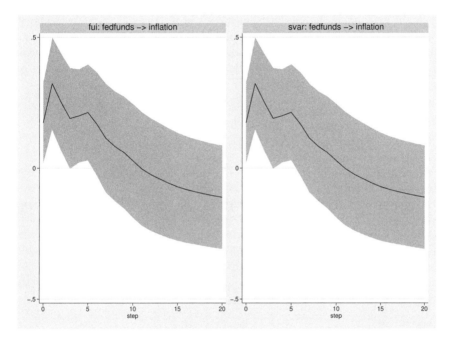

Figure 9.13. Comparison of recursive VAR and SVAR results

Check any of the other IRF-related statistics. They match.

Clearly, if we believe in a particular recursive ordering, we will just impose it. It's too much trouble using **svar** to get the same result. Also note that calculating IRF-related statistics works just the same for SVARs as for recursive VARs. In fact, all the postestimation commands that work with the **var** and **varbasic** commands work pretty much the same with **svar**.

Let's see if we can puzzle out a little more interesting SVAR example. Let's impose a simplified version of the Taylor rule on the data. The Taylor rule can be thought of either as a description of Federal Reserve policy during the Greenspan era or as a policy rule the Fed should follow. Following Stock and Watson (2001), we can write the rule in terms of our three variables as

$$r_t = r^* + 1.5\left(\overline{\pi}_t - \pi^*\right) - 1.25\left(\overline{u}_t - u^*\right)$$

where r_t is the Federal funds rate, r^* is the target real rate of interest, $\bar{\pi}_t$ and $\bar{u}_t$ are the average values of the inflation and unemployment rates over the most recent four quarters, and π^* and u^* are the target rates of inflation and unemployment.[59] As a policy guide, the Taylor rule suggests raising the Federal funds rate when inflation is above its target rate and lowering the funds rate when unemployment is above its target rate. This version of the Taylor rule is backward looking—the federal funds rate reacts to deviations of inflation and unemployment from their target levels—so we will estimate our baseline VAR with an `inflation` $\rightarrow$ `unrate` $\rightarrow$ `fedfunds` ordering.[60]

How might we impose the Taylor rule as a set of restrictions on our original VAR?[61] First, because the target rates in our simplified example are independent of time, we can ignore them. They will be absorbed by the vector of constant terms. Second, we can use Stata's `constraint` command to impose the feedback rule on the four-quarter averages of inflation and unemployment.[62]

```
. constraint 1 [fedfunds]L.inflation  = 1.5/4
. constraint 2 [fedfunds]L2.inflation = 1.5/4
. constraint 3 [fedfunds]L3.inflation = 1.5/4
. constraint 4 [fedfunds]L4.inflation = 1.5/4
. constraint 5 [fedfunds]L.unrate  = -1.25/4
. constraint 6 [fedfunds]L2.unrate = -1.25/4
. constraint 7 [fedfunds]L3.unrate = -1.25/4
. constraint 8 [fedfunds]L4.unrate = -1.25/4
```

Finally, we constrain the **A** matrix to prevent any influence of contemporaneous inflation or unemployment on the Federal funds rate.

```
. matrix A = [1,.,0\.,1,0\0,0,1]
. matrix list A
symmetric A[3,3]
     c1  c2  c3
r1    1
r2    .   1
r3    0   0   1
```

---

59. Taylor's rule is specified in terms of the difference between actual and potential GDP; however, it can be rewritten in terms of unemployment. In addition, the three target rates usually are allowed to vary over time. We have simplified the treatment here.
60. We can also specify a forward-looking Taylor rule, where the Federal funds rate is set based on the Federal Reserve's expectation of future inflation and unemployment. Stock and Watson (2001) compare estimates of the backward-looking and forward-looking Taylor rules.
61. This is far from the only way to impose the Taylor rule, and we make no assertion that it is the most apt. It does generate intriguing results, and, more importantly, demonstrates some more of the capabilities of Stata's `svar` command.
62. These commands constrain the coefficients on the lags of, say, inflation to be equal, and they set the magnitude of the response of the funds rate to changes in inflation. Because each lagged quarter has equal weight in the four-quarter average, we divide by four the Taylor rule assumption of a coefficient of 1.5.

Now to reestimate the VAR with these restrictions.[63,64]

```
. svar inflation unrate fedfunds if date<=tq(2002q1), lags(1 2 3 4) aeq(A)
> varconst(1/8) noislog
Estimating short-run parameters
Iteration 0:    log likelihood = -1815.8149
Iteration 1:    log likelihood = -1421.4442
Iteration 2:    log likelihood = -1317.5914
Iteration 3:    log likelihood =  -1264.722
Iteration 4:    log likelihood = -1262.7122
Iteration 5:    log likelihood = -1262.6708
Iteration 6:    log likelihood = -1262.6706
Iteration 7:    log likelihood = -1262.6706
Structural vector autoregression
 ( 1)   [/A]1_1 = 1
 ( 2)   [/A]1_3 = 0
 ( 3)   [/A]2_2 = 1
 ( 4)   [/A]2_3 = 0
 ( 5)   [/A]3_1 = 0
 ( 6)   [/A]3_2 = 0
 ( 7)   [/A]3_3 = 1
 ( 8)   [/B]1_1 = 1
 ( 9)   [/B]1_2 = 0
 (10)   [/B]1_3 = 0
 (11)   [/B]2_1 = 0
 (12)   [/B]2_2 = 1
 (13)   [/B]2_3 = 0
 (14)   [/B]3_1 = 0
 (15)   [/B]3_2 = 0
 (16)   [/B]3_3 = 1
```

	Coef.	Std. Err.	z	P>|z|	[95% Conf. Interval]	
/A						
1_1	1	(constrained)				
2_1	1.078809	.0434153	24.85	0.000	.9937164	1.163901
3_1	0	(constrained)				
1_2	3.578666	.1227797	29.15	0.000	3.338023	3.81931
2_2	1	(constrained)				
3_2	0	(constrained)				
1_3	0	(constrained)				
2_3	0	(constrained)				
3_3	1	(constrained)				

---

63. The noislog option suppresses the iteration log for the seemingly unrelated regression algorithm that is required by the addition of the constraints on the lag coefficients in the funds rate equation.
64. Because we placed restrictions on the **A** matrix but did not specify the **B** matrix, Stata set **B** = **I**.

```
/B
        1_1  |        1  (constrained)
        2_1  |        0  (constrained)
        3_1  |        0  (constrained)
        1_2  |        0  (constrained)
        2_2  |        1  (constrained)
        3_2  |        0  (constrained)
        1_3  |        0  (constrained)
        2_3  |        0  (constrained)
        3_3  |        1  (constrained)
```

LR test of identifying restrictions: chi2(4) = 1301   Prob > chi2 = 0.000

In this example, the likelihood-ratio test rejects the identifying restrictions. There can be multiple reasons for this rejection—the assumption of constant target rates for inflation and unemployment, an oversimplified version of the Taylor rule—but one likely contributing factor is the span of the estimation sample, which covers multiple Federal Reserve policy regimes.

For now, let's ignore the rejection of the restrictions and focus on the impact of these restrictions on the dynamic behavior implied by this version of the Taylor rule. Figure 9.14 compares the OIRFs (for the `fui`-ordered recursive VAR) and structural IRFs (for the SVAR estimate of the Taylor rule) of a random shock to the Federal funds rate on the inflation and unemployment rates. In both models, increases in the funds rate increase the unemployment rate initially, but the unemployment rate decreases after three years in the Taylor rule estimates. Similarly, the impact of monetary policy on inflation appears to differ in the out years, where the recursive model estimates continuing negative responses of inflation, while the responses drift back to 0 in the Taylor rule SVAR.

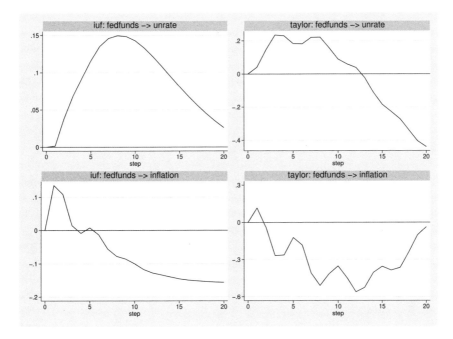

Figure 9.14. Impact of monetary policy in a recursive VAR and an SVAR

These results should not be taken too seriously yet. For one thing, they are very sensitive to the sample period. To highlight this issue, we reestimated this SVAR for our out-of-sample period, the last 10 years of data in this dataset. Figure 9.15 compares the in-sample estimates of monetary policy impact with the estimated impact during the out-of-sample period by using the identical specification of the Taylor rule restrictions. The graphs labeled `later` display the results for the out-of-sample period.

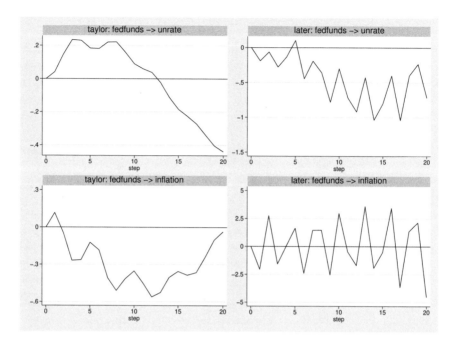

Figure 9.15. An example of the sensitivity of estimates to the sample period

9.4.2   Examples of a long-run SVAR

As we noted above, we can impose restrictions on the long-run impact of impulses by specifying elements of the **C** matrix in the equation

$$\mathbf{y}_t = \mathbf{C}\boldsymbol{\nu}_t$$

For instance, in a two-equation VAR of the Federal funds and unemployment rates, we might want to impose the assumption of long-run monetary neutrality by specifying that random impulses to the funds rate have no long-run impact on unemployment.

```
. use ${ITSUS_DATA}/varexample, clear
. matrix C = [.,.\0,.]
. matrix list C
C[2,2]
     c1  c2
r1   .   .
r2   0   .
. svar fedfunds unrate if date<=tq(2002q1), lags(1 2 3 4) lreq(C) nolog
Estimating long-run parameters
Structural vector autoregression
 ( 1)  [/C]2_1 = 0
Sample:  1961:1 - 2002:1                Number of obs     =        165
Exactly identified model               Log likelihood    = -190.4849
```

		Coef.	Std. Err.	z	P>|z|	[95% Conf. Interval]	
/C							
	1_1	5.799341	.3192432	18.17	0.000	5.173636	6.425046
	2_1	0	(constrained)				
	1_2	21.34239	1.258622	16.96	0.000	18.87554	23.80924
	2_2	10.97861	.6043524	18.17	0.000	9.794098	12.16312

This model is exactly identified, so it cannot provide a test of the assumption of monetary neutrality.

The syntax of the `svar` command for long-run SVARs is

svar *depvarlist* [*if*] [*in*], { <u>lrcon</u>straints(*constraints_lr*) <u>lreq</u>(*matrix_lreq*)
<u>lrcns</u>(*matrix_lrcns*) } [*long_run_options*]

The three critical options—`lrconstraints()`, `lreq()`, and `lrcns()`—are analogous to the short-run versions.

Let's estimate a slightly more complicated long-run SVAR where we impose long-run restrictions on our three-variable system. For illustration purposes, we impose some strong—and perhaps not entirely plausible—restrictions: neither inflation nor unemployment shocks have long-run impacts on the funds rate; neither funds rate nor inflation rate shocks have long-run impacts on unemployment; and unemployment shocks have no long-run impact on inflation (although funds rate shocks are allowed to have a long-run impact on inflation).

```
. matrix C = [.,0,0\0,.,0\.,0,.]

. matrix list C

C[3,3]
    c1  c2  c3
r1   .   0   0
r2   0   .   0
r3   .   0   .

. svar fedfunds unrate inflation if date<=tq(2002q1), lags(1 2 3 4) lreq(C)
> nolog
Estimating long-run parameters

Structural vector autoregression

 ( 1)  [/C]1_2 = 0
 ( 2)  [/C]1_3 = 0
 ( 3)  [/C]2_1 = 0
 ( 4)  [/C]2_3 = 0
 ( 5)  [/C]3_2 = 0
```

Sample: 1961:1 - 2002:1	Number of obs	=	165
Overidentified model	Log likelihood	=	-540.757

		Coef.	Std. Err.	z	P>|z|	[95% Conf. Interval]
/C							
	1_1	11.31427	.6228301	18.17	0.000	10.09355	12.535
	2_1	0	(constrained)				
	3_1	4.937364	.4918359	10.04	0.000	3.973384	5.901345
	1_2	0	(constrained)				
	2_2	5.40296	.2974231	18.17	0.000	4.820021	5.985898
	3_2	0	(constrained)				
	1_3	0	(constrained)				
	2_3	0	(constrained)				
	3_3	5.265466	.2898544	18.17	0.000	4.697362	5.83357

LR test of identifying restrictions: chi2(2) = 266.3   Prob > chi2 = 0.000

The likelihood-ratio test rejects these identifying restrictions. Let's try to weaken the restrictions a bit. We continue to impose long-run neutrality between monetary policy and unemployment and between inflation and unemployment, but we remove the other restrictions.

```
. matrix C = [.,0,.\0,.,0\.,0,.]
. matrix list C
symmetric C[3,3]
     c1  c2  c3
r1   .
r2   0   .
r3   .   0   .
. svar fedfunds unrate inflation if date<=tq(2002q1), lags(1 2 3 4) lreq(C)
> nolog
With the current starting values, the constraints are not sufficient for
    identification
The constraints placed on C  are
  1999:  [/C]1_2 = 0
  1998:  [/C]2_1 = 0
  1997:  [/C]2_3 = 0
  1996:  [/C]3_2 = 0
These constraints place 4 independent constraints on C
The order condition requires at least 3 constraints
r(498);

end of do-file

r(498);
```

What just happened? We have enough independent constraints to satisfy the order condition for identification, but we failed to satisfy the rank condition. If you are familiar with the rank and order conditions for the identification of traditional SEMs, you may be a little confused by some of the output you get from the **svar** command. The rank condition for SVARs is very different from, and less intuitive than, the SEM version.[65] Just be aware that you may run afoul of this requirement when using either the long-run or the short-run versions of the **svar** command.

9.5 Points to remember

- A VAR is a straightforward extension of a univariate autoregressive model to an autoregression for a vector of dependent variables. Without any additional assumptions or restrictions, a VAR is simply a time-series representation of a reduced-form econometric model.

- Reduced-form VARs were introduced as a challenge to traditional SEMs that often relied on strong assumptions for identification. Low-dimensional, reduced-form VARs often produced superior forecasts to more elaborate SEMs. The ease of generating dynamic forecasts from a VAR is one of the principal attractions of this approach.

- Without further assumptions, reduced-form VARs are limited in their ability to answer questions about the directions of causation and the impact of alternative

---

65. The derivation and explanation of this condition is beyond the scope of our discussion. You can see the formula for the rank condition in the *Methods and formulas* section of the description of the **svar** command in the *Time-Series Reference Manual* ([TS]). The explanation of the rank condition can be found in Amisano and Giannini (1997).

policy rules. Recursive VARs and SVARs provide methods for addressing the limitations but at the cost of introducing sometimes-strong identifying assumptions.

- Cross correlations and Granger causality tests can be helpful in characterizing the temporal sequence of influences in a VAR. Stronger statements about causation and timing require either a recursive VAR or an SVAR.

- A recursive VAR can be regarded as an exactly identified SVAR where the identifying assumption is a specification of the order of influence of the variables. Impulses (that is, random shocks) are assumed to influence first one dependent variable, then another, in a domino fashion. For some VARs with a small number of dependent variables, it may be possible to support this type of assumption from economic theory. Because of its simplicity and ease of implementation, recursive VARs have sometimes been used where the theoretical justification is questionable.

- Once an order has been chosen, the IRFs and FEVDs provide detailed information about timing and causation.

- Different choices of orderings frequently generate markedly different views of the interactions among the variables and of the impacts of alternative policy rules. The only solution available to the researcher is to ground orderings and restrictions in persuasive theoretical arguments.

- A short-run SVAR combines identifying restrictions of various types—preferably derived from theory—to identify the contemporaneous relationships among the dependent variables. Because a VAR is written without any explicit contemporaneous relationships among the dependent variables, those parameters are embedded in the covariance matrix of the random disturbances. The restrictions in an SVAR make it possible to decompose the covariance matrix in a way that solves for those parameters.

- A long-run SVAR places restrictions directly on the long-run impacts of the orthogonalized disturbances. As always, interpretation depends on the ability to convincingly link the unobservable orthogonal disturbances to a tangible concept.

- Although a VAR can be mapped to an SEM and vice versa, identification in an SVAR is approached differently than in an SEM. Restrictions can apply either to the short-run relationships in the VAR or to the long-run relationships between random disturbances and dependent variables. At the current state of the art, it is not possible to combine short-run and long-run identifying restrictions.

9.6   Looking ahead

We are approaching the finish line (at least for this book). The next topic, cointegration, reraises two topics that we have skirted so far. First, how can we model nonstationary variables? So far, we have induced stationarity by differencing. However, once we have multiple nonstationary variables, differencing is not always an appropriate solution. Second, what is the role of trends in time-series models? We talked about trends in the early chapters when we discussed filters and forecasts, but we dropped the topic when we began to focus on stationarity as the linchpin of our approach. If we are going to model nonstationary variables, we have to think more carefully about trends. The next chapter takes up these questions.

Stata commands and features discussed

fcast compute ([TS] **fcast compute**): Compute dynamic forecasts of dependent variables after var, svar, or vec; section 9.2.3

fcast graph ([TS] **fcast graph**): Graph forecasts of variables computed by fcast compute; section 9.2.3

irf cgraph ([TS] **irf cgraph**): Combine graphs of IRFs, dynamic-multiplier functions, and FEVDs; section 9.3.2

irf create ([TS] **irf create**): Obtain IRFs, dynamic-multiplier functions, and FEVDs; section 9.3.2

irf ctable ([TS] **irf ctable**): Combine tables of IRFs, dynamic-multiplier functions, and FEVDs; section 9.3.2

irf graph ([TS] **irf graph**): Graph IRFs, dynamic-multiplier functions, and FEVDs; section 9.3.2

irf table ([TS] **irf table**): Create tables of IRFs, dynamic-multiplier functions, and FEVDs; section 9.3.2

predict ([R] **predict**): Obtain predictions, residuals, etc., after estimation; section 9.2.3

svar ([TS] **var svar**): Structural vector autoregressive models; section 9.4

var ([TS] **var**): Vector autoregressive models; section 9.3.2

varbasic ([TS] **varbasic**): Fit a simple VAR and graph IRFs and FEVDs; section 9.2.1

vargranger ([TS] **vargranger**): Perform pairwise Granger causality tests after var or svar: section 9.3.2

varlmar ([TS] **varlmar**): Perform Lagrange multiplier test for residual correlation after var or svar; section 9.2.2

varnorm ([TS] **varnorm**): Test for normally distributed disturbances after var or svar; section 9.2.2

varsoc ([TS] **varsoc**): Obtain lag-order selection statistics for VARs and vector error-correction models; section 9.2.1

varstable ([TS] **varstable**): Check the stability condition of VAR or SVAR estimates; section 9.2.2

varwle ([TS] **varwle**): Obtain Wald lag-exclusion statistics after `var` or `svar`; section 9.2.2

xcorr ([TS] **xcorr**): Cross-correlogram for bivariate time series; section 9.3.1

10 Models of nonstationary time series

Chapter map

10.1 Trends and unit roots. Trend-stationary and unit-root models—two alternative views of nonstationary variables. Implications of trend stationarity and unit roots.

10.2 Testing for unit roots. Tests based on the Dickey–Fuller approach: augmented Dickey–Fuller, generalized least squares (GLS)-based, and Phillips–Perron tests.

10.3 Cointegration: Looking for a long-term relationship. Why nonstationary variables may obey a long-term equilibrium relationship.

10.4 Cointegrating relationships and VECMs. How cointegration changes a vector autoregression (VAR) to a vector error-correction model (VECM). Why differencing to induce stationarity leads to misspecification when there are cointegrating relationships. Linear and quadratic trends in VECMs.

10.5 From intuition to VECM: An example. An example of mapping our intuition about an equilibrium economic relationship to a VECM. How to check the reasonableness of our fitted model.

10.6 Points to remember. Trend-stationary and unit-root processes. Tests for unit roots. VAR and VECM. The difficulty in specifying the complete model.

10.7 Looking ahead. Summary and advanced topics.

Forecasting is at the heart of time-series analysis. Above all, we want to anticipate the future. Influencing the future is important, but even contemplating that exercise assumes we have some way of envisioning the future.

Given the importance of forecasting, it is curious that we have focused in chapters 5–9 almost exclusively on models for stationary variables. After all, practical univariate models for stationary variables provide useful information for only a few periods ahead. Indeed the signature of a stationary time series is the rapid decay of the autocorrelation function—in other words, the rapid decay of the association between the present and the future. The influence of present-day disturbances can persist a bit longer in VARs, where a shock to one variable sets off sympathetic vibrations in related variables and subsequent feedback loops. Nonetheless, the essence of stationarity is the more-or-less rapid dissipation of disturbances.

Many of the time series we track in practice are nonstationary: real output, population, and the like. These series often are dominated by a trend that constitutes the most significant contributor to their future values. Thus an understanding of trends is the key to longer-horizon forecasts.

All the way back in chapter 3, we decomposed time series into four parts: trend, cycle, seasonality, and random noise. Since then, we have focused mainly on techniques for modeling the cyclical component and taking account of the uncertainty introduced by the random noise. We have treated the trend as a nuisance factor, a nonstationary component to eliminate prior to modeling the stationary remainder.[1] In this chapter, we correct that neglect of trends. We begin by discussing two alternative views of trends in a univariate model. Next we consider the relationships between multiple trending variables. In some instances, groups of nonstationary variables obey one or more long-term relationships and the deviations from these relationships can be stationary. In such a case, we call these variables cointegrated, and we use a VECM to fit their relationships. Many of the tools and concepts we developed in the previous chapter's treatment of VARs can be extended to cointegrated variables and VECMs.

10.1   Trends and unit roots

"The trend is your friend"—Wall Street saying

U.S. national output provides a useful example of a nonstationary time series. In chapter 7, to model real output as an autoregressive moving-average (ARMA) process, we transformed it to a stationary variable. We took the first difference of the natural log of real gross domestic product (GDP) as a stationary measure of real growth. We forecasted real growth from some candidate ARMA models, then compared these forecasts with the growth forecasts implied by both a linear trend model and a Holt–Winters smooth—two approaches we applied directly to the nonstationary log of real GDP. We commented on some of the similarities and differences among these forecasts, but we did not provide a formal, statistical framework for assessing these approaches. Now we are in a position to correct that oversight.[2]

Recall from chapter 6 that the general linear process for a stationary variable—the fundamental representation of our time-series models—can be written as

$$y_t = \mu + \psi(L)\epsilon_t$$

where ϵ_t is white noise and

$$\psi(L) \equiv 1 + \psi_1 + \psi_2 + \cdots$$

such that the series

$$\sum_{i=0}^{\infty} \psi_i^2 < \infty$$

---

1. And we have ignored the seasonal component almost completely, a practice we will continue. A full treatment of seasonality requires another book.
2. This discussion follows Hamilton (1994, chap. 15).

Let's look closely at two nonstationary processes that represent extensions of the general linear process. For the first extension, we add a deterministic, linear time trend to create a *trend-stationary* process:

$$y_t = \mu + \delta t + \psi(L)\epsilon_t \tag{10.1}$$

This new process has a time-varying mean, $Ey_t = \mu + \delta t$, so it cannot be stationary. However, the process is stationary around the trend—that is, $y_t - \mu - \delta t$ is stationary— hence, the name "trend-stationary process".[3]

A second example of a nonstationary process is the unit-root process

$$(1 - L)y_t = \delta + \psi(L)\epsilon_t \tag{10.2}$$

We used this process to model real GDP in chapter 7. Because an inspection of the autocorrelation functions verified that log GDP is nonstationary while real growth is stationary, we fitted ARMA models for

$$(1 - L) \log \text{ GDP}_t = \delta + \psi(L)\epsilon_t$$

Why do we call this a unit-root process? Recall from our discussion in chapter 6 of the ARMA model

$$\phi(L)y_t = \theta(L)\epsilon_t$$

that y_t is stationary if all the roots of $\phi(z)$ lie outside the unit circle. In the unit-root process above, $\phi(z) \equiv 1 - z$, and the one and only root lies on the unit circle, hence, "unit root".[4] We also describe this type of process as an integrated process.[5] This terminology is derived from an analogy with calculus. In a discrete-time process, a first-difference is analogous to a derivative, and the inverse of a derivative is an integral.

We consciously chose to use the symbol δ in the expressions for both the trend-stationary (10.1) and unit-root processes (10.2). In both processes, δ is the typical one-period increment in y_t. Also, if we first-difference the trend-stationary process, we obtain

$$(1 - L)y_t = \delta + (1 - L)\psi(L)\epsilon_t$$

So δ plays the same role in both processes.

---

3. Technically, we do not have to subtract μ from y_t to induce stationarity, but we look for any excuse to eliminate annoying constants.

4. More generally, if we can rewrite the process

$$(1 - \phi_1 L - \phi_2 L^2 - \phi_3 L^3 - \cdots)y_t = \theta(L)\epsilon_t$$

as

$$(1 - L)(1 - \varphi_1 L - \varphi_2 L^2 - \cdots)y_t = \theta(L)\epsilon_t$$

we have a unit-root process. And, of course, there may be more than one unit root; in other words, we may have

$$(1 - L)^k (1 - \varphi_1^* L - \varphi_2^* L^2 - \cdots)y_t = \theta(L)\epsilon_t$$

5. This is the same use of "integrated" as in autoregressive integrated moving-average model.

This last equation may give the impression that first-differencing is the appropriate treatment for both unit-root and trend-stationary processes. That impression turns out to be mistaken. First-differencing a trend-stationary process introduces a unit root in the moving-average component of the process. As a result, the moving-average component is no longer invertible. The appropriate treatment for a trend-stationary process is detrending, that is, subtracting the time-varying mean $\mu + \delta t$ from y_t.

Similarly, it would be a mistake to detrend a unit-root process. To see this point, we simplify the unit-root process by setting $\psi(L) \equiv 1$. This assumption produces a random walk with drift

$$y_t = y_{t-1} + \delta + \epsilon_t$$

If we detrend this process, we get

$$y_t - \delta t = y_0 + \sum_{i=1}^{t} \epsilon_i \equiv y_0 + u_t$$

This detrended process has a constant mean (y_0); however, the variance of u_t is $t\sigma^2$, which increases with t.

Now we are in a quandary. Returning to our example of real GDP, should we model this time series as a trend-stationary process or as a unit-root process? If we choose the wrong model, we will choose the wrong method for inducing stationarity. Before we attempt to answer this difficult question, let's look at some of the other differences between these two types of nonstationary processes.

Trend-stationary and unit-root processes have important differences in their forecasts, their forecast errors, and their implications for the impact of a random shock to the process. Let's start with the differences in the forecasts of y_{t+s} made at time t—that is, an s-period-ahead forecast—and the associated forecast errors. For a trend-stationary process, we add the deterministic component to the forecast of the stationary stochastic component

$$\widehat{y}_{t+s|t} = \mu + \delta(t + s) + \psi_s \epsilon_t + \psi_{s+1} \epsilon_{t-1} + \cdots$$

Note the subscripts on the ψ and ϵ terms. At time t, the forecast of all future ϵ terms is 0; thus the first s terms of the moving-average component are set to 0 in the forecast. Because the series $\sum_{i=0}^{\infty} \psi_i$ is required to be finite, the ψ_i terms must eventually become vanishingly small. As a consequence, the forecast $\widehat{y}_{t+s|t}$ converges to $\mu + \delta(t + s)$ as s grows larger. In other words, deviations from trend are projected to evaporate eventually. The output loss from a recession is recovered in the future. Similarly, the extra output from an expansion is eventually "paid back".

The forecast error for the trend-stationary process is a ψ-weighted sum of the s disturbances that occur after time t:

$$y_{t+s} - \widehat{y}_{t+s|t} = \epsilon_{t+s} + \psi_1 \epsilon_{t+s-1} + \cdots + \psi_{s-1} \epsilon_{t+1}$$

Thus the mean squared error of the forecast is

$$E\left(y_{t+s} - \widehat{y}_{t+s|t}\right) = \left(1 + \psi_1^2 + \psi_2^2 + \cdots + \psi_{s-1}^2\right)\sigma^2$$

As s grows larger, this squared error approaches

$$\sigma^2 \sum_{i=0}^{\infty} \psi_i^2$$

which is just the unconditional variance of $\psi(L)\epsilon_t$.

For the unit-root process, we can use the stationary first-difference process to help derive the s-period-ahead forecast. The s-period-ahead forecast of the first difference is

$$\Delta\widehat{y}_{t+s|s} = \delta + \psi_s\epsilon_t + \psi_{s+1}\epsilon_{t-1} + \cdots$$

Notice that the unit-root process also follows a trending process with slope δ, but the intercept changes permanently each period to y_t. Random disturbances have permanent impacts that are embedded in y_t. Output loss in recessions is never expected to be recovered, and expansions represent good fortune that need never be reversed.

We can characterize these statements more formally in terms of the impulse–response functions, which quantify the persistence of disturbances. For the trend-stationary process (10.1), the impact s periods ahead of a shock today is

$$\frac{\partial y_{t+s}}{\partial\epsilon_t} = \psi_s$$

Because the ψ-weights are square summable, this impact grows smaller and smaller as s goes to infinity, that is,

$$\lim_{s\to\infty} \frac{\partial y_{t+s}}{\partial\epsilon_t} = 0$$

One-period shocks have no permanent impact on a trend-stationary variable.

For a unit-root process (10.2), the impact of today's shock on y_{t+s} can be shown to be

$$\frac{\partial y_{t+s}}{\partial\epsilon_t} = 1 + \psi_1 + \cdots + \psi_s$$

Thus the long-run impact of ϵ_t is

$$\lim_{s\to\infty} \frac{\partial y_{t+s}}{\partial\epsilon_t} = 1 + \psi_1 + \psi_2 + \cdots = \psi(1)$$

Given these differences in long-run behavior, it seems to be very important to determine whether a nonstationary variable follows a trend-stationary process or a unit-root process. Unfortunately, we cannot in general make that determination in a finite sample. It can be shown that if a variable follows a unit-root process, there exists a stationary process that cannot be distinguished from the true unit process for any given sample size. Conversely, if the true process is stationary, there exists a unit-root process that cannot be rejected for any finite sample.[6]

At the risk of oversimplifying, we can motivate this result by considering the familiar question of fitting any model parameter—say, a coefficient, β, in a classical linear regression model. The ordinary least-squares (OLS) estimator, $\hat{\beta}$ converges to β in an infinite sample. However, in any finite sample, there are an infinite number of values close to β that cannot be rejected on the basis of the sample. Analogously, a univariate process is stationary if all the roots of $\phi(L)$ lie outside the unit circle, while, for a unit-root process, one of the roots lies on the unit circle. But in any finite sample, it is impossible to distinguish between a unit root and a root that is very, very close to unity.

Before you become discouraged, let's put this conundrum in perspective. The observational equivalence of stationary and unit-root processes is a reminder of the extreme flexibility of time-series models. We have seen in previous chapters the ability of multiple ARMA specifications to provide roughly equivalent fits to a real-world process. The emphasis on parsimony in model selection is rooted in this realization. The simplest alternative among viable models generally will be the most tractable.

Once we are willing to restrict the candidate models to a reasonable set, the problem of observational equivalence becomes much smaller. We can, in a large enough sample, discriminate convincingly between specific stationary and unit-root alternatives. Moreover, the question of the "permanent" (that is, long-run) impact of an individual shock is somewhat beside the point. None of us are going to live long enough to observe the permanent impact of anything. But we can estimate bounds on the impact of a shock at some meaningful horizon, that is, on the magnitude of $\partial y_{t+s}/\partial \epsilon_t$ for a reasonable value of s. If the question is, say, the impact on GDP two years from now of a change in the Federal funds rate today, the tools at hand can help.

From this point on, we are going to focus on unit-root processes and leave trend-stationary processes behind. We do this for two reasons. First, it turns out that the familiar OLS t and F tests have the same asymptotic distribution in the trend-stationary case as in a static regression.[7] Analyzing unit-root models, on the other hand, requires some new tools. Second, the concept of cointegration is based on relationships among unit-root processes. We begin in the next section with some tests for unit roots.

10.2 Testing for unit roots

We are going to present several approaches that have been proposed for testing the hypothesis that a series contains a unit root. However, after the previous section, the language used to describe hypothesis tests for unit roots may seem a bit out of place.

---

6. Hamilton (1994, chap. 15) provides a handy summary of the literature on this question. The observational equivalence of unit-root and stationary processes refers to general stationary process, not just trend-stationary processes.

7. Although the proof of that statement is challenging. See Hamilton (1994, chap. 16).

Formal hypothesis tests are based on the notion of a "true" model and a formula that can help us decide between two competing hypotheses—the null and the alternative— about that true model. But as we discussed in the section above, the extreme flexibility of time-series models makes it difficult to discriminate definitively between alternative specifications for a time-series variable.

In fact, the notion of a true model can be a bit of a red herring in time-series analysis. Much of time-series modeling is atheoretic; it concentrates on finding a specification that captures the observable dynamic behavior of a series without requiring a theoretical explanation for that behavior. Of course, time-series models can be grounded in theory, and exogenous influences can be incorporated in these models, but these features are not required. Many time-series models are best regarded as providing useful approximations to complex, and perhaps not fully understood, real-world phenomena—approximations that help us to predict future realizations and to estimate the impact of disturbances or interventions in the process. Accordingly, you should think of unit-root tests as a way to decide if a unit-root specification provides a reasonable approximation for the variable of interest, not as a step in revealing the true model for a process.

With these caveats, let's look at some alternative tests. In all the tests we consider, the null hypothesis states that the process determining our time-series variable contains a unit root. Specifically, the null assumes the true model is

$$y_t = \alpha + y_{t-1} + u_t$$

The model of real GDP growth we explored in chapter 7 fits this specification. Our specification was

$$\log \text{real GDP}_t = \alpha + \log \text{real GDP}_{t-1} + u_t$$

where α, the drift term, represented the average rate of real growth and u_t incorporated the ARMA dynamics that we identified and estimated.

If we subtract y_{t-1} from both sides of the equation above, we have

$$\Delta y_t = \alpha + u_t$$

Thus if we estimated the regression

$$\Delta y_t = \alpha + \beta y_{t-1} + u_t$$

we should find that $\widehat{\beta} = 0$ under the null hypothesis. Note, however, that the u_t are likely to be serially correlated, so the plain OLS standard errors probably are incorrect.

The augmented Dickey–Fuller test uses OLS to fit the slightly different equation[8]

$$\Delta y_t = \alpha + \beta y_{t-1} + \delta t + \zeta_1 \Delta y_{t-1} + \cdots + \zeta_k \Delta y_{t-k} + \epsilon_t$$

The addition of the lagged differences in this regression soaks up the serial correlation so that now u_t will be identically and independently distributed. Rather than relying on

---

8. The "augmentation" here is the addition of the lagged differences to the right-hand side of the equation.

a first-round test for serial correlation in u_t, the augmented Dickey–Fuller test simply includes a reasonable number of lagged values to Δy_t to soak up any serial correlation.[9] This specification also allows for a trend, δt.

If y_t followed a first-order autoregressive process instead of a unit-root process, then

$$y_t = \rho y_{t-1} + u_t$$

where we have suppressed α and δt for simplicity. In this case,

$$\Delta y_t = (\rho - 1)y_{t-1} + u_t$$

Stationarity requires $|\rho| < 1$, so $(\rho - 1) < 0$ under this alternative specification. Because $\beta \equiv \rho - 1$, Dickey and Fuller (1979) developed critical values for testing $\widehat{\beta}/\widehat{\sigma}_\beta = 0$ against the alternative $\widehat{\beta}/\widehat{\sigma}_\beta < 0$. While this test statistic looks like a standard t statistic, it does not have an asymptotic t distribution.

The `dfuller` command calculates the augmented Dickey–Fuller test (the easy part—it's just an OLS regression) and compares the test statistic with the tabled critical values developed by Dickey and Fuller (the hard part). The syntax is

`dfuller` *varname* $\big[\,if\,\big]$ $\big[\,in\,\big]$ $\big[$, no̲nconstant tr̲end dr̲ift regress la̲gs(#) $\big]$

`lags(#)` specifies the number of lags of Δy_t to include in the regression. If you specify the `regress` option, the Dickey–Fuller regression will be displayed. Normally, it is omitted. The other three options—`noconstant`, `drift`, and `trend`—indicate whether drift and trend terms should be included.

Aside from the number of lags, there are four possible specifications of the Dickey–Fuller regression.

Case	Null hypothesis	Restrictions	Option
1	No drift, no trend	$\alpha = 0, \delta = 0$	`noconstant`
2	No drift, no trend	$\delta = 0$	default
3	Drift	$\delta = 0$	`drift`
4	Drift and trend	none	`trend`

At first glance, it looks as if case 2 and case 3 are identical; the only restriction is $\delta = 0$. The difference is the assumption about α. In both cases, the regression includes a constant term, which estimates α, but in case 2, we assume the population value of α is 0. This assumption affects the distribution of the test statistic and, hence, the critical values calculated by Dickey and Fuller.

---

9. Choosing the appropriate number of lags depends largely on guesswork. In practice, it makes sense to try several alternatives to see if the significance of the test statistic is sensitive to small changes in the lag length.

Let's use the augmented Dickey–Fuller test to confirm the specification we used in chapter 7 for the time-series process for real GDP growth.

```
. use ${ITSUS_DATA}/quarterly, clear
(Quarterly U.S. GDP, annualized, BEA, as of 4/27/2012)
. generate lgdp = log(gdp2005)
. dfuller lgdp, lags(4) reg
Augmented Dickey-Fuller test for unit root         Number of obs   =       256
                           ──────── Interpolated Dickey-Fuller ────────
                     Test      1% Critical      5% Critical     10% Critical
                 Statistic        Value            Value            Value

 Z(t)              -2.000         -3.460           -2.880           -2.570

MacKinnon approximate p-value for Z(t) = 0.2866
```

D.lgdp	Coef.	Std. Err.	t	P>|t|	[95% Conf. Interval]	
lgdp						
L1.	-.0019467	.0009734	-2.00	0.047	-.0038638	-.0000296
LD.	.3292161	.0628588	5.24	0.000	.2054158	.4530165
L2D.	.13482	.0658649	2.05	0.042	.0050992	.2645408
L3D.	-.093799	.0657394	-1.43	0.155	-.2232726	.0356746
L4D.	-.0900451	.0624493	-1.44	0.151	-.213039	.0329488
_cons	.0224022	.0085882	2.61	0.010	.0054878	.0393166

After some experimentation, we decided to include four lags of Δy_t in the Dickey–Fuller regression. In this case, the result of the test is not highly sensitive to choice of lag length.[10] The table at the top of the output displays the test statistic (-2.000) and interpolated estimates (from the Dickey–Fuller calculations) of the 1%, 5%, and 10% critical values. The test statistic is larger algebraically (that is, closer to 0) than any of the displayed critical values. We accept the null hypothesis of a unit root.

We included the regression output to confirm that the Dickey–Fuller test statistic is the regression estimate of the t statistic of the coefficient on L.lgdp, -2.00 in this example. However, the test statistic does not follow a t distribution. Notice that the p-value in the regression output is 0.047, while the MacKinnon approximate p-value of the Dickey–Fuller test is 0.2866.

---

10. The choice of lag length tends to be as much art as science. A common rule of thumb is to make an initial guess based on the frequency of the data: four lags for quarterly data, twelve (if possible) for monthly, and so on. Another approach—a sort of informal stepwise approach that is more in line with classical statistical theory—is to choose a number of lags that you are confident is somewhat larger than needed (although there is no guidance on how to achieve that confidence). Then you trim the longest lags, one by one, until the longest remaining lag is statistically significant. In any event, it's generally a good idea to try some alternatives to make sure the results are not highly sensitive to your choice of lag length.

We have not specified the strongest test of our specification. We used the default specification, where α is estimated but the critical values are calculated under the assumption that the population value of α is 0. But α represents the long-run growth rate of real GDP, clearly a value greater than 0.

Let's correct our error by indicating that a nonzero drift term should be in the model.

```
. dfuller lgdp, lags(4) drift
Augmented Dickey-Fuller test for unit root          Number of obs   =        256
```

	Test Statistic	1% Critical Value	Z(t) has t-distribution 5% Critical Value	10% Critical Value
Z(t)	-2.000	-2.341	-1.651	-1.285

```
p-value for Z(t) = 0.0233
```

Now we have a puzzle. The value of the test statistic has not changed, because the specification of the Dickey–Fuller regression has not changed. However, the critical values have changed, and we now reject the null hypothesis. Time to seek a second (and third) opinion.

Stata offers two other tests based on the Dickey–Fuller approach but offering some improvements. `dfgls` calculates a modified Dickey–Fuller test proposed by Elliott, Rothenberg, and Stock (1996). In this approach, the time series undergoes a GLS transformation prior to the estimation of the Dickey–Fuller regression.[11] Studies have shown that this test has significantly greater power than the traditional augmented Dickey–Fuller test.

The syntax is

`dfgls` *varname* $\big[$ *if* $\big]$ $\big[$ *in* $\big]$ $\big[$, <u>m</u>axlag(*#*) <u>not</u>rend ers $\big]$

The `ers` option specifies that `dfgls` should present alternative critical values tabulated by Elliott, Rothenberg, and Stock (1996).

---

11. See the Stata manual for details.

Let's see what this test says about the log of real GDP.

```
. dfgls lgdp, maxlag(4) notrend

DF-GLS for lgdp                                  Number of obs =    256

                  DF-GLS mu      1% Critical     5% Critical     10% Critical
       [lags]   Test Statistic      Value           Value           Value

         4          3.087          -2.580          -2.002          -1.687
         3          3.198          -2.580          -2.006          -1.690
         2          3.147          -2.580          -2.009          -1.693
         1          3.939          -2.580          -2.013          -1.696

Opt Lag (Ng-Perron seq t) =  2 with RMSE  .0095169
Min SC   = -9.244386 at lag  2 with RMSE  .0095169
Min MAIC = -9.188314 at lag  2 with RMSE  .0095169
```

The output of `dfgls` includes three estimates of the optimal number of lags of Δy_t to include. In this example, all three methods agree that two lags are best. However, the null hypothesis is accepted at any of the lag lengths. Indeed the test statistic is positive, thus larger than any of the critical values.

The `pperron` command calculates two test statistics, Z_ρ and Z_τ, proposed by Phillips and Perron (1988). The Phillips–Perron test uses Newey–West standard errors to deal with the serial correlation of u_t. The syntax is

pperron *varname* $\big[$ *if* $\big]$ $\big[$ *in* $\big]$ $\big[$, <u>no</u>constant <u>tr</u>end <u>regress</u> <u>lags</u>(#) $\big]$

```
     Phillips-Perron test for unit root           Number of obs   =      260
                                                  Newey-West lags =        4

                                    ──────── Interpolated Dickey-Fuller ────────
                        Test        1% Critical     5% Critical     10% Critical
                     Statistic         Value           Value           Value

     Z(rho)           -0.681         -20.308         -14.000         -11.200
     Z(t)             -1.977          -3.459          -2.880          -2.570

MacKinnon approximate p-value for Z(t) = 0.2966
```

Both test statistics support acceptance of the null hypothesis.

10.3 Cointegration: Looking for a long-term relationship

There are so many ways to motivate the subject of cointegration that it's difficult to know where to start. We could talk about "balancing" equations, so the order of integration on both sides is equal. We could talk about the role of cointegration in recognizing that many VARs specified in terms of first-differenced variables may be misspecified. Or we could talk about cointegration as a formal way of addressing some of the questions around trends that we discussed in the beginning of this chapter.

We are not going to start with any of these approaches, as useful and important as they are. Instead we are going to start with some nonstatistical intuition about the meaning of nonstationary variables. From there, we will segue to the math of cointegration. Finally, we will see some of the hurdles in moving from our intuition about the real world to a reasonable model of cointegrated variables.

By this point in the book, you should be convinced that real GDP has a unit root, but what does that mean? Speaking just for myself, I do not believe that the log of real GDP is literally unbounded. At any point in time, resources—land, labor, capital—are finite, and I have a sneaking, Malthusian suspicion that while increases in productivity may generate currently unimaginable increases in output per capita, infinity is still a long way off. However, over any reasonable horizon we care to contemplate—a year, a decade, a century—modeling the log of real GDP as a unit-root process is a useful approach for inference and prediction, as long as we do not take it too literally.

One reason the unit-root formulation is useful is that it makes us think about what factors might pin down the time path of real output over the long run. At bottom, the answer is simple,

$$\text{Output} = \text{Resources} \times \text{Productivity}$$

If, for the sake of discussion, we assume some sort of fixed relationship between the categories of resources (land, labor, capital), we can use labor as an index and write

$$\text{Output} = \frac{\text{Labor}}{\text{Unit labor requirement}}$$

or

$$\log(\text{Output}) = \log(\text{Labor}) - \log(\text{Unit labor requirement})$$

Real output can increase only if there is growth in the available labor force or an increase in labor productivity (decrease in the unit labor requirement). This formulation encapsulates a lot of debate about economic policy. A nation seeking to increase economic welfare can encourage growth in the labor force by increasing economic incentives to enter the labor force; liberals might propose subsidies for day care and health care, while conservatives might advocate reductions in the minimum wage, unemployment insurance, and welfare. Nations with a long planning horizon may try to influence the birth rate directly (very controversial). A nation also can adopt policies that stimulate productivity growth, investment tax credits for instance.

The equation above is a tautology—productivity (or its reciprocal, unit labor requirement) is the unobservable residual that guarantees the equality of both sides of the equation. Economists and others add identifying assumptions derived from theory to convert this to an observable statistical relationship, say, something like

$$\log(\text{Output}_t) = \log(\text{Labor}_t) - \log(\text{Unit labor requirement}_t) + \epsilon_t$$

where a lot of work goes into defining measures of output, labor, and the unit labor requirement. For our purposes, the important thing to notice is that at least one quantity on the right-hand side of this equation must contain a unit root to balance the left-hand side.[12] Labor is a likely candidate; it tends to grow with population with variations driven mainly by events, like the baby boom, which skew the demographic distribution. Labor productivity also has trended upward over time.

To the extent that we are measuring a relationship that is not a tautology, there can be persistent deviations from the long-run relationship between output, labor, and productivity. During wartime, labor-force participation may spike as older workers return to the labor force and female labor participation increases to replace young men called away to military service. During recessions, discouraged workers may be sidelined for an extended period, and labor productivity may surge as the diminished labor force is supported by an ample capital base. These events will be measured as extended periods of either positive or negative values of ϵ_t. Over time, however, output will return to its "normal" or equilibrium relationship with labor and labor productivity. In other words, ϵ_t is likely to be stationary, albeit with significant autocorrelation.

This characterization—that nonstationary variables may obey a long-run relationship with each other whose residual is stationary—is the central intuition of cointegration. Now let's see how to formalize this intuition.

10.4 Cointegrating relationships and VECMs

A cointegrating relationship is a stationary linear combination of two or more nonstationary variables. Because this relationship requires at least two variables, we are immediately in the world of chapter 9, and we will find a lot of overlap between the techniques used in that chapter and the techniques used in this one.

Let's start small, with just two variables:

$$y_t = \alpha + \beta x_t + \epsilon_t$$

If y_t and x_t both are covariance stationary, then we can use classical regression to quantify their relationship.[13] If y_t is nonstationary, either x_t or ϵ_t, or both, must be nonstationary. If ϵ_t is stationary, then

$$y_t - \alpha - \beta x_t$$

is stationary, and we call $y_t - \alpha - \beta x_t$ the cointegrating relationship between y_t and x_t.

There can be only one cointegrating relationship between these two variables. To see this point, imagine that there is a second, distinct cointegrating relationship, say,

$$y_t - \delta - \gamma x_t$$

---

12. This constraint is not a mere technical requirement. The key implication of the unit process in output is the projection of unbounded future values. If the labor supply and the unit labor requirement were both stationary, their expected future values would be bounded (in probability), making the equation nonsensical.

13. To capture the dynamic properties of y_t or ϵ_t, we may choose to use an ARMA with exogenous inputs model, that is, an autoregressive integrated moving-average model with explanatory variables in the measurement equation.

But that means that over the long-run,

$$\alpha + \beta x_t = \delta + \gamma x_t$$

y_t cannot have two distinct long-run equilibrium paths. In general, K nonstationary variables can have at most $K - 1$ cointegrating relationships.

Why not just first-difference y_t and x_t and estimate

$$\Delta y_t = \beta \Delta x_t + \Delta \epsilon_t$$

After all, that approach worked for fitting univariate time-series models. It turns out that this equation is misspecified when y_t and x_t are cointegrated, but it will take us a couple of paragraphs to make that point.

Another potential pitfall: What if y_t and x_t are independent unit-root processes? That is, what if y_t and x_t are not cointegrated and $\beta = 0$? Granger and Newbold (1974) showed that

$$y_t = \alpha + \beta x_t + \epsilon_t$$

is a spurious regression; that is, classical least squares will almost always reject the (true) hypothesis $\beta = 0$.

While an example is not a substitute for a proof, let's illustrate this point about spurious regressions by generating two independent random walks, then regress one on the other.

```
. clear
. set obs 2000
number of observations (_N) was 0, now 2000
. set seed 47
. generate e1 = rnormal()
. generate rw1 = e1 in f
(1,999 missing values generated)
. replace rw1 = rw1[_n-1] + e1 in 2/l
(1,999 real changes made)
. set seed 23
. generate e2 = rnormal()
. generate rw2 = e2 in f
(1,999 missing values generated)
. replace rw2 = rw2[_n-1] + e2 in 2/l
(1,999 real changes made)
```

```
. regress rw1 rw2
      Source │       SS           df       MS            Number of obs   =      2,000
─────────────┼──────────────────────────────────        F(1, 1998)      =     516.51
       Model │  424128.599         1  424128.599         Prob > F        =     0.0000
    Residual │  1640633.37     1,998  821.137825         R-squared       =     0.2054
─────────────┼──────────────────────────────────        Adj R-squared   =     0.2050
       Total │  2064761.97     1,999  1032.89744         Root MSE        =     28.656

─────────────┼──────────────────────────────────────────────────────────────────────
         rw1 │     Coef.   Std. Err.      t    P>|t|     [95% Conf. Interval]
─────────────┼──────────────────────────────────────────────────────────────────────
         rw2 │   1.10833   .0487673    22.73   0.000     1.01269     1.20397
       _cons │  -3.895474  1.654338    -2.35   0.019    -7.139883   -.6510645
```

Even though we generated these random walks independently, the t statistic on $\beta_{\mathtt{rw2}}$ is 22.7, highly significant. Moreover, the R^2 statistic is 0.21.

Now we have a dilemma. If we estimate the equation in levels, we run the risk of estimating a spurious regression, but if we take first-differences of y_t and x_t, we run the risk of estimating a misspecified equation. What we need is some way of assessing whether there is a cointegrating relationship. If we are analyzing more than two nonstationary variables, we also need to assess how many cointegrating relationships there are.

Again let's start with a small example to anchor our intuition. Assume we have two unit-root variables, y_t and z_t, that happen to be cointegrated. Assume further that they obey the relationship

$$y_t = \alpha_1 y_{t-1} + \gamma_0 z_t + \gamma_1 z_{t-1} + \epsilon_t$$

Subtract y_{t-1} from both sides of this equation and simultaneously add and subtract z_{t-1} from the right-hand side. After rearranging terms, we can rewrite the equation as

$$\Delta y_t = (\alpha_1 - 1)y_{t-1} + \gamma_0 \Delta z_t + (\gamma_0 + \gamma_1)z_{t-1} + \epsilon_t \qquad (10.3)$$

Let's collect terms a little differently

$$\Delta y_t = \gamma_0 \Delta z_t - \lambda(y_{t-1} - \theta z_{t-1}) + \epsilon_t$$

where $\lambda \equiv (1 - \alpha_1)$ and

$$\theta \equiv \frac{\gamma_0 + \gamma_1}{\alpha_1 - 1}$$

Equation 10.3 is an error-correction model (ECM), the fundamental representation used to characterize the relationships among cointegrated variables.[14] The cointegrating relationship in this example is $y_{t-1} - \theta z_{t-1}$. The errors in the ECM are the nonzero values

---

14. If there were more lagged values of y_t and x_t in the original relationship, our rewriting would end up with the ECM

$$A(L)\Delta y_t = \Gamma(L)\Delta z_t - \lambda(y_{t-1} - \theta z_{t-1}) + \epsilon_t$$

of $y_{t-1} - \theta z_{t-1}$, that is, the short-term deviations from the long-term cointegrating relationship. λ quantifies the speed of adjustment (the correction) to these errors.

This characterization of the adjustment process ties back to the initial intuition we used to motivate the concept of cointegration. The cointegrating relationship describes the long-term relationship that links the levels of the nonstationary variables, x_t and z_t. Deviations from this relationship represent disequilibria that cannot persist indefinitely, and the ECM helps us analyze how this two-equation system returns to equilibrium.

Note that all the terms in the ECM—Δy_t, Δz_t, the cointegrating relationship, and ϵ_t—are stationary; the equation balances. This ECM can be fitted by OLS by using Stata's `regress` command. We can use the `nlcom` command to back out an estimate of θ.[15] Alternatively, we can use the `nl` command to obtain estimates of all the parameters in one step.

Things get a little more complicated when there are more than two variables (and, hence, potentially more than one cointegrating relationship). For the next step, let's consider a second-order VAR for a K-element vector $\mathbf{y}_t$ where all the elements contain unit roots.

$$\mathbf{y}_t = \boldsymbol{\mu} + \boldsymbol{\Phi}_1 \mathbf{y}_{t-1} + \boldsymbol{\Phi}_2 \mathbf{y}_{t-2} + \epsilon_t$$

We can subtract $\mathbf{y}_{t-1}$ from both sides of this equation and rearrange terms to obtain

$$\Delta \mathbf{y}_t = \boldsymbol{\mu} + (\boldsymbol{\Phi}_1 + \boldsymbol{\Phi}_2 - \mathbf{I})\mathbf{y}_{t-1} - \boldsymbol{\Phi}_2 \Delta \mathbf{y}_{t-1} + \epsilon_t$$

With a bit more algebra, we can show that we can rewrite a general, pth-order VAR for a K-element vector $\mathbf{y}_t$ as a VECM

$$\Delta \mathbf{y}_t = \boldsymbol{\mu} + \boldsymbol{\Pi} \mathbf{y}_{t-1} + \sum_{i=1}^{p-1} \boldsymbol{\Gamma}_i \Delta \mathbf{y}_{t-i} + \epsilon_t$$

where

$$\boldsymbol{\Pi} = \sum_{j=1}^{p} \boldsymbol{\Phi}_j - \mathbf{I}$$

and

$$\boldsymbol{\Gamma}_i = - \sum_{j=i+1}^{p} \boldsymbol{\Phi}_j$$

Engle and Granger (1987) showed that $\boldsymbol{\Pi}$ has rank $r, 0 \leq r < K$, where r is the number of linearly independent cointegrating relationships among the elements of $\mathbf{y}_t$. This result explains why we cannot just estimate a VAR in $\Delta \mathbf{y}_t$ when there are cointegrating relationships. We would be omitting the term $\boldsymbol{\Pi} \mathbf{y}_{t-1}$.[16]

---

15. Recall we followed a similar strategy in chapter 5.
16. If there are no cointegrating relationships among the elements of $\mathbf{y}_t$, the rank of $\boldsymbol{\Pi}$ is 0, and the $\boldsymbol{\Pi}\mathbf{y}_{t-1}$ term does not appear. Alternatively, if all the elements of $\mathbf{y}_t$ are stationary instead of unit-root processes, $\boldsymbol{\Pi}$ has full rank.

Because $\mathbf{\Pi}$ has rank r, we can decompose $\mathbf{\Pi}$ into $\boldsymbol{\alpha\beta}'$, where $\boldsymbol{\alpha}$ and $\boldsymbol{\beta}$ both are $K \times r$ matrices with rank r. Thus we can rewrite our VECM as

$$\Delta\mathbf{y}_t = \boldsymbol{\mu} + \boldsymbol{\delta}t + \boldsymbol{\alpha\beta}'\mathbf{y}_{t-1} + \sum_{i=1}^{p-1}\boldsymbol{\Gamma}_i\Delta\mathbf{y}_{t-i} + \boldsymbol{\epsilon}_t$$

We have snuck in a $K \times 1$ trend term, $\boldsymbol{\delta}t$; we will get to that in a minute. First, consider identifying the parameters in $\boldsymbol{\alpha\beta}'$. We can replace $(\boldsymbol{\alpha}, \boldsymbol{\beta})$ with $(\boldsymbol{\alpha}\mathbf{Q}, \boldsymbol{\beta}\mathbf{Q}^{-1})$, where $\mathbf{Q}$ is any nonsingular $r \times r$ matrix, without changing our VECM. Consequently, some sort of restrictions have to be imposed to obtain identification. In the rest of this chapter, we will rely on some identifying conventions proposed by Johansen, but Stata will let you impose your own restrictions if you like.

10.4.1   Deterministic components in the VECM

What is that trend term doing in our VECM? Well, remember that a constant in an equation for the first-difference of a variable represents a linear trend in the level of the variable, that is,

$$y_t = \kappa + \lambda t \implies \Delta y_t = \lambda$$

Similarly, a quadratic time trend in the level equation for a variable represents a linear time trend in the first-difference equation

$$y_t = \kappa + \lambda t + \omega t^2 \implies \Delta y_t = \lambda + 2\omega t - \omega$$

Thus, because our VECM is written in terms of $\Delta\mathbf{y}_t$, the deterministic components $\boldsymbol{\mu} + \boldsymbol{\delta}t$ represent linear and quadratic trends in $\mathbf{y}_t$.

But wait—we can make this even more complicated. Let's decompose these terms into trends in $\mathbf{y}_t$ and separate trends in the cointegrating equations. We can write

$$\boldsymbol{\mu} \equiv \boldsymbol{\alpha\nu} + \boldsymbol{\gamma}$$

and

$$\boldsymbol{\delta}t \equiv \boldsymbol{\alpha\rho}t + \boldsymbol{\tau}t$$

where $\boldsymbol{\nu}$ and $\boldsymbol{\rho}$ are $r \times 1$ vectors and $\boldsymbol{\gamma}$ and $\boldsymbol{\tau}$ are $K \times 1$ vectors.[17] Then we can rewrite the VECM as

$$\Delta\mathbf{y}_t = \boldsymbol{\gamma} + \boldsymbol{\tau}t + \boldsymbol{\alpha}(\boldsymbol{\beta}'\mathbf{y}_{t-1} + \boldsymbol{\nu} + \boldsymbol{\rho}t) + \sum_{i=1}^{p-1}\boldsymbol{\Gamma}_i\Delta\mathbf{y}_{t-i} + \boldsymbol{\epsilon}_t$$

---

17. These vectors are defined so that $\boldsymbol{\gamma}$ is orthogonal to $\boldsymbol{\alpha\nu}$ and $\boldsymbol{\tau}$ is orthogonal to $\boldsymbol{\alpha\rho}$.

The cointegrating equations represent the equilibrium relationships among the levels of the nonstationary variables in this multiple equation model, just the same as the single cointegrating relationship represented the equilibrium relationship between the levels of the two nonstationary variables in our illustrative ECM above. The deterministic terms in the cointegrating equations, $\nu + \rho t$, represent the means and linear trends of those relationships, but not a quadratic trend.

This form is very flexible—perhaps too flexible. We have to specify what kinds of trends enter our equations before we can test for the number of cointegrating equations or fit the VECM. Moreover, as we shall see below, our assumptions about these trend terms are not innocuous. Graphical analysis of the variables may provide some clues, but as usual, some guidance from theory is likely to be most useful.

In testing and estimation, Stata allows for five cases:

Case 1: Unrestricted trend. All four parameters are estimated. We are assuming there are quadratic trends in the level of $\mathbf{y}_t$ and the cointegrating equations are trend stationary.

Case 2: Restricted trend, $\tau = 0$. $\mathbf{y}_t$ includes linear, but not quadratic, trends. As in Case 1, the cointegrating equations are trend stationary.

Case 3: Unrestricted constant, $\tau = 0$ and $\rho = 0$. The observable variables follow linear trends, but the cointegrating equations are stationary around constant means.

Case 4: Restricted constant, $\tau = 0$, $\rho = 0$, and $\gamma = 0$. There are no trends in $\mathbf{y}_t$, but the cointegrating equations are stationary around a constant mean.

Case 5: No trend, $\tau = 0$, $\rho = 0$, $\gamma = 0$, and $\nu = 0$. No nonzero means or trends.

10.5   From intuition to VECM: An example

Now it's time to look at an example to illustrate both the powerful tools that Stata provides for fitting and analyzing VECMs and, more important, the types of issues that you will have to resolve in order to translate your intuitions about a time-series relationship into a persuasive model.

Imagine we are interested in questions related to regional labor mobility and wage equalization. In particular, we want to see if wages in the construction industries equalize across political boundaries in the Washington, DC metro area. We suspect there are likely to be short-run differences in wages for similar construction jobs in adjoining areas. For example, the license requirements in DC to practice a particular trade may be different from the requirements in Montgomery County, Maryland, or Arlington County, Virginia. For union jobs, there may be some hurdles to moving from one local to another. In the long run, no significant disparity in wages for essentially the same job should persist.[18] If the incentive is great enough, workers will acquire the needed licenses, serve the required apprenticeships, pay additional union dues, commute a long distance, or, in the extreme, migrate within the region.

Let's formalize this economic intuition. Let w_{it} be the wage in the construction trades in the ith region in period t. For now, we will assume w_{it} is nonstationary, but we will test that hypothesis in a minute. We suspect that, in equilibrium,

$$\beta_i w_{it} = \beta_j w_{jt} + \nu + \rho t$$

or perhaps

$$\beta_i \log w_{it} = \beta_j \log w_{jt} + \nu + \rho t$$

if regions i and j are reasonably close. The constant ν represents any difference in wages that might persist in equilibrium due to differences in income tax rates or some similar irreducible wedge. Alternatively, any persistent wedge might be captured by $\beta_i \neq 1$. For instance, an equilibrium relationship of the form

$$1.1 w_{it} = w_{jt}$$

would indicate that construction wages in region i will remain 10% higher in equilibrium than wages in region j. A nonzero value of ρ would indicate a trend in the equilibrium wage difference. This sort of trend is a little harder to motivate, but it could arise if there is a trend in, say, tax differences that persists in a systematic way over a substantial amount of time. Thus, in the notation of the previous section, we are interested in fitting the VECM

$$\Delta \mathbf{w}_t = \boldsymbol{\gamma} + \boldsymbol{\tau} t + \boldsymbol{\alpha}(\boldsymbol{\beta}' \mathbf{w}_t - 1 + \boldsymbol{\nu} + \boldsymbol{\rho} t) + \sum_{i=1}^{p-1} \boldsymbol{\Gamma} \Delta \mathbf{w}_{t-i} + \epsilon_t$$

where $\mathbf{w}_t$ is a vector of wages in each region in period t; that is, the ith element of $\mathbf{w}_t$ is the construction wage in the ith region, w_{it}.

Ideally, we would like to have granular time-series data on wages for specific trades—painting, carpentry, plumbing, etc.—within narrow political jurisdictions—cities, counties, etc. Those sorts of data are a little tough to find, so we start with a much coarser view—aggregate construction wages in DC, Maryland (MD), and Virginia (VA).[19] Intuitively, the forces promoting wage equalization are stronger when the regions are not only adjacent but also small. By measuring wages across the entire DC–MD–VA area, we may not be capturing the wage equalization process in the DC metro area very accurately. On the other hand, the metro area represents the dominant concentration

---

18. More precisely, any equilibrium wedge between the wage for a specific job in adjacent areas should be related to institutional differences that sustain differences in the after-tax-and-other-costs wage and the value of the marginal product of the worker.

19. The data for this example were taken from the FRED database provided by the Federal Reserve Bank of St. Louis (https://fred.stlouisfed.org/). The data in this online database change as sources update their series. The construction wage series were updated as of 22 June 2011. For some reason, the construction wage data for DC are missing for the year 2005. We filled in the missing values by linearly interpolating between the values for 2004:4 and 2006:1.

of population in the region. According to the 2010 Census, there are approximately 14.4 million people in the combined DC–MD–VA area. The most recent Census Bureau estimate puts the population in the DC metro area at around 5.5 million people or 38% of the three-region total.[20]

Let's start as usual by looking at the data. Figure 10.1 displays the quarterly data for construction wages in DC, MD, and VA from 1990:1 through 2011:1.

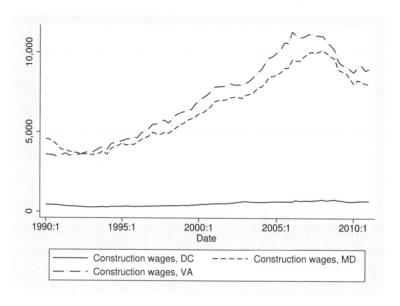

Figure 10.1. Construction wages in DC–MA–VA, millions of dollars

Construction wages in MD and VA appear to be closely correlated and show evidence of nonstationarity. The small size of DC makes it difficult to compare the time series for DC with the other two; however, all three series are highly correlated.

```
. use ${ITSUS_DATA}/conwages, clear
. correlate conwgdc conwgmd conwgva
(obs=85)
```

	conwgdc	conwgmd	conwgva
conwgdc	1.0000		
conwgmd	0.9694	1.0000	
conwgva	0.9380	0.9895	1.0000

---

20. The DC metropolitan statistical area includes some portions of West Virginia.

Figure 10.1 highlights our first challenge. We have a hypothesis about wage equalization, but these series do not measure hourly or weekly wages. Instead they measure aggregate wage income, the total dollars paid as wages in the construction industries each quarter.[21] Unfortunately, we do not have readily available data on hours worked in construction in each quarter in these three areas.[22]

We have chosen to proceed by converting our construction wage (or, more precisely, construction wage income) data to construction wage indices. We normalize the data so that the value of the index is equal to 100 in the first quarter of 1990—the first period we observe—in each of the three regions.

```
. generate dcwage = 100 * conwgdc/conwgdc[1]
. generate mdwage = 100 * conwgmd/conwgmd[1]
. generate vawage = 100 * conwgva/conwgva[1]
. label variable dc "DC"
. label variable md "MD"
. label variable va "VA"
. format dcwage mdwage vawage %6.0fc
```

---

21. These data are published by the Bureau of Economic Analysis, a division of the Department of Commerce, using the North American Industry Classification System. See https://www.census.gov/programs-surveys/economic-census/guidance/understanding-naics.html for more information about the North American Industry Classification System.

22. The Bureau of Labor Statistics publishes detailed industry-level employment and wage data in its Quarterly Census of Employment and Wages (https://www.bls.gov/cew/), but for some reason, data on construction employment in Washington, DC, end in 2004. There are additional challenges to this line of attack. For one thing, we are assuming that the mix of trades in construction employment is roughly constant across the three regions. After trying several approaches, we landed on the wage index method we adopt here. Not perfect, but it appears to work tolerably well. More importantly, it provides a good step-by-step illustration of using Stata's tools for fitting a VECM.

```
. tsline dcwage mdwage vawage
```

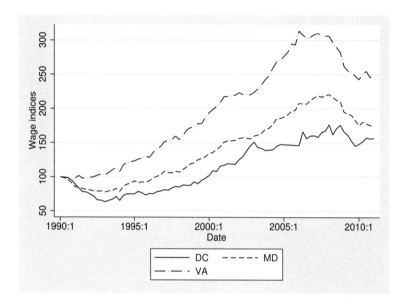

Figure 10.2. Construction wages in DC–MA–VA, 1990:1=100

   With this rescaling, the qualitative features of the wages in MD and VA are roughly the same as in figure 10.1, although wages in VA appear to grow substantially faster than in MD or DC. The rescaling also makes the DC time series easier to compare with the other two. The general shape of the DC data is similar to that of the MD and VA data, although construction wages in DC appear to plateau from 2003 Q1 through 2008 Q1, while wages in MD and VA increase sharply during that period.

VA is a right-to-work state,[23] while MD and DC are not. Perhaps the faster growth in the VA wage index reflects faster percentage growth from a lower base. Alternatively, these differences may reflect differences in the growth rates of population and the regional economies. Persistent differences in the growth of wages also may reflect differences over time in the share of compensation paid as benefits and in differences in working conditions affecting worker safety, for instance.[24]

We will follow a step-by-step process to fit a VECM for these data. We will

1. confirm that the construction wage indices contain a unit root (ac and dfgls);

2. identify the number of lags to include in the VECM (varsoc);

3. identify the number of cointegrating relationships (vecrank);

4. fit a VECM (vec);

5. test for stability and white-noise residuals (vecstable, veclmar, and vecnorm);

6. review the model implications for reasonableness (lots of stuff).

Let's get started.

Step 1: Confirm the unit root

Up to now, we have assumed that the wage indices are nonstationary, an essential assumption for the variables in a VECM. Let's see if the data are consistent with that assumption. We will look at the data for VA, but the results are similar for DC and MD.

Figure 10.3 displays the autocorrelations of the VA wage data. While these autocorrelations do not provide a formal test for a unit root, they appear consistent with that

---

23. Section 14(b) of the Taft–Hartley Act permits states to pass laws requiring an open shop for union jobs. In an open shop, workers cannot be required to join the union or to pay dues as a nonmember in a union job category. Unions have argued that wages and benefits tend to be lower in right-to-work states. There is some empirical research supporting that hypothesis, although, because right-to-work states also tend to have other pro-business policies, it is difficult to attribute observed differences solely to the right-to-work laws.

24. Can we use these data to explore our hypothesis? Maybe. As we noted above, the hypothesis of wage equalization is strongest when applied to hourly compensation (including all economic compensation such as performance bonuses, benefits, on-the-job training, etc.) for a specific trade (such as electrician or plumber) in a restricted area (for example, DC, Arlington County, VA, and Montgomery County, MD). As is frequently the case in empirical research, we are working with less-than-ideal proxies for the concepts addressed by our theory. I did explore normalizing the wage income series by measures of construction employment. The Bureau of Labor Statistics publishes a Quarterly Census of Employment and Wages, which provides some of the desired information, but for some reason, the series on construction employment in DC is unavailable after 2004. I also fit a VECM with the construction wage income series normalized by the labor force in each area. Surprisingly, the results were very good—easily interpreted and generally consistent with the wage equalization hypothesis. However, this approach requires a maintained hypothesis that the share of construction employment in the labor force in each area is roughly constant. This hypothesis seemed a little tenuous (even to me). In sum, the example below can be thought of as a sort of exploratory time-series analysis. To the extent that the results are encouraging, they should serve as a spur to search out data more closely related to the phenomena of interest.

hypothesis. The autocorrelation function falls off approximately linearly rather than exhibiting either exponential decay or a sudden drop to 0.

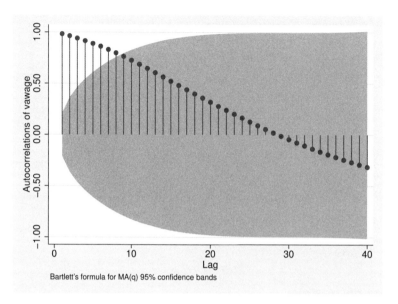

Figure 10.3. Autocorrelations of quarterly construction wages per capita in VA

The `dfgls` command we introduced in section 10.2 provides more formal support for the hypothesis of a unit root in construction wages.

```
. dfgls vawage

DF-GLS for vawage                                       Number of obs =      73
Maxlag = 11 chosen by Schwert criterion

                  DF-GLS tau     1% Critical     5% Critical    10% Critical
        [lags]   Test Statistic     Value           Value           Value

          11       -2.264          -3.637          -2.747          -2.473
          10       -2.350          -3.637          -2.785          -2.510
           9       -2.120          -3.637          -2.823          -2.546
           8       -2.173          -3.637          -2.860          -2.582
           7       -2.159          -3.637          -2.897          -2.616
           6       -1.997          -3.637          -2.932          -2.649
           5       -2.011          -3.637          -2.965          -2.680
           4       -2.327          -3.637          -2.996          -2.709
           3       -1.785          -3.637          -3.025          -2.735
           2       -1.134          -3.637          -3.051          -2.759
           1       -0.717          -3.637          -3.074          -2.780

Opt Lag (Ng-Perron seq t) =  4 with RMSE  4.976939
Min SC   =  3.496996 at lag  3 with RMSE  5.108733
Min MAIC =  3.448081 at lag  3 with RMSE  5.108733
```

The test statistic is not smaller than any of the critical values at any of the lags, so we accept the null hypothesis of a unit root.

Step 2: Identify the number of lags

In converting the VAR representation of our variables to the VECM representation, we have reduced the number of lags by one. That is, we have gone from

$$\mathbf{y}_t = \boldsymbol{\mu} + \sum_{i}^{p} \boldsymbol{\Phi}_i \mathbf{y}_{t-i} + \boldsymbol{\epsilon}_t$$

with p lags of $\mathbf{y}_t$, to

$$\Delta \mathbf{y}_t = \boldsymbol{\gamma} + \boldsymbol{\tau} t + \boldsymbol{\alpha}(\boldsymbol{\beta}' \mathbf{y}_{t-1} + \boldsymbol{\nu} + \boldsymbol{\rho} t) + \sum_{i=1}^{p-1} \boldsymbol{\Gamma}_i \Delta \mathbf{y}_{t-i} + \boldsymbol{\epsilon}_t$$

with $p - 1$ lags of $\Delta \mathbf{y}_t$.

It turns out that we can use the same methods we used in chapter 9 to identify the number of lags in a VAR to determine the number of lags in a VECM. In fact, we use the same Stata command, `varsoc`, to determine the number of lags in the VAR form of the model. We will not repeat the description we gave in chapter 9 of the test statistics. For each method, Stata places an asterisk by the test statistic associated with the recommended lag length.

```
. varsoc dcwage mdwage vawage
   Selection-order criteria
   Sample:  1991:1 - 2011:1                     Number of obs      =        81

  | lag |     LL       LR      df    p      FPE      AIC      HQIC      SBIC
  |-----+-------------------------------------------------------------------
  |  0  | -983.999                         7.7e+06  24.3703  24.4059   24.459
  |  1  | -637.128  693.74    9  0.000  1834.37*  16.0278*  16.1702*  16.3826*
  |  2  | -632.057  10.141    9  0.339  2023.41   16.1249  16.3739  16.7457
  |  3  | -623.309  17.496*   9  0.041  2041.14   16.1311  16.4869  17.0179
  |  4  | -618.368  9.8829    9  0.360  2266.69   16.2313  16.6939  17.3842

   Endogenous:  dcwage mdwage vawage
   Exogenous:   _cons
```

The likelihood-ratio test suggests three lags, while the other four test statistics indicate that a single lag should be sufficient. With a single lag in the VAR form of the model, the VECM has the form

$$\Delta \mathbf{w}_t = \boldsymbol{\gamma} + \boldsymbol{\tau} t + \boldsymbol{\alpha}(\boldsymbol{\beta}' \mathbf{w}_{t-1} + \boldsymbol{\nu} + \boldsymbol{\rho} t) + \boldsymbol{\epsilon}_t$$

These tests can be sensitive to the maximum lag length considered, so, to be sure, we rerun these tests with a maximum of eight lags.

```
. varsoc dcwage mdwage vawage, maxlag(8)
    Selection-order criteria
    Sample:  1992:1 - 2011:1                        Number of obs     =        77
```

lag	LL	LR	df	p	FPE	AIC	HQIC	SBIC
0	-928.691				6.5e+06	24.1998	24.2363	24.2911
1	-606.477	644.43	9	0.000	1902.62	16.0643	16.2104*	16.4296*
2	-600.322	12.31	9	0.196	2051.05	16.1382	16.3939	16.7774
3	-587.863	24.917	9	0.003	1880.3*	16.0484*	16.4137	16.9616
4	-583.116	9.4937	9	0.393	2111.23	16.1589	16.6337	17.346
5	-580.694	4.844	9	0.848	2526.31	16.3297	16.9141	17.7908
6	-574.279	12.831	9	0.170	2736.44	16.3968	17.0908	18.1319
7	-568.734	11.089	9	0.270	3047.4	16.4866	17.2902	18.4956
8	-558.028	21.412*	9	0.011	2986.33	16.4423	17.3554	18.7252

```
    Endogenous:  dcwage mdwage vawage
    Exogenous:   _cons
```

The final prediction error (FPE) and Akaike's information criterion (AIC) statistics now favor three lags, and the likelihood-ratio test suggests eight lags. At this point, it's not clear how to proceed. Lütkepohl (2005) has shown that the Hannan and Quinn's information criterion (HQIC) and Schwarz's Bayesian information criterion (SBIC) statistics provide consistent estimates of p, the true lag length, while the FPE and AIC statistics overestimate p even in infinite samples. Of course, none of that tells us what to rely on in this very finite sample. We will experiment with the number of lags in some of the following steps to see if it makes a difference in the results.[25]

Step 3: Identify the number of cointegrating relationships

Recall from section 10.4 that the number of linearly independent cointegrating relationships, r, lies between 0 and $K - 1$, where K is the number of dependent variables in $\mathbf{y}_t$. Furthermore, the rank of $\boldsymbol{\alpha\beta}'$ is r. Thus an estimate of the rank of $\boldsymbol{\alpha\beta}'$ is also an estimate of the number of cointegrating relationships.

The `vecrank` command provides three different approaches that can help identify r. By default, Stata calculates and displays a trace statistic proposed by Johansen (1995). The statistic is

$$-T \sum_{i=r+1}^{K} \ln\left(1 - \widehat{\lambda}_i\right)$$

---

25. These tests are very sensitive, making it difficult to choose a lag length with confidence. For example, if we leave the maximum lag at 4 but suppress the constant term (with the `noconstant` option), the likelihood-ratio tests suggest 4 lags, while the other statistics agree on 1 lag. And if we set the maximum number of lags to 24, all the test statistics except the FPE suggest 24 lags. With these data, the FPE can only be calculated for up to 19 lags, and the minimum final prediction error occurs with 18 lags.

where λ_i is the ith eigenvalue of $\boldsymbol{\alpha\beta'}$. The null hypothesis is that there are no more than r cointegrating relationships against the alternative that there are more than r cointegrating relationships. Under the null, all the λ's in the trace statistic are 0, so the trace statistic also is 0 under the null.

Tests for the number of cointegrating relationships, like tests for lag length, involve multiple tests—are there two relationships, are there three? and so on. This leads to a familiar problem: the nominal size of a single test is not the true size of the test when multiple hypotheses are examined. For the trace test, Johansen (1995) developed a method that has accurate nominal coverage despite the multiple tests. We start with the null that $r = 0$. If that null is rejected, we test $r = 1$. We continue until the null is accepted.

The other two approaches provided by **vecrank** are the maximum eigenvalue test and an assortment of information criteria. In the maximum eigenvalue approach, we perform a likelihood-ratio test of the null hypothesis of exactly r cointegrating relationships versus the alternative hypothesis of exactly $r + 1$ relationships. This approach does not resolve the multiple test issue; nominal sizes of the tests will not correspond to the actual size of the test when multiple tests are conducted. The information-criterion tests identify the value of r that minimizes the SBIC, HQIC, and AIC criteria. The information-criterion tests avoid the multiple-testing issue. As in testing for lag length, SBIC and HQIC provide consistent estimates of r.

Let's see what sort of guidance **vecrank** gives us in our model of construction wages. The syntax of the **vecrank** command is

vecrank *depvar* $\big[\,if\,\big]$ $\big[\,in\,\big]$ $\big[$, <u>lags</u>(*#*) <u>trend</u>(*trend_type*) <u>notr</u>ace <u>max</u> <u>ic</u>

  other_options $\big]$

Unless you specify the **notrace** option, Stata displays the trace statistic. The maximum eigenvalue (**max**) and information-criterion (**ic**) tests are reported only if you ask for them.

Notice that we have to choose a trend specification, one of the five cases we listed in section 10.4. The allowable trend types are

Case	Restrictions	*trend_type*
1: Unrestricted trend	none	<u>t</u>rend
2: Restricted trend	$\tau = 0$	<u>r</u>trend
3: Unrestricted constant	$\tau = \rho = 0$	<u>c</u>onstant
4: Restricted constant	$\tau = \rho = \gamma = 0$	<u>rc</u>onstant
5: No trend	$\tau = \rho = \gamma = \nu = 0$	<u>n</u>one

How do we choose among these five specifications? In the best case, we have a firm opinion derived from theory for choosing. In reality, we often are choosing without much of any guidance from theory. Moreover, the results of our `vecrank` can be sensitive to our choice.

In our example of construction wages, we indicated some situations that might motivate a constant (ν) or possibly a trend term (ρ) in the cointegrating relationship, but these explanations represent our speculation, not a theory or institutional facts. We do not have a strong reason to choose one specification over another. And we have not considered at all whether $\mathbf{w}_t$ incorporates a linear (γ) or quadratic (τ) trend. We will just have to try some alternatives and see what happens. Let's start by seeing what the default behavior of the command produces.

```
. vecrank dcwage mdwage vawage
                         Johansen tests for cointegration
 Trend: constant                                    Number of obs =      83
 Sample:  1990:3 - 2011:1                                    Lags =       2
```

					5%
maximum				trace	critical
rank	parms	LL	eigenvalue	statistic	value
0	12	-678.52116	.	60.1980	29.68
1	17	-657.40906	0.39874	17.9738	15.41
2	20	-649.22589	0.17896	1.6074*	3.76
3	21	-648.42217	0.01918		

`vecrank` reports the type of trend—by default, Stata chooses `constant` (Case 3: Unrestricted constant)—the number of observations, the date coverage of the sample, the number of lags (Stata chose two lags because we did not specify anything) and a table of tests. Each row of the table represents one hypothesis test. For each test, `vecrank` displays the maximum rank under the null, the number of parameters that Stata estimated, the log likelihood, the rth eigenvalue, the trace statistic, and a 5% critical value for the trace statistic. For instance, the row with maximum rank 1 represents a test of the null that our candidate VECM contains no more than one cointegrating relationship (rank of $\alpha\beta' \leq 1$) versus the alternative that there is more than one cointegrating relationship. (With three variables, two is the maximum number of cointegrating relationships.) The trace statistic for this row is 17.9738, which is greater than the 5% critical value of 15.41.[26]

---

26. You can verify that the trace statistic is calculated as

$$-T \sum_{i=r+1}^{K} \ln(1 - \widehat{\lambda}_i)$$

where the $\widehat{\lambda}_i$ are listed in the eigenvalue column. For example, the trace statistic for the rank 0 row is

$$60.1980 = -83 \times [\ln(1 - 0.01918) + \ln(1 - 0.17896) + \ln(1 - 0.39874)]$$

With this specification, the null hypothesis that $r \leq 2$ is accepted (trace statistic $=$ 1.6074 with a 5% critical value of 3.76). Let's see what the other test statistics suggest.

```
. vecrank dcwage mdwage vawage, max ic
                    Johansen tests for cointegration
Trend: constant                                 Number of obs =      83
Sample:  1990:3 - 2011:1                             Lags =        2
```

					5%
maximum				trace	critical
rank	parms	LL	eigenvalue	statistic	value
0	12	-678.52116	.	60.1980	29.68
1	17	-657.40906	0.39874	17.9738	15.41
2	20	-649.22589	0.17896	1.6074*	3.76
3	21	-648.42217	0.01918		

					5%
maximum				max	critical
rank	parms	LL	eigenvalue	statistic	value
0	12	-678.52116	.	42.2242	20.97
1	17	-657.40906	0.39874	16.3663	14.07
2	20	-649.22589	0.17896	1.6074	3.76
3	21	-648.42217	0.01918		

maximum						
rank	parms	LL	eigenvalue	SBIC	HQIC	AIC
0	12	-678.52116		16.98878	16.77956	16.63906
1	17	-657.40906	0.39874	16.74625	16.44985	16.25082
2	20	-649.22589	0.17896	16.70878*	16.36008*	16.12593
3	21	-648.42217	0.01918	16.74265	16.37652	16.13065

With rare unanimity, these results all suggest two cointegrating relationships. Do not, however, let this example of agreement among test statistics give you a false sense of confidence. The test statistics are sensitive to the choice of trend specifications and lag length. For instance, if we retain the **Trend: constant** specification but include only one lag, the trace statistic favors $r = 3$, an impossible choice.[27] Other combinations of trend specification and lag length suggest $r = 1$.

---

27. The VECM specification requires $r \leq K$. If $r = K$, then we have a model of three unit-root variables that are not cointegrated, which we can estimate as an ordinary VAR in first-differences, that is,

$$\Delta \mathbf{w}_t = \sum_i^p \mathbf{\Phi}_i \Delta \mathbf{w}_{t-i} + \boldsymbol{\epsilon}_t$$

We have dropped the constant term as a consequence of taking the first-differences of $\mathbf{w}_t$. However, we can reintroduce a constant and a trend term in this VAR if we believe $\mathbf{w}_t$ follows a linear or quadratic trend.

We are on familiar, if inconvenient, ground. With nonexperimental data, like the construction wage data in our example, the data rarely speak for themselves, and our statistical tools, as sophisticated as they are, rarely provide unambiguous guidance. A tempting mistake in this situation is to select from this plethora of specifications the one that generates estimates most favorable to our hypothesis. After all, we believe our hypothesis is reasonable; thus it is only human nature to regard as reasonable any specification that supports our prior opinion. We compound our error if we report only the results that accord with our hypothesis.

Unfortunately, our obligation is to challenge our hypothesis as rigorously as possible. In this situation, we should fit several specifications to the data to determine how sensitive the estimates of the relationship of interest are to these variations in model form. If the estimates are robust to variations in specification, our confidence in our hypothesis increases. If not—well, no one said empirical research is easy.

Step 4: Fit a VECM

We have as much statistical guidance about specification as we are going to get. It's time to fit a VECM. Actually, we are going to fit several VECMs. Because we are uncertain about the number of lags and the constant and trend terms, we will start with an encompassing model—the largest (that is, most fully parameterized) model we are willing to consider—then we will see how far the data will allow us to trim the specification. The first model we fit will contain three lags (in the VAR form of the model), two cointegrating relationships, and the `trend(trend)` specification; that is, it will include terms for τ, ρ, γ, and ν.

Stepping aside from the statistical test results, we have intuitive reasons for thinking there might be two cointegrating relationships. The pace of wage equalization depends on the impediments workers face in changing work locations: differing income tax rates, licensing hurdles, apprenticeship periods, and costs (some nonpecuniary) related to the distance between workplaces. The totality of those barriers determines the rate at which wage gaps are closed. There is no reason to believe that the barriers (and hence, the speed of adjustment) between DC and MD are the same as the barriers between MD and VA.

With three regions and two cointegrating relationships, we can write these relationships in multiple equivalent ways. For instance, we can express the cointegrating expression in our VECM in terms of a cointegrating relationship between wages in DC, $w_{\mathrm{D},t}$, and VA, $w_{\mathrm{V},t}$,

$$\alpha_{D1}(w_{\mathrm{D},t} - \beta_{13}w_{\mathrm{V},t} + \nu_1 + \rho_1 t)$$

and a similar relationship between wages in MD, $w_{\mathrm{M},t}$, and VA,

$$\alpha_{M2}(w_{\mathrm{M},t} - \beta_{23}w_{\mathrm{V},t} + \nu_2 + \rho_2 t)$$

Consider, for the moment, the first cointegrating relationship. If construction wages in DC are higher than equilibrium compared with VA wages, then

$$K_{DV} \equiv (w_{D,t} - \beta_{13}w_{V,t} + \nu_1 + \rho_1 t) > 0$$

To return DC and VA wages to their equilibrium relationship, we require $-1 < \alpha_{D1} < 0$.[28] In the succeeding quarter, $w_{D,t}$ will increase by $\alpha_{D1}K_{DV}$. In other words, α_{D1} determines the speed of adjustment to equilibrium. In each quarter, $100\alpha_{D1}$ percent of the disequilibrium is eliminated.[29]

With this formulation of the cointegrating relationships, our initial VECM has the form

$$\begin{bmatrix} \Delta w_{D,t} \\ \Delta w_{M,t} \\ \Delta w_{V,t} \end{bmatrix} = \begin{bmatrix} \gamma_D \\ \gamma_M \\ \gamma_V \end{bmatrix} + \begin{bmatrix} \tau_D \\ \tau_M \\ \tau_V \end{bmatrix} t \qquad (10.4)$$
$$+ \sum_{i=1}^{2} \begin{bmatrix} \Delta w_{D,t-i} \\ \Delta w_{M,t-i} \\ \Delta w_{V,t-i} \end{bmatrix} + \begin{bmatrix} \alpha_{D1} & 0 \\ 0 & \alpha_{M2} \\ \alpha_{V1} & \alpha_{V2} \end{bmatrix}$$
$$\left\{ \begin{bmatrix} 1 & 0 & -\beta_{13} \\ 0 & 1 & -\beta_{23} \end{bmatrix} \begin{bmatrix} w_{D,t-1} \\ w_{M,t-1} \\ w_{V,t-1} \end{bmatrix} + \begin{bmatrix} \nu_1 \\ \nu_2 \end{bmatrix} + \begin{bmatrix} \rho_1 \\ \rho_2 \end{bmatrix} t \right\}$$

The equation for $\Delta w_{D,t}$ includes the first cointegrating relationship (the one that defines an equilibrium relationship between wages in DC and MD) but not the second cointegrating relationship (the one that defines an equilibrium between wages in MD and VA). The equation for $\Delta w_{M,t}$ includes the second cointegrating relationship but not the first. And the equation for $\Delta w_{V,t}$ includes both cointegrating relationships.

Because $\boldsymbol{\alpha\beta'}$ is less than full rank, we need to apply restrictions to identify the elements of these two matrices. If there are r cointegrating relationships, then we need at least r^2 restrictions for identification. Johansen (1995) proposed a scheme for identification that has become the default (and is the default applied by Stata). We could ask Stata to impose the constraints we have written above, but let's see what the default restrictions produce first.

The **vec** command fits VECMs, just as the **var** command fits VARs. And just as we encountered a host of VAR-related commands that began with **var** ..., we will see several **vec** ... commands before we are done. (We have already used **vecrank**.)

The syntax for **vec** is

vec *varlist* $\big[$ *if* $\big]$ $\big[$ *in* $\big]$ $\big[$, <u>r</u>ank(*#*) <u>lags</u>(*#*) <u>t</u>rend(*trend_type*) *other_options* $\big]$

---

28. This restriction is a characteristic of this normalization, not a general restriction on the elements of $\boldsymbol{\alpha}$.

29. Of course, the path of adjustment to equilibrium also is affected by the dynamic properties of the VAR terms in $\Delta\mathbf{w}_t$.

The rank() option specifies r, the rank of $\boldsymbol{\alpha\beta'}$, and the number of cointegrating relationships. The other options are familiar from previously introduced commands.

So here goes. Let's see what we get.

```
. vec dcwage mdwage vawage, rank(2) lags(3) trend(trend)
Vector error-correction model
Sample:  1990:4 - 2011:1              Number of obs    =         82
                                      AIC              =    16.1159
Log likelihood = -628.7517            HQIC             =   16.49297
Det(Sigma_ml)  =  917.6133            SBIC             =    17.0551
```

Equation	Parms	RMSE	R-sq	chi2	P>chi2
D_dcwage	10	4.5171	0.3198	33.85087	0.0002
D_mdwage	10	2.87895	0.5368	83.4269	0.0000
D_vawage	10	4.91914	0.3934	46.69483	0.0000

	Coef.	Std. Err.	z	P>|z|	[95% Conf. Interval]	
D_dcwage						
_ce1						
L1.	-.2698322	.0924401	-2.92	0.004	-.4510114	-.0886529
_ce2						
L1.	.1047753	.1891919	0.55	0.580	-.266034	.4755845
dcwage						
LD.	-.040737	.1219412	-0.33	0.738	-.2797373	.1982634
L2D.	.1761336	.1171884	1.50	0.133	-.0535515	.4058186
mdwage						
LD.	-.2675864	.2820195	-0.95	0.343	-.8203345	.2851617
L2D.	-.1807267	.2766111	-0.65	0.514	-.7228745	.3614212
vawage						
LD.	.2755834	.186576	1.48	0.140	-.0900988	.6412657
L2D.	-.0296447	.1866936	-0.16	0.874	-.3955574	.336268
_trend	-.0013943	.0247345	-0.06	0.955	-.0498729	.0470843
_cons	.1731448	1.252224	0.14	0.890	-2.281168	2.627458

D_mdwage						
_ce1						
L1.	.126876	.0589162	2.15	0.031	.0114024	.2423497
_ce2						
L1.	-.7079289	.1205804	-5.87	0.000	-.9442622	-.4715956
dcwage						
LD.	-.1356118	.0777186	-1.74	0.081	-.2879373	.0167138
L2D.	-.1453265	.0746894	-1.95	0.052	-.291715	.001062
mdwage						
LD.	.1116314	.1797436	0.62	0.535	-.2406596	.4639224
L2D.	.1260702	.1762966	0.72	0.475	-.2194648	.4716053
vawage						
LD.	-.1735084	.1189132	-1.46	0.145	-.406574	.0595572
L2D.	-.1407472	.1189882	-1.18	0.237	-.3739598	.0924653
_trend	.0052978	.0157644	0.34	0.737	-.0255998	.0361954
_cons	-.7182236	.798098	-0.90	0.368	-2.282467	.8460197
D_vawage						
_ce1						
L1.	.212845	.1006675	2.11	0.034	.0155403	.4101497
_ce2						
L1.	-.7910777	.2060305	-3.84	0.000	-1.19489	-.3872654
dcwage						
LD.	-.321446	.1327943	-2.42	0.015	-.581718	-.0611739
L2D.	-.2809364	.1276185	-2.20	0.028	-.5310641	-.0308088
mdwage						
LD.	.3634736	.30712	1.18	0.237	-.2384706	.9654178
L2D.	-.0197474	.3012303	-0.07	0.948	-.6101479	.5706531
vawage						
LD.	-.2436385	.2031818	-1.20	0.230	-.6418675	.1545905
L2D.	.0654525	.2033098	0.32	0.748	-.3330274	.4639325
_trend	-.0049256	.0269359	-0.18	0.855	-.057719	.0478678
_cons	.418861	1.363675	0.31	0.759	-2.253893	3.091615

Cointegrating equations

Equation	Parms	chi2	P>chi2
_ce1	1	34.95274	0.0000
_ce2	1	631.1734	0.0000

```
Identification:  beta is exactly identified
                 Johansen normalization restrictions imposed
```

beta	Coef.	Std. Err.	z	P>|z|	[95% Conf. Interval]	
_ce1						
dcwage	1	.	.	.	.	.
mdwage	0	(omitted)				
vawage	-.5358217	.0906316	-5.91	0.000	-.7134564	-.3581871
_trend	.0073648	.	.	.	.	.
_cons	-11.75511	.	.	.	.	.
_ce2						
dcwage	0	(omitted)				
mdwage	1	.	.	.	.	.
vawage	-.7232581	.0287885	-25.12	0.000	-.7796825	-.6668337
_trend	.1734272	.	.	.	.	.
_cons	-8.578372	.	.	.	.	.

```
. estimates store trendlags3
```

That is a lot of output, and normally we would omit most of it. However, it's worth looking at some of the details this time. But before we dig into these details, notice that we have stored the estimates (actually, a lot of information associated with these estimates). Storing the estimates makes it easier to test this specification against nested alternatives. More on that later.

Let's start by orienting ourselves. (Because the vec output covers more than one page, you will have to flip back and forth between this explanation and the output. Sorry.) The top section of the output—the part before the coefficient estimates—is organized just like the output of the var command, so we will not discuss it.

The next section of the output, which contains the coefficient estimates, is the longest, again similar to the output of the var command. However, there are a few new terms in the equation. The parameters that control the rate of adjustment to disequilibrium, that is, the elements of α, appear before the other parameters. For example, the table of coefficients for the $\Delta w_{D,t}$ equation begins with

	Coef.	Std. Err.	z	P>|z|	[95% Conf. Interval]	
D_dcwage						
_ce1						
L1.	-.2698322	.0924401	-2.92	0.004	-.4510114	-.0886529
_ce2						
L1.	.1047753	.1891919	0.55	0.580	-.266034	.4755845

The coefficient that Stata calls `L1._ce1` corresponds to α_{D1}, the speed of adjustment to nonzero (that is, disequilibrium) values of the first cointegrating relationship.[30] Similarly, `L1._ce2` corresponds to α_{D2}. Note that the p-values of these coefficients are consistent with our normalizations $\alpha_{D1} \neq 0$ and $\alpha_{D2} = 0$. Furthermore, $-1 < \alpha_{D1} < 0$, as required.

The last two coefficients in the table for $\Delta w_{D,t}$ are

_trend	-.0013943	.0247345	-0.06	0.955	-.0498729	.0470843
_cons	.1731448	1.252224	0.14	0.890	-2.281168	2.627458

These coefficients correspond to τ_D and γ_D, the quadratic and linear trend terms, respectively, in the $w_{D,t}$ equation. Given their p-values, they do not appear essential.

Before we discuss other components of the output, take a look at the estimates of the elements of $\boldsymbol{\alpha}$ in the other two equations in the system. Both cointegrating relationships are significant at the 5% level in the $\Delta w_{V,t}$ equation, consistent with the normalization in (10.5). However, both cointegrating relationships also are significant at the 5% level in the $\Delta w_{V,t}$ equations, contradicting our assumption that $\alpha_{M1} = 0$. A nonzero value of α_{M1} will complicate somewhat the interpretation of the estimates. Note also that the estimates of τ_M, τ_V, γ_M, and γ_V all are insignificant at the 5% level.

Page down to the end of the coefficient estimates, and you will find information about the cointegrating relationships. The first table in this section is analogous to the table at the top of the output that displays the significance of each of the three equations in the VECM. In this section, the table displays the significance of each of the cointegrating equations. Note that only one parameter is estimated in each equation. This reflects the Johansen restrictions applied by default to identify the elements of $\boldsymbol{\alpha}$ and $\boldsymbol{\beta}$. (Note that this default scheme exactly identifies $\boldsymbol{\beta}$.)

The final table reports the estimated (and restricted) coefficients in the cointegrating equations. The Stata term `_cons` corresponds to $\boldsymbol{\nu}$, the constant in the cointegrating relationship, and the term `_trend` corresponds to $\boldsymbol{\rho}$, the trend term. Thus, in our notation, we can write the first cointegrating relationship as

$$w_{D,t} - 0.54 w_{V,t} - 11.8 + 0.007 t$$

and the second cointegrating relationship as

$$w_{M,t} - 0.72 w_{V,t} - 8.58 + 0.17 t$$

Ignoring the constant and trend terms for the moment, these relationships suggest that in equilibrium, the index of construction wages in DC is 54% of the index in VA, and

---

30. Read the label `L1._ce1` as the coefficient on the first cointegrating equation (hence, `_ce1`), which contains one-period lags of the levels of the dependent variables (hence, `L1.`). Stata calls the cointegrating relationships

$$\boldsymbol{\beta}' \mathbf{w}_{t-1} + \boldsymbol{\nu} + \boldsymbol{\rho} t$$

the cointegrating equations because, in equilibrium,

$$0 = \boldsymbol{\beta}' \mathbf{w}_{t-1} + \boldsymbol{\nu} + \boldsymbol{\rho} t$$

the MD index is 72% of the VA index. These equilibrium orderings—a VA index higher than the MD index, which is higher than the DC index—are consistent with the relative levels we saw in figure 10.2.[31]

Because our data are indices and not hourly wages, we cannot translate them directly into cross-region comparisons of wage levels. However, some rough approximations suggest that construction wages in the base period (1990:1) were highest in MD and lowest in VA, so perhaps these equilibrium relationships among the indices suggest some convergence of wage levels over time.

Do we need all the parameters in this specification to provide an adequate description of the dynamic behavior of construction wages in DC, MD, and VA? Probably not. For one thing, we are asking a lot of a modest amount of data. With this specification, we are estimating 34 parameters from 82 observations. That is not quite 2 1/2 observations per parameter. As a rough rule of thumb, I like to have at least five observations per estimated parameter, if possible. Let's see how much we can trim this specification.

For instance, are the linear and quadratic trend terms in the equations for $\Delta \mathbf{w}_t$ adding much explanatory power? None of these coefficients are individually significant. Let's test whether the quadratic trend terms are jointly significant.

```
. test [D_dcwage]_trend [D_mdwage]_trend [D_vawage]_trend
 ( 1)  [D_dcwage]_trend = 0
 ( 2)  [D_mdwage]_trend = 0
 ( 3)  [D_vawage]_trend = 0
            chi2( 3) =     0.73
          Prob > chi2 =    0.8672
```

These results suggest we safely can accept the null hypothesis that $\tau = 0$.

There is another way to assess whether we need to include the τt term in our VECM. We will fit a nested model, a VECM with three lags (in the VAR form of the model), two cointegrating relationships, but no τt term. Then we will perform a likelihood-ratio test of the nested model (the null hypothesis) against the encompassing model (the alternative). For this purpose, we are not interested in the output from the `vec` command, so we will precede it with the `quietly` qualifier.

```
. quietly vec dcwage mdwage vawage, rank(2) lags(3) trend(rtrend)
. estimates store rtrendlags3
. lrtest trendlags3 rtrendlags3
Likelihood-ratio test                          LR chi2(1)  =      0.39
(Assumption: rtrendlags3 nested in trendlags3)  Prob > chi2 =    0.5333
```

The answer has not changed: we can accept the nested model.

---

31. The indices diverge immediately, suggesting that cross-region construction wages were not at equilibrium in the first quarter of 1990.

Note that the p-value is different here than it was in the χ^2 test above. In the previous test, we held all other parameters constant, while in the likelihood-ratio test, we reestimated all the parameters, hence, the change in the significance level.

That was a start, but perhaps we can eliminate more terms. Let's test the null hypothesis that $\rho = 0$, that is, the hypothesis that we can eliminate the trend term in the cointegrating relationships.

```
. quietly vec dcwage mdwage vawage, rank(2) lags(3) trend(constant)

. estimates store constantlags3

. lrtest rtrendlags3 constantlags3
Likelihood-ratio test                              LR chi2(2)  =      6.32
(Assumption: constantlags3 nested in rtrendlags3)  Prob > chi2 =    0.0425
```

The null is rejected at the 5% level but not at the 1% level. How seriously should we worry about this rejection? Let's look at the individual coefficient estimates.

```
Johansen normalization restrictions imposed
```

beta	Coef.	Std. Err.	z	P>|z|	[95% Conf. Interval]	
_ce1						
dcwage	1	.	.	.	.	.
mdwage	0	(omitted)				
vawage	-.5469229	.0909488	-6.01	0.000	-.7251791	-.3686666
_trend	.0099875	.2716107	0.04	0.971	-.5223596	.5423346
_cons	-9.540815	.	.	.	.	.
_ce2						
dcwage	0	(omitted)				
mdwage	1	.	.	.	.	.
vawage	-.7248064	.0288334	-25.14	0.000	-.7813187	-.668294
_trend	.1607105	.0861083	1.87	0.062	-.0080587	.3294798
_cons	-7.530798	.	.	.	.	.

Neither coefficient is individually significant. Because there is not an intuitive explanation for the alternative hypothesis $\rho \neq 0$ and because the p-value of the likelihood-ratio test is not far from the 5% threshold, we will accept the null.

Next up is the test of the null hypothesis that $\gamma = 0$, that is, the hypothesis that there are no trends in the levels of the wage indices.

```
. quietly vec dcwage mdwage vawage, rank(2) lags(3) trend(rconstant)

. estimates store rconstantlags3

. lrtest constantlags3 rconstantlags3
Likelihood-ratio test                               LR chi2(1)  =      3.24
(Assumption: rconstantlags3 nested in constantlags3) Prob > chi2 =    0.0716
```

We can accept this null at the 5% level.

Only one `trend()`-related test left. Let's see if there is a nonzero mean (ν) in the cointegration relationships.

```
. quietly vec dcwage mdwage vawage, rank(2) lags(3) trend(none)
. estimates store nonelags3
. lrtest rconstantlags3 nonelags3
Likelihood-ratio test                          LR chi2(2)  =      18.10
(Assumption: nonelags3 nested in rconstantlags3)  Prob > chi2 =     0.0001
```

The data forcefully reject the null.

We have managed to eliminate τ, ρ, and γ (but not ν). We are down to 30 parameters.[32] Can we trim more? What about the lags? Each lag accounts for nine coefficients (three coefficients in each of three equations).

```
. quietly vec dcwage mdwage vawage if date>=tq(1990:4), rank(2) lags(2)
> trend(rconstant)
. estimates store rconstantlags2
. lrtest rconstantlags3 rconstantlags2
Likelihood-ratio test                          LR chi2(9)  =      16.41
(Assumption: rconstantlags2 nested in rconstantlags3) Prob > chi2 =     0.0587
. quietly vec dcwage mdwage vawage if date>=tq(1990:4), rank(2) lags(1)
> trend(rconstant)
. estimates store rconstantlags1
. lrtest rconstantlags2 rconstantlags1
Likelihood-ratio test                          LR chi2(9)  =      12.67
(Assumption: rconstantlags1 nested in rconstantlags2) Prob > chi2 =     0.1779
```

It appears we can trim the model to one lag.[33,34] The trimmed model has 12 model degrees of freedom. Let's look at the estimates.

```
. vec dcwage mdwage vawage if date>=tq(1990:4), rank(2) lags(1) trend(rconstant)
Vector error-correction model
Sample:  1990:4 - 2011:1              Number of obs    =        82
                                      AIC              =  16.05537
Log likelihood = -648.2701            HQIC             =   16.1732
Det(Sigma_ml)  =  1477.098            SBIC             =  16.34887
```

Equation	Parms	RMSE	R-sq	chi2	P>chi2
D_dcwage	2	4.56783	0.2271	23.21885	0.0000
D_mdwage	2	2.91137	0.4736	71.08517	0.0000
D_vawage	2	4.94908	0.3178	36.79699	0.0000

---

32. We have eliminated more than four parameters. Why are we still estimating 30 parameters? Remember that the Johansen normalization pins down some of the undetermined parameters; thus eliminating parameters does not produce a one-for-one reduction in the model degrees of freedom.
33. Note the restriction on the estimation sample. The likelihood-ratio test requires that the restricted model be completely nested in the encompassing model, including the estimation sample.
34. There must be at least one lag in the VAR form of the model to allow for the first-differences of the dependent variables and lagged levels in the cointegrating relationships in the VECM specification.

	Coef.	Std. Err.	z	P>|z|	[95% Conf. Interval]	
D_dcwage						
_ce1						
L1.	-.1350438	.0498343	-2.71	0.007	-.2327173	-.0373703
_ce2						
L1.	-.1258783	.1055188	-1.19	0.233	-.3326914	.0809347
D_mdwage						
_ce1						
L1.	.0339051	.0317626	1.07	0.286	-.0283483	.0961586
_ce2						
L1.	-.4691332	.0672538	-6.98	0.000	-.6009481	-.3373182
D_vawage						
_ce1						
L1.	.1399138	.0539937	2.59	0.010	.034088	.2457395
_ce2						
L1.	-.666114	.1143258	-5.83	0.000	-.8901884	-.4420395

Cointegrating equations

Equation	Parms	chi2	P>chi2
_ce1	1	276.266	0.0000
_ce2	1	4428.437	0.0000

Identification: beta is exactly identified

Johansen normalization restrictions imposed

beta	Coef.	Std. Err.	z	P>|z|	[95% Conf. Interval]	
_ce1						
dcwage	1	.	.	.	.	.
mdwage	0	(omitted)				
vawage	-.5788087	.0348234	-16.62	0.000	-.6470614	-.5105561
_cons	.9315457	7.352195	0.13	0.899	-13.47849	15.34158
_ce2						
dcwage	6.94e-18	.	.	.	.	.
mdwage	1	.	.	.	.	.
vawage	-.6666528	.0100178	-66.55	0.000	-.6862875	-.6470182
_cons	-11.48179	2.115048	-5.43	0.000	-15.62721	-7.336374

Notice that $\widehat{\alpha}$ now appears to obey our normalizations: $\alpha_{D2} = \alpha_{M1} = 0$.

In this final specification, the speed of adjustment parameter in the $\Delta w_{D,t}$ equation is -0.135, or about roughly 15% in absolute value. Thus the half-life of a disequilibrium between DC and VA wage indices is about six quarters. The speed of adjustment is faster in the $\Delta w_{M,t}$ equation. The half-life of a disequilibrium between VA and MD wage indices is about two quarters.[35] The equilibrium relationship between construction wage indices in DC and VA is[36]

$$\text{Wage index in DC} = 0.58 \times \text{Wage index in VA}$$

and the equilibrium relationship between MD and VA wage indices is

$$\text{Wage index in MD} = 11.5 + 0.67 \times \text{Wage index in VA}$$

which implies the long-term MD/DC relationship

$$\text{Wage index in MD} = 11.5 + 1.15 \times \text{Wage index in DC}$$

Step 5: Test for stability and white-noise residuals

As with any model, after estimation, we look for trouble. Is the model stable? Is there autocorrelation in the residuals. Stata provides VECM analogs of the commands used to test VARs. We will apply these `vec ...` commands without detailed comment.

```
. vecstable
```

Eigenvalue stability condition

Eigenvalue	Modulus
1	1
.8794532 + .07937285i	.883028
.8794532 - .07937285i	.883028

The VECM specification imposes a unit modulus.

```
. veclmar
```

Lagrange-multiplier test

lag	chi2	df	Prob > chi2
1	3.4711	9	0.94266
2	7.6575	9	0.56900

H0: no autocorrelation at lag order

---

35. Both cointegrating relationships have significant coefficients in the equation for $\Delta w_{M,t}$. By combining these terms, we can obtain a single equilibrium relationship for construction wages in MD relative to construction wages in both DC and VA. This relationship is a bit more complicated to interpret because it collapses a three-way relationship into a single equation.
36. For readability, I have taken advantage of the statistical insignificance of ν_1 to set it to 0 here.

```
. vecnorm
```

Jarque-Bera test

Equation		chi2	df	Prob > chi2
D_dcwage		1.848	2	0.39697
D_mdwage		2.365	2	0.30652
D_vawage		4.293	2	0.11691
ALL		8.505	6	0.20336

Skewness test

Equation	Skewness	chi2	df	Prob > chi2
D_dcwage	.36149	1.786	1	0.18142
D_mdwage	-.30823	1.298	1	0.25450
D_vawage	.55836	4.261	1	0.03900
ALL		7.345	3	0.06167

Kurtosis test

Equation	Kurtosis	chi2	df	Prob > chi2
D_dcwage	3.1346	0.062	1	0.80358
D_mdwage	3.5587	1.067	1	0.30173
D_vawage	3.0967	0.032	1	0.85815
ALL		1.160	3	0.76253

The VECM specification imposes a unit root. Aside from that unit root, there is no evidence of instability. Nor is there evidence of autocorrelated errors. The evidence of nonnormality is minimal: the null is rejected for the $\Delta w_{V,t}$ equation by the skewness test. In all other cases, it is accepted.

Step 6: Review the model implications for reasonableness

As with VARs, VECMs generate a host of implications by which their reasonableness can be gauged: impulse–response functions, dynamic multipliers, forecasts. There is almost too much information to absorb. VECMs extend this list with information about the cointegrating relationships. These relationships are required to be stationary. Any indications of nonstationarity in the estimated cointegrating relationships raise questions about the adequacy of the model.

Let's start by taking a look at the estimated cointegrating relationships in our VECM for construction wage indices. The workhorse `predict` command can generate these estimates for us.

```
. predict ce1, ce equation(_ce1)
. label variable ce1 "CE for DC/VA"
. predict ce2, ce equation(_ce2)
. label variable ce2 "CE for MD/VA"
```

. tsline ce1 ce2, yline(0) lpattern(solid longdash_dot) ytitle("Disequilibria")

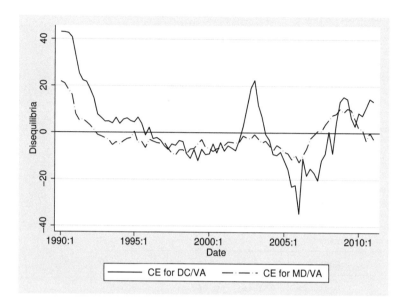

Figure 10.4. Estimated cointegrating relationships

Figure 10.4 displays the estimated cointegrating relationships. As expected, these series are positively autocorrelated, an implication of the values of $\boldsymbol{\alpha}_{D1}$ and $\boldsymbol{\alpha}_{M2}$. The only curious feature is the explosive oscillation in the first cointegrating relation in the mid-2000s. This swing in the cointegrating relationship reflects the widening of the gap between wage indices in VA and DC that occurred during that period. (Recall that we noted that pattern in our discussion of figure 10.2.) From these data, something appears to have suppressed construction wage growth in DC in the mid-2000s. The patterns in MD and VA are similar; DC is the outlier. This pattern merits further investigation to determine whether it reflects real events in the DC economy or, alternatively, some kind of defect in our data.

While the cointegrating relationships appear stationary in figure 10.4, it would be nice to have some quantitative support. It's not clear that we can apply the same statistical tests to these estimated relationships that we used to check for unit roots in the observable data. Nonetheless, we will sneak a peek at the autocorrelation functions for heuristic guidance.

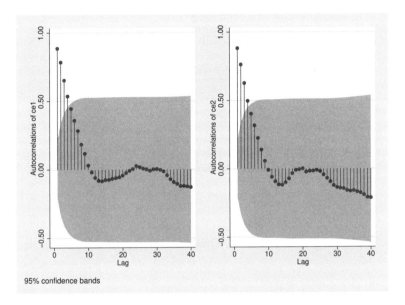

Figure 10.5. Autocorrelation functions of the cointegrating relationships

There does not appear to be anything suspicious in figure 10.5.

Stata uses the same impulse–response function commands (the `irf` family of commands) and forecast commands (the `fcast` family) for VECMs as for VARs. This reuse saves us the trouble of learning a new set of commands.[37]

---

37. There are some subtle differences in the ways these commands handle VARs and VECMs—see the *Time-Series Reference Manual* ([TS]) for details.

Figure 10.6 displays selected orthogonalized impulse–response functions (OIRFs) from our VECM. For readability, the y-scales are independent to make the qualitative features of the OIRFs more readable.

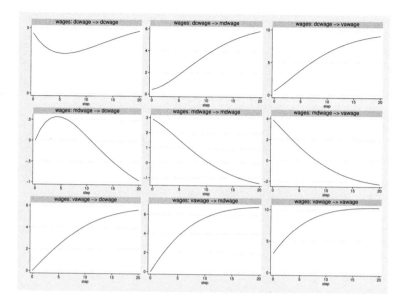

Figure 10.6. Orthogonalized impulse–response functions

In contrast to a stationary VAR, the OIRF in a VECM need not, although they may, die out over time. Impulses that die out are termed transitory. Those that do not are termed permanent. In this model, none of the OIRF die out completely, although the impacts of impulses to wages in MD tend to decay slowly. The impacts of DC impulses decline initially, then increase sooner and more sharply in MD and VA than in DC itself. And impulses in the VA equation have an immediate and permanent impact on wages in all three regions.

Perhaps the most practical application of our VECM is forecasting changes in wage indices in these three regions.[38] Let's see what the model has to say about the eight quarters after the end of our sample.

```
. fcast compute O_, step(8)
. fcast graph O_dcwage O_mdwage O_vawage
```

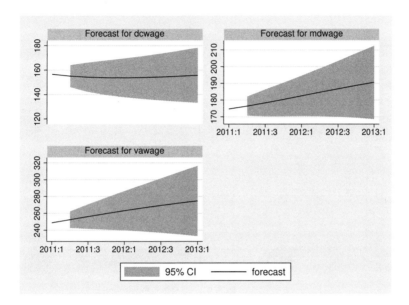

Figure 10.7. Projected construction wage indices in DC, MD, and VA

Figure 10.7 displays forecasts for the period 2011 Q2 through 2013 Q1. According to the model, the historical ranking of wage indices—VA highest, DC lowest—is expected to be unchanged. Wage growth trends are expected to differ across the three regions. The wage index in DC is expected to decline initially but end the two-year period almost unchanged (down -0.5%). The index in MD is expected to increase 9% over the two-year period, and the VA index is expected to increase just a bit more (10%).

---

38. The translation to changes in wages is complicated for the reasons described in detail earlier in this section.

Are these projected trends reasonable? Well, remember that the last estimated values of both cointegrating relationships are positive, suggesting that wage gaps should begin to narrow between DC and VA and between MD and VA to restore equilibrium. Figure 10.8 appends the cointegrating relationship for DC and VA implied by these forecasts to the within-sample estimate of this relationship. As expected, the cointegrating relationship is projected to return to equilibrium over the forecast horizon. The model projects a similar trend for the second cointegrating relationship.[39]

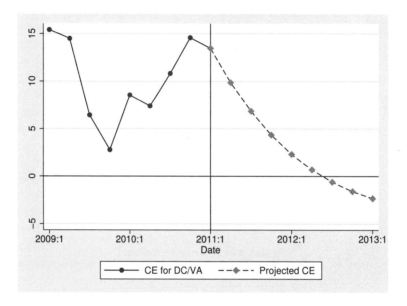

Figure 10.8. Evolution of the cointegrating relationship

These results indicate that our VECM is internally consistent, but they do not give us a sense of the accuracy and reliability of the forecasts. We will not repeat the comparisons we provided in chapter 9 using the `varbench` program (although that would be an excellent idea). Instead we will present a graph that highlights how cautious we should be in relying on the forecasts of this particular model.

---

39. However, the return to equilibrium is slower over this period for the second cointegrating relationship.

We will generate two sets of forecasts that overlap with the estimation sample, and we will compare them with what actually occurred (during the sample period) and with each other (out of sample). The first forecast uses wage index observations through the end of 2005 and constructs forecasts from that point on. The second forecast incorporates wage index observations through the end of 2009.

Figure 10.9 displays this comparison for the construction wage index in DC. The forecast that begins in 2006 tracks some fairly volatile actuals reasonably well through the end of 2008, but once the recession hits (an event that was not foreseen at the end of 2005), this forecast is far too optimistic. The forecast that starts in 2010 parts ways with the actuals immediately; it misses the recovery completely. By the end of the forecast horizon in the fourth quarter of 2013, the optimistic forecast is 31% higher than the later, more pessimistic forecast.

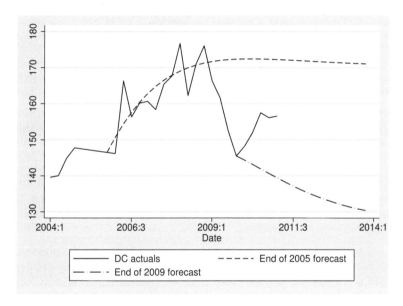

Figure 10.9. Comparison of two forecasts of the construction wage index in DC

This example provides a useful reminder of the limits of our forecasting abilities. Even with sophisticated tools such as the VECM, our statistical "headlights" often do not illuminate very much of the road ahead.

10.6   Points to remember

- By definition, disturbances to stationary time series dissipate quickly. Forecasting beyond a short horizon depends primarily on trends (for trend-stationary variables) or unit roots (for unit-root—that is, difference-stationary—variables).

- Trend-stationary and unit-root processes imply distinctly different long-run behavior. Shocks to trend-stationary processes imply the series will "catch up" over time to return to the trend line. In contrast, shocks to unit-root processes are permanent.

- It is not possible, in general, to determine whether a nonstationary series is a trend-stationary or a unit-root process. We can always find a unit-root process that resembles a trend-stationary process arbitrarily closely (and vice versa). As a practical matter, though, we can distinguish between a typical trend-stationary and unit-root process over a reasonable horizon.

- Multiple approaches have been proposed to test for unit roots. Arguments for one test over another depend on large-sample properties. In practice, researchers often run more than one test to help them assess the weight of the evidence for a unit root.

- Nonstationary time series may nonetheless obey a stationary equilibrium condition that relates two or more of the nonstationary variables. Such variables are said to be cointegrated.

- When nonstationary time series are cointegrated, the VAR model described in chapter 9 can be recast as a VECM. The VECM relates the first-differences of the nonstationary time series to their past first-differences (similar to a VAR model) and to lagged deviations from the cointegrating equations, the equilibrium relationships between the variables.

- A VECM may include linear and quadratic trend terms in the observable values. It also may include a constant and a linear trend term in the cointegrating relationships. These variations make it more challenging to specify the complete model appropriately. The multiple tests for lag length and for the number of cointegrating relationships frequently leave the researcher without clear guidance. As always, choices based on theoretical arguments are desirable.

- In addition to statistical tests of model adequacy, the researcher should review the implications of the fitted VECM to ensure they make sense. The plethora of implications of VECMs can make this an involved task.

10.7   Looking ahead

We are done for now. We have provided what we hope is a coherent introduction to time-series analysis and to the ways Stata can help you carry out that analysis. With these tools, you now are equipped to undertake serious research on time-series topics.

In the next, brief chapter, we will summarize what we hope we have communicated and indicate some of the more advanced topics (and Stata tools) that await your further exploration.

Stata commands and features discussed

constraint ([R] **constraint**): Define and list constraints; section 10.2

dfgls ([TS] **dfgls**): Dickey–Fuller–GLS unit-root test; section 10.2, section 10.5

dfuller ([TS] **dfuller**): Augmented Dickey–Fuller unit-root test; section 10.2

estimates ([R] **estimates**): Save and manipulate estimation results; section 10.2

fcast compute ([TS] **fcast compute**): Compute dynamic forecasts of dependent variables after var, svar, or vec; section 10.5

fcast graph ([TS] **fcast graph**): Graph forecasts of variables computed by fcast compute; section 10.5

irf create ([TS] **irf create**): Obtain impulse–response functions, dynamic-multiplier functions, and forecast-error variance decompositions; section 10.5

irf graph ([TS] **irf graph**): Graph impulse–response functions, dynamic-multiplier functions, and forecast-error variance decompositions; section 10.5

lrtest ([R] **lrtest**): Likelihood-ratio tests after estimation; section 10.2

pperron ([TS] **pperron**): Phillips–Perron unit-root test; section 10.2

predict ([R] **predict**): Obtain predictions, residuals, etc., after estimation; section 10.5

vec ([TS] **vec**): VECM; section 10.5

veclmar ([TS] **veclmar**): Perform Lagrange multiplier test for residual correlation after var or svar; section 10.5

vecnorm ([TS] **vecnorm**): Test for normally distributed disturbances after var or svar; section 10.5

vecrank ([TS] **vecrank**): Fit the cointegrating rank of a VECM; section 10.5

varsoc ([TS] **varsoc**): Obtain lag-order selection statistics for VARs and VECMs; section 10.5

vecstable ([TS] **vecstable**): Check the stability condition of VECM estimates; section 10.5

11 Closing observations

Chapter map

11.1 Making sense of it all. Final points to remember.

11.2 What did we miss? Time-series topics to study next. Stata time-series features we did not cover—some advanced, some just useful.

11.3 Farewell. Possible errors and gratitude.

11.1 Making sense of it all

Welcome to the finish line. Take a deep breath and relax. You have covered the core concepts of time-series analysis, and you have learned how to use Stata to apply those tools to your data. I hope you developed an appreciation of the power of these tools to make sophisticated inferences. I also hope you appreciate the limits of the tools—situations where additional assumptions or theory are required to resolve ambiguity and examples where careless use of time-series tools can sometimes lead us astray.

While this book focused on explaining time-series analysis and the Stata tools for that analysis, there are some overarching messages running through the previous chapters—some specific to time-series analysis and some more generally applicable. Here are some of the key ideas I tried to convey, a final list of points to remember.

Points to remember

- Look at your data. A close analysis and understanding of one's data comprises 90% of any successful time-series analysis, particularly an analysis of new or unfamiliar data. Be sure you understand how the data are defined. Call the people responsible for compiling the data (you may be surprised what you discover). Are your data seasonally adjusted? Think hard about whether your data actually measure the phenomenon you are studying. And look at your data from different perspectives—not just univariate statistics and correlations, but lots of graphs.[1]

- Time series are assumed to be composed of four unobservable components: trend, seasonal, cycle, and random noise. Time-series analysis attempts to decompose a time series into these components and to characterize their separate behavior.

---

1. The opening pages of Tufte's classic *The Visual Display of Quantitative Information* (2001) present a compelling example.

- Time-series analysis often is described as reducing a time series to white noise, that is, identifying, estimating, and extracting all the systematic components (trend, seasonal, and cycle) until only a white noise residual is left.

- Filters and smoothers aim to reduce the contribution of random noise to a series so that its systematic components can be more easily apprehended.

- If your only interest is in short-term forecasts, simple filter- and smoother-based methods often work about as well—sometimes even better—than more sophisticated approaches. More complex approaches shine when the goal is analysis, especially counterfactual policy analysis. Match the complexity of the tool to the task at hand.

- Forecast accuracy decays rapidly with forecast horizon.

- Autocorrelated disturbances arise naturally, often from the choice of observation frequency.

- There is not a single approach for all cases of autocorrelated disturbances in a regression. You have to analyze the situation and match the "cure" to the "disease".

- **Very important:** Acceptable estimation strategies can provide materially different estimates, even when these strategies are asymptotically equivalent. The econometrics literature provides limited guidance for choosing among these strategies. It is the responsibility of the researcher to investigate whether the estimates from a given method make theoretical and practical sense.

- All the time-series models you are likely to encounter resolve eventually to a stochastic linear difference equation. Accordingly, standard difference equation methods can reveal the dynamic properties of time series.

- Stochastic difference equations provide flexible fits to observed time series—at times too flexible. Often several specifications fit almost equally well.

- The concept of stationarity allows us to divide and conquer. The transitory dynamics of a time series are contained in its stationary component, while its long-term behavior is contained in its nonstationary (trend or unit-root) component.

- Vector autoregression (VAR) became popular as a corrective to the too-ready adoption of "incredible" a priori assumptions. However, strong assumptions frequently have to be incorporated to produce definite answers to research questions. The reliability of these answers depends crucially on the theoretical soundness of these assumptions.

11.2   What did we miss?

The title of this book is *Introduction to Time Series Using Stata*, not *Advanced Time Series Analysis*. We have covered the basics, hopefully in enough detail and with clear enough explanations that you can carry out meaningful time-series analysis of your own data. And do not underestimate the power of these "basics". The material covered in the previous chapters takes you tolerably close to the frontiers of our knowledge.

Nonetheless, there are important topics in time-series analysis that we could not cover in the span of this book. In addition, there are useful time-series tools in Stata that we could not include. In this section, we briefly indicate some of the items we missed to provide you with a road map to future study.

11.2.1 Advanced time-series topics

In no particular order, here are some additional topics in time-series statistics.[2]

Bayesian analysis. One consistent message of the last two chapters is the frequent difficulty obtaining precise estimates of parameters and impacts. Frequently, the standard errors of these estimates are distressingly large.

Bayesian analysis of time-series models provides a flexible method for refining estimates and potentially reducing standard errors of the estimates by introducing prior information. Approaches like structural vector autoregression apply prior beliefs in an "all-or-none" fashion. The data either accept or reject the restrictions applied. The Bayesian approach mixes prior information with data with weights that reflect both the strength of the prior beliefs and the definiteness of the data evidence.

The Kalman filter and state-space models. We briefly discussed the state-space representation of autoregressive moving-average (ARMA) models in chapter 7 when we explained how the `arima` command displays its results. However, the state-space approach is very general and has much broader applicability than just the `arima` command. An essential element of the state-space model is the Kalman filter, an essential method for sequentially updating linear projections. The Kalman filter and state-space representation underlie dynamic factor and unobserved-components models (UCMs), which we mention below.

Spectral analysis. In this book, we analyzed time series in the time domain, that is, as sequences of observations at discrete intervals of time. The fundamental representation of a time-domain model is the linear filter

$$y_t = \mu + \sum_{j=0}^{\infty} \psi_j \epsilon_{t-j}$$

---

2. Hamilton (1994) contains useful discussions of most of these topics. Another source you should not overlook is the Internet. Many contributors to the field post working papers and lecture notes on the Internet. More than once during the writing of this book, I have found it useful to scan some of these materials.

An alternative approach models time series in the frequency domain, that is, as weighted sums of periodic functions of the form

$$y_t = \mu + \int_0^\pi \alpha(\omega) \cos(\omega t) d\omega + \int_0^\pi \delta(\omega) \sin(\omega t) d\omega \qquad (11.1)$$

where ω represents a particular frequency. This spectral analysis decomposes y_t into cycles of different frequencies with the goal of determining which cycles contribute most to the evolution of y_t.

It turns out that there is a close correspondence between time-domain and frequency-domain models. The spectral representation theorem demonstrates that any covariance-stationary model in the time domain can be expressed in the form of (11.1). This relationship proves to be a great convenience because some analyses and proofs turn out to be easier to complete in the frequency domain.

Continuous time models. Spectral analysis is one way of representing time series observed at discrete intervals as the outcome of a continuous process (in this case, continuous in frequency). An important class of truly continuous time processes are described by Brownian motion or diffusion models. These models form the foundation of much of mathematical finance, and Wall Street depends on models of this type to value interest-rate-sensitive assets and derivatives.

Regime changes. We have touched upon the topic of regime change at points in this book. Autoregressive conditional heteroskedasticity and generalized autoregressive conditional heteroskedastic models represent a successful method of finessing the problem, a method of accounting for periods of qualitatively different behavior within a single model with fixed parameters. A more fundamental example is the Stock and Watson (2001) macroeconomic model we used in chapter 9 to illustrate VAR models. Over the sample period 1960:1 through 2004:4 (and the out-of-sample period that runs through 2012:1), there arguably have been several discretely different Federal Reserve policy regimes. These differences in monetary policy rules are likely to change the dynamic relationships we attempt to estimate in the VAR. The inclusion of data from different policy environments will generate estimates that represent a muddled average of different underlying processes.

Time-series models of regime change attempt to account for this type of change. Typically, a traditional time-series model is combined with a Markov process that describes the probability of switching regimes, characterized by discrete changes in parameters, error distributions, or some other aspect of the underlying process.

ARMA and vector ARMA with exogenous inputs. The gist of time-series models is found in their representation of the dynamic behavior of endogenous variables. As a result, we have focused narrowly on this aspect of time-series analysis. However, in real-world data analysis, we also are interested in the impact of exogenous variables on the evolution of the endogenous series. Including exogenous variables in the ARMA model

$$\phi(L)y_t = \mu + \theta(L)\epsilon_t$$

produces the ARMA with exogenous inputs model

$$\phi(L)y_t = \mu + \mathbf{\Gamma}(L)\mathbf{X}_t + \theta(L)\epsilon_t$$

Similarly, adding exogenous variables to the VAR model produces the VARMAX model. While we did not discuss these features, the Stata commands for fitting autoregressive integrated moving-average and VAR models also allow for the incorporation of exogenous variables.[3]

Policy analysis. Both models of regime change and models with exogenous variables can be used to analyze the historical and hypothetical impacts of policy interventions. Structural vector autoregressions also are intended to provide insight into these types of questions. A full treatment of policy analysis turns on some profound questions of statistical inference and of economic theory. An adequate treatment would require a book of its own.

Seasonal models. We touched very briefly on seasonal variations in chapters 3, 6, and 7, but seasonal analysis and adjustment is a topic that merits a much more complete discussion.

11.2.2 Additional Stata time-series features

Below we list the Stata commands we did not discuss in this book.

Data management tools and utilities

- Stata's date formats and date functions can handle business calendars, that is, calendars that account for business holidays such as Thanksgiving without generating gaps in the time series. Type `help datetime business calendars` for details.

- `rolling` allows for rolling estimation, in other words, a sequence of estimates of a model using a sample window that "rolls forward" an observation at a time. The `rolling` command provides a model-free method to examine time-varying relationships. Many Stata estimation commands, such as `regress`, can be combined with `rolling`.

- `import haver` imports data from the online Haver Analytics database (Stata for Windows only). You must be a subscriber to the Haver Analytics database.

- `tsappend` adds observations to a time-series dataset, fills in the date variable appropriately, and updates the `tsset` information.

- `tsfill` fills in gaps in a time-series dataset with missing values for all the variables except the date variable.

- `tsreport` displays the number of gaps in a time-series dataset.

---

3. In other settings, ARMA with exogenous inputs models are described as transfer function models.

- `tsrevar` is a specialized command that provides a convenient way of handling time-series variable lists—variable lists such as

      L(1/4).gnp D.fedfunds

  inside a Stata program.

- `import fred` imports data from the Federal Reserve Economic Data (FRED) into Stata. `import fred` supports data on FRED as well as historical vintage data on Archival FRED (ALFRED).

Univariate models

- `arfima` fits the parameters of autoregressive fractionally integrated moving-average models. We analyzed the rapid dissipation of shocks in autoregressive integrated moving-average models in chapter 7 and highlighted in chapter 10 the implication that long-term forecasts in these models depend almost exclusively on the nonstationary component—the trend or unit root. We describe this characteristic by saying that stationary ARMA models are short-memory processes. However, if we replace the unit-root integration $(1 - L)y_t$ with fractional integration $(1 - L)^d y_t$, where $-0.5 < d < 0.5$, we can model a stationary process with a long memory.

- `cumsp` plots the cumulative sample spectral distribution function of a time series.

- `pergram` plots the log-standardized periodogram of a time series.

- `psdensity` uses the estimated parameters from a prior `arfima`, `arima`, or `ucm` command to estimate the spectral density of a stationary time series.

- `tsfilter` filters a time series, extracting selected periodicities. Because these filters are characterized by the periodicities they pass through, they represent frequency domain tools. These filters have been used by macroeconomists to analyze business-cycle-related questions, but the tools are completely general.

- `ucm` fits UCMs that decompose a time series into its four components: trend, seasonal, cycle, and random disturbance. Exogenous variables can be included in a UCM.

- `estat sbcusum` performs a test of whether the coefficients in a time-series regression fit with `regress` are stable over time. The test statistic is constructed from the cumulative sum of either the recursive residuals or the ordinary least-squares residuals.

- `estat sbknown` performs a Wald or a likelihood-ratio test of whether the coefficients in a time-series regression fit with `regress` or `ivregress 2sls` vary over the periods defined by known break dates.

- `estat sbsingle` performs a test of whether the coefficients in a time-series regression fit with `regress` or `ivregress 2sls` vary over the periods defined by an unknown break date.

- `mswitch` fits dynamic regression models that exhibit different dynamics across unobserved states using state-dependent parameters to accommodate structural breaks or other multiple-state phenomena. These models are known as Markov-switching models because the transitions between the unobserved states follow a Markov chain.

- `threshold` extends linear regression to allow coefficients to differ across regions. Those regions are identified by a threshold variable being above or below a threshold value. The model may have multiple thresholds, and you can either specify a known number of thresholds or let threshold find that number for you through information criteria.

- `estat acplot` plots the estimated autocorrelation and autocovariance functions of a stationary process using the parameters of a previously fit parametric model with the `arima` or `arfima` command.

Multivariate models

- `dfactor` fits dynamic factor models, flexible models for multivariate time series. In dynamic factor models, unobservable factors are assumed to have a vector autoregressive structure.

- `mgarch` fits multivariate generalized autoregressive conditional heteroskedastic models.

- `sspace` fits general state-space models. Many time-series models are special cases of state-space models.

- `forecast` suite of commands computes dynamic forecasts by solving multiple-equation models. Equations can be stochastic relationships fit using estimation commands such as `regress`, `ivregress`, `var`, or `reg3`; or they can be nonstochastic relationships, called identities, that express one variable as a deterministic function of other variables. The `forecast` commands can also be used to obtain dynamic forecasts in single-equation models.

- `dsge` and `dsgenl` fit linear and nonlinear dynamic stochastic general equilibrium (DSGE) models. DSGE models are systems of equations derived by economic theory, where expectations of future values may have an effect on the current values. The calculated parameters are often explicitly interpretable in terms of economic theory.

11.3   Farewell

I hope you found this book useful and tolerably readable. If you see errors in the book or you feel areas are confusing or could be improved, please contact the folks at Stata Press, who will pass on the information to me. Thanks for your time and attention.

Go forth and do good work.

References

Amisano, G., and C. Giannini. 1997. *Topics in Structural VAR Econometrics*. 2nd, revised, and enlarged ed. Heidelberg: Springer.

Bartlett, M. S. 1946. On the theoretical specification and sampling properties of autocorrelated time-series. *Journal of the Royal Statistical Society (Supplement)* 8: 27–41. https://doi.org/10.2307/2983611.

Bollerslev, T. 1986. Generalized autoregressive conditional heteroskedasticity. *Journal of Econometrics* 31: 307–327. https://doi.org/10.1016/0304-4076(86)90063-1.

———. 1988. On the correlation structure for the generalized autoregressive conditional heteroskedastic process. *Journal of Time Series Analysis* 9: 121–131. https://doi.org/10.1111/j.1467-9892.1988.tb00459.x.

Bollerslev, T., R. Y. Chou, and K. F. Kroner. 1992. ARCH modeling in finance: A review of the theory and empirical evidence. *Journal of Econometrics* 52: 5–59. https://doi.org/10.1016/0304-4076(92)90064-X.

Box, G. E. P., and G. M. Jenkins. 1976. *Times Series Analysis: Forecasting and Control*. 2nd ed. Hoboken, NJ: Wiley.

Box, G. E. P., G. M. Jenkins, and G. C. Reinsel. 2008. *Time Series Analysis: Forecasting and Control*. 4th ed. Hoboken, NJ: Wiley.

Cleves, M., W. W. Gould, and Y. V. Marchenko. 2016. *An Introduction to Survival Analysis Using Stata*. Rev. 3rd ed. College Station, TX: Stata Press.

Diaconis, P. 1985. Theories of data analysis: From magical thinking through classical statistics. In *Exploring Data Tables, Trends, and Shapes*, ed. D. C. Hoaglin, F. Mosteller, and J. W. Tukey, 1–36. New York: Wiley. https://doi.org/10.1002/9781118150702.ch1.

Dickey, D. A., and W. A. Fuller. 1979. Distribution of the estimators for autoregressive time series with a unit root. *Journal of the American Statistical Association* 74: 427–431. https://doi.org/10.1080/01621459.1979.10482531.

Ehrenberg, A. S. C. 1977. Rudiments of numeracy. *Journal of the Royal Statistical Society, Series A* 140: 277–297. https://doi.org/10.2307/2344922.

Elliott, G., T. J. Rothenberg, and J. H. Stock. 1996. Efficient tests for an autoregressive unit root. *Econometrica* 64: 813–836. https://doi.org/10.2307/2171846.

Engle, R. F. 1982. Autoregressive conditional heteroskedasticity with estimates of the variance of United Kingdom inflation. *Econometrica* 50: 987–1008. https://doi.org/10.2307/1912773.

Engle, R. F., and C. W. J. Granger. 1987. Co-integration and error correction: Representation, estimation, and testing. *Econometrica* 55: 251–276. https://doi.org/10.2307/1913236.

Engle, R. F., D. M. Lilien, and R. P. Robins. 1987. Estimating time varying risk premia in the term structure: The Arch-M model. *Econometrica* 55: 391–407. https://doi.org/10.2307/1913242.

Granger, C. W. J. 1980. *Forecasting in Business and Economics*. New York: Academic Press.

Granger, C. W. J., and P. Newbold. 1974. Spurious regressions in econometrics. *Journal of Econometrics* 2: 111–120. https://doi.org/10.1016/0304-4076(74)90034-7.

———. 1989. *Forecasting in Business and Economics*. 2nd ed. Bingley, UK: Emerald.

Greene, W. H. 2018. *Econometric Analysis*. 8th ed. Upper Saddle River, NJ: Prentice Hall.

Hamilton, J. D. 1994. *Times Series Analysis*. Princeton, NJ: Princeton University Press.

Hendry, D. F. 1995. *Dynamic Econometrics*. Oxford: Oxford University Press.

Hildreth, C., and J. Y. Lu. 1960. Demand relations with autocorrelated disturbances. Reprinted in *Agricultural Experiment Station Technical Bulletin*, No. 276. East Lansing, MI: Michigan State University Press.

Holt, C. C. 1957. Forecasting seasonal and trends by exponentially weighted moving averages. Memorandum 52, Office of Naval Research.

Johansen, S. 1995. *Likelihood-Based Inference in Cointegrated Vector Autoregressive Models*. Oxford: Oxford University Press.

Judge, G. G., W. E. Griffiths, R. C. Hill, H. Lütkepohl, and T.-C. Lee. 1985. *The Theory and Practice of Econometrics*. 2nd ed. New York: Wiley.

Judge, G. G., R. C. Hill, W. E. Griffiths, H. Lütkepohl, and T.-C. Lee. 1988. *Introduction to the Theory and Practice of Econometrics*. 2nd ed. New York: Wiley.

Long, J. S., and J. Freese. 2014. *Regression Models for Categorical Dependent Variables Using Stata*. 3rd ed. College Station, TX: Stata Press.

Lucas, R. E., Jr. 1976. Econometric policy evaluation: A critique. *Carnegie–Rochester Conference Series on Public Policy* 1: 19–46. https://doi.org/10.1016/S0167-2231(76)80003-6.

Lütkepohl, H. 2005. *New Introduction to Multiple Time Series Analysis*. New York: Springer.

Mitchell, M. N. 2012. *A Visual Guide to Stata Graphics*. 3rd ed. College Station, TX: Stata Press.

Nelson, D. B. 1991. Conditional heteroskedasticity in asset returns: A new approach. *Econometrica* 59: 347–370. https://doi.org/10.2307/2938260.

Phillips, P. C. B., and P. Perron. 1988. Testing for a unit root in time series regression. *Biometrika* 75: 335–346. https://doi.org/10.2307/2336182.

Sargent, T. J. 1976. The observational equivalence of natural and unnatural rate theories of macroeconomics. *Journal of Political Economy* 84: 631–640. https://doi.org/10.1086/260465.

Shaman, P., and R. A. Stine. 1988. The bias of autoregressive coefficient estimators. *Journal of the American Statistical Association* 83: 842–848. https://doi.org/10.1080/01621459.1988.10478672.

Sims, C. A. 1980. Macroeconomics and reality. *Econometrica* 48: 1–48. https://doi.org/10.2307/1912017.

Stock, J. H., and M. W. Watson. 2001. Vector autoregressions. *Journal of Economic Perspectives* 15: 101–115. https://doi.org/10.1257/jep.15.4.101.

———. 2019. *Introduction to Econometrics*. 4th ed. New York: Pearson.

Tufte, E. R. 2001. *The Visual Display of Quantitative Information*. 2nd ed. Cheshire, CT: Graphics Press.

Tukey, J. W. 1977. *Exploratory Data Analysis*. Reading, MA: Addison–Wesley.

Wainer, H. 1984. How to display data badly. *American Statistician* 38: 137–147. https://doi.org/10.2307/2683253.

Winters, P. R. 1960. Forecasting sales by exponentially weighted moving averages. *Management Science* 6: 324–342. https://doi.org/10.1287/mnsc.6.3.324.

Author index

A

Amisano, G. 373

B

Bartlett, M. S. 232
Bollerslev, T. 280, 286
Box, G. E. P. 71, 209, 230, 254, 263, 330

C

Chou, R. Y. 286
Cleves, M. 143

D

Diaconis, P. xxv
Dickey, D. A. 384

E

Ehrenberg, A. S. C. xxv
Elliott, G. 386
Engle, R. F. 277, 281, 296, 392

F

Freese, J. 143
Fuller, W. A. 384

G

Giannini, C. 373
Gould, W. 143
Granger, C. W. J. . . . 122, 128, 218, 390, 392
Greene, W. H. 78
Griffiths, W. E. 71
Gutierrez, R. G. 143

H

Hamilton, J. D. 71, 209, 210, 230, 259, 263, 285, 287, 293, 309, 378, 382, 429

Hendry, D. F. 227
Hildreth, C. 187
Hill, R. C. 71
Holt, C. C. 131

J

Jenkins, G. M. . . . 71, 209, 230, 254, 263, 330
Johansen, S. 402, 403, 407
Judge, G. G. 71

K

Kroner, K. F. 286

L

Lütkepohl, H. 71, 307, 343, 402
Lee, T.-C. 71
Lilien, D. M. 296
Long, J. S. 143
Lu, J. Y. 187
Lucas, Jr., R. E. 358

M

Marchenko, Y. V. 143
Mitchell, M. N. 38, 319

N

Nelson, D. B. 293
Newbold, P. 218, 390

P

Perron, P. 387
Phillips, P. C. B. 387

R

Reinsel, G. C. 71
Robins, R. P. 296
Rothenberg, T. J. 386

S

Sargent, T. J. 339
Shaman, P. 184
Sims, C. A. 302, 339, 340, 358
Stine, R. A. 184
Stock, J. H. 71, 303, 304, 306, 307,
            313, 328, 338, 365, 366, 386,
            430

T

Tufte, E. R. 427
Tukey, J. W. 86, 100, 118

W

Wainer, H. xxv
Watson, M. W. . . 71, 303, 304, 306, 307,
            313, 328, 338, 365, 366, 430
Winters, P. R. 131

Subject index

A

ac command 231
AIC..*see* Akaike's information criterion
Akaike's information criterion......306
alternative hypothesis...............73
analytic weights 18
ARCH....*see* autoregressive conditional heteroskedasticity
arch command...............281–284
arfima command..................432
ARIMA ... *see* autoregressive integrated moving average
arima command...................243
ARMA(p,q).........................82
ARMA models .... *see* autoregressive moving-average models
autocorrelated 77
    disturbances..............167–200
autocorrelation, ... 77, 80–82, 168–172
    coefficients....................230
    testing for 176–177
autoregression 80
autoregressive conditional heteroskedasticity
    extensions to the model 298
    in-mean......................296
    model 277–285
    test for.......................280
autoregressive integrated moving average......................228
    model........................230
autoregressive moving-average models........82–83, 205–208
aweight .... *see* analytic weights

B

backtesting........................164
Bayesian analysis..................429
Box–Jenkins approach 226–227
Breusch–Godfrey test.............178

C

Cholesky decomposition 340
Cochrane–Orcutt 187
coeflegend option 247
cointegrating equations 394
cointegrating relationships 389–394
cointegration...................387–389
command name.....................15
common factor 250
constraint command 55
constraints.........................55
continuous time models...........430
correlate command 52
count command 49
covariance
    stationarity 209
    stationary.....................81
cross-correlation functions 331
cumsp command 432
current directory...................29
cycle 93, 95–98

D

D .... *see* difference operator
data
    integrity 90
    revision 92
date...............................62
date and time formats 64
date() function 65–66
Datetime business calendars 431

describe command 2, 12, 28
detrending 229
dfactor command 433
dfgls command 386–387
dfuller command 384–386
Dickey–Fuller test, ... 384–386
     augmented 383
difference
     equations 81
     operator 69, 204
differencing 229
display command 28, 65
double-exponential moving averages ...
     .... 122, 130–131
drop command 25
dsge command 433
dsgenl command 433
Durbin's alternative test 176
Durbin–Watson test 176
dynamic forecasts 146, 316

E

ECM .... *see* error-correction model
EGARCH .... *see* exponen-
     tial generalized autoregressive
     conditional heteroskedasticity
Elliott, Rothenberg, and Stock test ....
     386–387
error-correction model 391
estat
     acplot command 433
     archlm commmand 280
     command 58–59
     durbinalt command 177
     ic command 59
     sbcusum command 432
     sbknown command 432
     sbsingle command 432
     summarize command 59
EWMA ...... *see* exponentially weighted
     moving averages
exit command 13
=*exp* 16
expectation 72
expected value 72

exponential generalized autoregressive
     conditional
     heteroskedasticity 293
exponentially weighted moving averages
     .... 122, 126
export
     dbase command 24
     delimited command 24
     excel command 24
     sasxport5 command 24
     sasxport8 command 24

F

F .... *see* lead operator
fcast
     compute command 322–325
     graph command 322–325
feasible generalized least squares ... 77–
     78, 178
     strategy 186–188
FEVDs ...... *see* forecast-error variance
     decompositions
FGLS ..... *see* feasible generalized least
     squares
final prediction error 306
first-order autocorrelation 173–174
fit 100
fitted values 56
foreach command 61
forecast 141
     horizon 144
forecast command 433
forecast-error variance
     decompositions 343
format command 64
forvalues command 61
FPE .... *see* final prediction error
frequency domain 430
frequency weights 18
fweight .... *see* frequency weights

G

GARCH .. *see* generalized autoregressive
     conditional heteroskedasticity
Gaussian white noise 80

general linear process 202–203
generalized autoregressive conditional
       heteroskedasticity 285–298
generate command 25
Granger causality 336–339
Granger causality tests 330
graph command 38
graph commands 38
Great Moderation 275

H
H_0 .... *see* null hypothesis
H_a .... *see* alternative hypothesis
Hannan and Quinn's information crite-
       rion 306
Hanning smoother 102
help command 10
heteroskedasticity 272
histogram 36
histogram command 37
Holt–Winters smoothers .. 123, 131–138
homoskedasticity 76
HQIC .... *see* Hannan and Quinn's
       information criterion
hypothesis tests 73–74

I
if qualifier 16
import
    dbase command 24
    delimited command 24
    excel command 24
    fed command 24
    fred command 432
    haver command 24, 431
    sas command 24
    sasxport5 command 24
    sasxport8 command 24
    spss command 24
importance weights 18
impulse–response function 214,
       344–358
in qualifier 16
infile command 23, 24
infix command 24

input command 22, 24
instrumental variables 77, 178
invertibility 208–210
invertible 210
irf
    cgraph command 357
    create command 344, 345
    graph command 344, 352–353
    table command 349–352
IRF ..... *see* impulse–response function
ITSUS_DATA macro 30
IV .... *see* instrumental variables
ivregress command 55
iweight .... *see* importance weights

K
K-step-ahead forecast 145
Kalman filter 429
keep command 25

L
L .... *see* time-series operators
label 27
label command 27
lag
    operator 203
    polynomials 203–205
lagging moving averages 126
lead operator 69, 204
linear regression 74–78
list command 28
loop 61
loss function 144
Lucas critique 358

M
mata command 61
matrix 61
matrix command 61
mdy() function 63, 65–66
mean 72
median 72
median smoothers 102
mgarch command 433
moments 72

moving-average component 82
`mswitch` command 432
multiple-equation models 78–79
multiplicative seasonality 137

N
naïve forecast 152
`newey` command 55, 182
Newey–West estimator 182
`nl` command 54–55
`nlcom` command 58
noise 100
nonlinear least squares 54
nonstationary variable 381
null hypothesis 73

O
`odbc` command 24
OIRF . . . . *see* orthogonalized
        impulse–response function
OLS . . . . *see* ordinary least squares
one-sided alternative 73
one-step-ahead 145
one-step-ahead forecasts 316
operators 18
ordinary least squares 74–77, 178
        strategy 182–183
orthogonalized impulse–response func-
        tion 341,
        420
        cumulative 356
`outfile` command 22, 24

P
p-values 74
`pac` command 234
panel dataset 31
parameter redundancy 250
parsimony 227
partial autocorrelation coefficients . . 234
`pergram` command 432
Phillips–Perron test 387
`pperron` command 387
`prais` command 55
Prais–Winsten 187

`predict` command 56, 283
prefix command 15
probability
        density functions 72
        distribution function 72
`psdensity` command 432
pseudorandom numbers 61
*p*th-order autocorrelation 173
`pwd` command 29
`pweight` . . . . *see* sampling weights

Q
Q statistic 254

R
random
        variables 72
        walk 72, 82
rational expectations 358
`real()` function 34
regime changes 430
`regress` command 53–54
residual 56, 100
RMSE . . . . . *see* root mean squared error
`rolling` command 431
root mean squared error 154

S
S . . . . . . *see* seasonal difference operator
sampling weights 18
`save` command 21
SBIC . . . . *see* Schwarz's Bayesian
        information criterion
`scatter` command 6, 40–43
Schwarz's Bayesian information
        criterion 306
`search` command 10
seasonal 98–100
        adjustment 91
        component 93
        difference operator 69, 204
        Holt–Winters 137–138
seemingly unrelated regressions model
        79

SEMs .... *see* structural econometric models
set seed command 61
signal 100
smoothers 100
span 102
spectral analysis 429–430
splitting 113
sspace command 433
state-space models 246, 429
static forecasts 146
stationarity 80–82, 208–210
stationary process 174
strictly stationary 81, 208
strpos() function 34
structural econometric models 358–361
structural impulse–response function 365
structural vector autoregression ... 303, 358–373
    long-run 360, 370–373
    short-run 360–370
substr() function 34
summarize command 3, 28, 51
svar command 361–373
SVARs .... *see* structural vector autoregression
syntax 15–20
sysuse command 2

T
table command 4, 52
tabulate command 4, 28, 50
Taylor rule 365–369
td() function 68
test command 57
test hypotheses 57
testparm command 58
threshold command 433
time 62
time and dates 62
time-series operators 68
time-varying volatility 272, 277–285
tm() function 68

transformation strategy 183–186
trend 93–95
    component 93
trend-stationary 379
tsappend command 431
tsfill command 431
tsfilter command 432
tsline command 48
tsreport command 431
tsrevar command 431
tsset command 48, 62
tssmooth
    dexponential command 130–131, 147, 150–151
    exponential command .. 126–130, 147, 149
    hwinters command.. 131–137, 147
    ma command 123–126
    nl command 102–121
    shwinters command 137–138, 147, 161
ttest command 5
twicing 103
two-sided alternative 73
twoway
    command 44
    connected command 45–48
type command 23
type I error 74
type II error 74

U
ucm command 432
unit root, ... 82, 378–382
    process 379
    testing for 382–387
univariate time-series models 145
use command 21

V
VAR .... *see* vector autoregression
var command 345
varbasic command 307–309
varbench command 327–329
vargranger command 338

variance 72
varlist 16
`varlmar` command 313
`varnorm` command 313–314
`varsoc` command 305–306, 401–402
`varstable` command 309–312
`varwle` command 314–315
`vec` command 407–423
`veclmar` command 416
VECMs ...... *see* vector error-correction
          model
`vecrank` command 402–405
`vecstable` command 416
vector autoregression 300–303
      recursive 303, 330
      reduced-form 302, 305
      stationarity 305, 309
vector error-correction model ..389–394

W
weak stationarity 81, 209
weight 18
weighted moving averages 122–125
white noise 80–82
`wntestb` command 256
`wntestq` command 255

X
`xcorr` command 331–333